K-12 Mathematics Education in Israel

Issues and Innovations

SERIES ON MATHEMATICS EDUCATION

Mathematics education is a field of active research in the past few decades. Plenty of important and valuable research results were published. The series of monographs is to capture those output in book form. The series is to serve as a record for the research done and to be used as references for further research. The themes/topics may include the new maths forms, modeling and applications, proof and proving, amongst several others.

Published

Vol. 13 *K-12 Mathematics Education in Israel: Issues and Innovations*
edited by Nitsa Movshovitz-Hadar

Vol. 12 *Special Secondary Schools for the Mathematically Talented: An International Panorama*
edited by B. R. Vogeli

Vol. 11 *Mathematics Education in Korea*
Volume 2: Contemporary Trends in Researches in Korea
edited by Jinho Kim, Inki Han, Mangoo Park and Joongkwoen Lee

Vol. 10 *Mathematics and Its Teaching in the Southern Americas: with An Introduction by Ubiratan D'Ambrosio*
edited by H. Rosario, P. Scott and B. R. Vogeli

Vol. 9 *Primary Mathematics Standards for Pre-Service Teachers in Chile: A Resource Book for Teachers and Educators*
by P. Felmer, R. Lewin, S. Martínez, C. Reyes, L. Varas, E. Chandía, P. Dartnell, A. López, C. Martínez, A. Mena, A. Ortíz, G. Schwarze and P. Zanocco

Vol. 8 *Mathematical Modelling: From Theory to Practice*
edited by N. H. Lee and D. K. E. Ng

Vol. 7 *Mathematics Education in Korea*
Volume 1: Curricular and Teaching and Learning Practices
edited by Jinho Kim, Inki Han, Joongkwoen Lee and Mangoo Park

Vol. 6 *How Chinese Teach Mathematics: Perspectives from Insiders*
edited by L. Fan, N.-Y. Wong, J. Cai and S. Li

Vol. 5 *Russian Mathematics Education: Programs and Practices*
edited by A. Karp and B. R. Vogeli

Vol. 4 *Russian Mathematics Education: History and World Significance*
edited by A. Karp and B. R. Vogeli

*For the complete list of titles in this series, please go to
http://www.worldscientific.com/series/sme

Series on Mathematics Education Vol. 13

K-12 MATHEMATICS EDUCATION IN ISRAEL

Issues and Innovations

Edited by

Nitsa Movshovitz-Hadar

Technion – Israel Institute of Technology, Israel

World Scientific

NEW JERSEY • LONDON • SINGAPORE • BEIJING • SHANGHAI • HONG KONG • TAIPEI • CHENNAI • TOKYO

Published by

World Scientific Publishing Co. Pte. Ltd.

5 Toh Tuck Link, Singapore 596224

USA office: 27 Warren Street, Suite 401-402, Hackensack, NJ 07601

UK office: 57 Shelton Street, Covent Garden, London WC2H 9HE

Library of Congress Cataloging-in-Publication Data
Names: Movshovitz-Hadar, Nitsa, editor.
Title: K-12 mathematics education in Israel : issues and innovations / edited by
Nitsa Movshovitz-Hadar (Technion, Israel Institute of Technology, Israel).
Other titles: Mathematics education in Israel
Description: New Jersey : World Scientific, 2018. |
Series: Series on mathematics education ; volume 13
Identifiers: LCCN 2017046285 | ISBN 9789813231184 (hc : alk. paper)
Subjects: LCSH: Mathematics--Study and teaching--Israel. |
Mathematics--Study and teaching--Israel--History.
Classification: LCC QA14.I75 K25 2018 | DDC 510.95694--dc23
LC record available at https://lccn.loc.gov/2017046285

British Library Cataloguing-in-Publication Data
A catalogue record for this book is available from the British Library.

For any available supplementary material, please visit
http://www.worldscientific.com/worldscibooks/10.1142/10741#t=suppl

Desk Editors: V. Vishnu Mohan/Tan Rok Ting

Typeset by Stallion Press
Email: enquiries@stallionpress.com

Printed in Singapore

List of Authors*

1. AISIK, Revital
2. AMICHAI, Shlomit
3. AMIT, Miriam (Miri)
4. ARAD, Zvi
5. ARCAVI, Abraham
6. ARTSTEIN, Zvi
7. ASHKENAZY, Ronit
8. ASHKENAZI, Sarit
9. AVISSAR-ZELDIS, Tamara
10. BARKAI, Ruthi
11. BERMAN, Abraham (Avi)
12. BITON, Yaniv
13. CASPI, Shai
14. CORRY, Leo
15. DANA-PICARD, Thierry (Noah)
16. DOLEV, Lea
17. EYLON, Bat-Sheva
18. EISENBERG, Eli
19. EFRAT, Osnat
20. ELBAUM-COHEN, Avital
21. ELIMELECH, Yael
22. EVEN, Ruhama
23. FARES, Muhana

*E-mail address and affiliation of the authors appear under the title of each chapter. Personal details about the authors appear at the end of each chapter.

24. FEILCHENFELD, David
25. FELLUS, Osnat
26. FRIED, Michael
27. FRIEDLANDER, Alex
28. FRIDMAN, Elisheva (Gerstein)
29. GAL, Reuven
30. GENDLER, Olga
31. GUREVITCH, Irina
32. HAZZAN, Orit
33. HERMON, Noga
34. HERSHKOVITZ, Sara
35. HERSHKOWITZ, Rina
36. HESS-GREEN, Rachel
37. HEYD-METZUYANIM, Einat
38. HOCH, Maureen
39. HURVITZ, Eli
40. ILANI, Lea
41. IMANUEL-NOY, Dalia
42. KARSENTY, Ronnie
43. KATSAP, Ada
44. KELLER, Nelly
45. KOICHU, Boris
46. LAVY, Ilana
47. LEVENSON, Esther
48. LEIKIN, Roza
49. LINCHEVSKI, Liora
50. LIVNEH, Drora
51. MARKOVITS, Zvia
52. MARK-ZIGDON, Nitza
53. MEVARECH, Zemira
54. MORGENSTERN, Yehuda
55. MOVSHOVITZ-HADAR, Nitsa
56. NACHLIELI, Talli
57. NERIA, Dorit
58. OBERMAN, John
59. PERL, Hannah

60. RASSLAN, Shaker
61. RAVIV, Dafna
62. ROBINSON, Naomi
63. RON, Gila
64. RON, Saleit
65. ROSENFELD, Sherman
66. SEGAL, Ruth
67. SEVER, Geula
68. SHARIF-RASSLAN, Amal
69. SHECHTER, Taly
70. SHRIKI, Atara
71. SILVERMAN, Boaz
72. SILVERMAN, Sarit
73. SION, Niza
74. SPEKTOR-LEVY, Ornit
75. STEINER, Aya
76. TABACH, Michal
77. TALMON, Varda
78. TIROSH, Dina
79. TSAMIR, Pessia
80. VAKNIN, Anna
81. WAGNER, Tili
82. WALLACH, Tali
83. WEISS, Dovi
84. YERUSHALMY, Michal
85. ZASLAVSKY, Orit
86. ZODIK, Iris

Editor's Preface

This book was prepared by the mathematics education community in Israel as a gesture towards the 70th anniversary of the State of Israel.

A Brief Background about the State of Israel

The State of Israel is located in the Middle East, on the southeastern shore of the Mediterranean Sea and the northern shore of the Red Sea, bordering with Egypt, Jordan, Lebanon, Syria, and The Palestinian Authority. It is smaller in area than at least 150 other countries in the world (22,072 km^2 or 8522 sq. mi, including the Golan Heights and East Jerusalem), and in population (over 839,000) it is smaller than at least 100 of them. The median age is 30.3 years. The Jewish population is about 75% of the total population (about 11% Haredi-Jews, namely extremely religious, and 64% non-Haredi Jews). The non-Jewish population consists of about 21% Arabs (Muslims, Christians, and Druze), and the rest are Armenians, Circassians, Samaritans, Baha'is, and others.

The official languages in Israel are Hebrew and Arabic. Hebrew is the primary language of Israel, and is spoken by most of the population either as native speakers or proficiently as a second language. Arabic, the country's other official language, is used mainly by Israel's Arab minority, which comprises about one-fifth of the population. English, spoken as a second language by the majority of the Israeli population, is used widely in official logos, road signs and product labels, and is a mandatory subject in the school system. Russian, spoken by the large immigrant population from the former Soviet Union, is also heavily used. Additional commonly spoken languages are Amharic and French.

Since its establishment in 1948, the population of Israel increased more than tenfold. Hundreds of thousands of immigrants from many countries and different cultures integrated in a relatively short span of time, creating a melting pot of customs and beliefs. A major role in this effort was played by giving high priority to educational and cultural frameworks and by allocating extensive resources to them. This helped in raising a scholarly generation that made the economy of the young state a success, putting it on par with the world's most developed countries.

The Israeli System of Education

The Free Compulsory Education Law of 1949 was one of the earliest laws accepted by the Israeli parliament (Knesset). It stated that the existing Jewish educational trends as well as the Arab educational system would be recognized and funded by the State. A wider scope State Education Law was accepted in 1953, amended several times in the later years, to warrant education in the principles of democracy, provided for all students (K-12). This basic law determines guiding principles with regard to the content and procedures of State education and provides for a six-day school week. State education is defined as education provided by the state on the basis of the curriculum approved by the Ministry of Education, under the supervision of the Ministry of Education, without attachment to a political party or other non-government organization. The law outlines the objectives of State education with regard to universal values; the values of Israel's society and heritage; remembrance of the Holocaust and heroism; development of the child's personality; abilities and creativity; various disciplines of knowledge; science and the arts; physical activity; and culture and recreation.

The Israeli education system consists of 12 years of schooling and one year of mandatory pre-school, starting at age 5, in two trends: "State education" and "State-religious education." It is separated into Hebrew and Arabic education by the main language used in the schools.

Israeli education system is centralized. The Ministry of Education governs, through its six geographical districts, and two country-wide administrative districts, the school budgets, their management, the curricula, and all the aspects of implementing it according to the central policy. Schooling comprises four stages: Preschool (ages 3–6), Primary School (commonly grades 1–6, in some cases grades 1–8), Middle School (grades 7–9), and High School (commonly grades 10–12, in some cases grades 9–12). The total number of schools in the Hebrew Education sector is about 3300, and about 780 in the Arabic Education sector.

High schools in Israel prepare students for the Israeli secondary-school graduation examinations (in Hebrew *Bagrut*). These examinations cover various academic disciplines, which are studied in units of one to five on an ascending scale of difficulty and scope. A secondary-school graduation certificate that includes scores, often determines acceptance into elite military units upon joining the Israel Defense Force (IDF) for a mandatory service, admission to academic institutions, and job prospects.

About the Book

As in many other countries in the world, mathematics is recognized by the public as one of the central school disciplines. It is taught from Kindergarten to 12th grade according to mandatory, centrally governed curricula (see details in the Appendix). The Ministry of Education provides supervision and guidance to teachers in implementing the mandatory curricula. (For details see Chapter 4.)

Mathematics teachers' preparation for K-10th grade takes place at 24 academic colleges of education. The preparation of mathematics teachers for 7–12th grade take place at the six research universities. A variety of innovative professional development programs for in service mathematics teachers are originated by the Ministry of Education as well as by the academic institutions (see Chapters 24–26, 29–34). Particular attention is devoted in these programs to improvement of the quality of the teaching force. For example, engineering graduates, High-Tech and IDF veterans are offered back to school scholarships to earn a mathematics teaching license (see Chapters 27 and 28).

The State of Israel has taken part in several international evaluations of mathematics achievements comparative studies, the results of which have had an impact on initiatives taken towards the improvement of mathematics education (see Chapter 40).

This book opens with a timeline of highlights in the development of the mathematics education system as a part of the entire system of education (see Chapter 1). Creating a Modern Hebrew language for mathematics was a challenge faced in the years before and after the establishment of the State of Israel. The story of the teaching of mathematics in Israel cannot be told without telling the story of the development of a Modern Hebrew lexicon for mathematics (see Chapter 35). A related issue is that of the Arabic speaking schools, a significant minority in the school system (see Chapter 7).

Another significant minority is the State-religious schools, where mathematics education takes place alongside religious studies and is separated by gender (see Chapter 5). A particularly big challenge in striving for excellence in mathematics education exists in ultra-orthodox Jewish community (see Chapter 6). At the other end stands an education system that belongs to the unique Jewish Socialist Kibbutz Movement. This movement established small peripheral settlements based on communal life starting in the early years before the State of Israel was established. The story of mathematics education in the Kibbutz Movement is included in Part V (see Chapter 39).

Issues, challenges and innovative programs related to mathematics education in grades K-6, and in grades 7–12, appear in Part II (Chapters 8–12) and Part III (Chapters 13–23) of this book, respectively. These chapters include issues of caring for struggling children as well as for mathematically talented ones, the use of technology-based tools for improving the teaching and the learning of school mathematics, curricular innovations and extracurricular activities for diverse student populations.

In recent years a State policy has taken place that encourages youth to take the highest possible level of school mathematics, in order to exhaust their potential capability, thus nurturing widening the excellence circles of the population. It was initiated and is widely supported by a philanthropic foundation (see Chapters 2 and 3). This policy yielded many innovative programs for students and teachers including programs for prospective teachers, reflected in several chapters of this book.

Several issues may be of general interest to the international community of mathematics education. Among them are the role of technology in mathematics education (see Chapters 36, 37 and also Chapters 12, 14, 15), interweaving ethnomathematics in the study of school mathematics (see Chapter 42), diagnosing mathematical learning disabilities (see Chapter 43), caring for the mathematically talented (see Chapter 16 for a comprehensive overview, and for specific programs see Chapters 13, 17, 18, 41).

The community of Israeli mathematics education researchers became quite reputable internationally. It is an active and vivid professional community that has a periodical journal (in Hebrew) and annually holds the Jerusalem conference on research in mathematics education in Israel (see Chapter 44). Many members of this community publish the results of their studies internationally. Some of which are reflected in this book which is devoted to school implementation and practice rather than to research (in particular, see Chapter 38).

At the end of the book the reader can find a glossary of terms related to mathematics education, acronyms and abbreviations as used in this book, and an appendix describing the mandatory mathematics curricula for K-12 updated in 2017.

Nitsa Movshovitz-Hadar

January 2018

Foreword

This book presents various aspects of the centrally administered system of mathematics education in Israel. It is a result of the professional cooperation between the academic professionals who specialize in research and development in mathematics education, the people holding administrative positions in the Ministry of Education in our country, and school educators who take care of the daily work with students, which I appreciate.

The fact that we work in collaboration and put great value on shared work interface, such as meticulous work on joint constant updating of the curriculum, structured feedback and work procedures, is in my opinion a significant factor in Israel's remarkable progress in the area of mathematics education.

This book sheds light on the past and present of mathematics education in Israel and attempts to predict parts of its future. I hope you will find reading it both enriching and enjoyable.

Wishing good luck to us all in our endeavor to improve mathematics education for all our students in the coming years, I am

Nerit Katz

Chief Superintendent of School Mathematics in Middle and High Schools
Israeli Ministry of Education
January 2018

Acknowledgments

When I took upon myself the task of editing this book I did not know how exciting, enriching, and fulfilling this task would be. It was a wonderful collaborative work with 86 participating authors, who also served as peer-reviewers, supporting the process in each and every step. Pearla Nesher, Ted Eisenberg, and Beba Shternberg also contributed to the review of the first drafts. Thanks are due to the organizers of the 5th Jerusalem conference for research in mathematics education which took place in February 2017, who accepted a round table proposal that enabled the gathering of 19 colleagues to consider the structure of the book and its partition into five parts based upon the abstracts provided by the authors. The deep thought invested by these 19 colleagues made the editing of the book much easier. The appendix is Johnny Oberman's contribution. He took upon himself the English translation for the various mathematics curricula and put them in a concise form which made the appendix clear and readable. Last but not least, my appreciation goes to Boaz Silverman who went through the final drafts and prepared a glossary which allowed a consistent and unified translation of the Hebrew terms across the various chapters. All these people were kind enough to bear with me for 20 months of pressure to complete their part on time so that the publisher can make it happen for the 70th anniversary of the State of Israel. Thank you all.

Nitsa Movshovitz-Hadar

March 2018

Contents

Part I

Issues and Innovations Related to the Structure of Mathematics Education in Israel

Chapter 1

Highlights in the Development of Education and Mathematics Education in the State of Israel: A Timeline

Michael N. Fried

Ben-Gurion University of the Negev, Israel
mfried@bgu.ac.il

Hannah Perl

Israel Ministry of Education, Israel
hannah.perl1948@gmail.com

Abraham Arcavi

Weizmann Institute of Science, Israel
abraham.arcavi@weizmann.ac.il

This chapter provides a timeline for what the authors consider to be the most significant and influential events in mathematics education from 1900 until today, with special emphasis on the years since the establishment of the State of Israel in 1948. Commentary corresponding to each date in the timeline is also given.[a]

1. Introduction

It is nearly impossible to reduce history to the neat sequence of events and dates that make up a timeline, as every historian knows. The moment one begins to investigate an "event", it becomes increasingly clear how

[a]More details about some parts of this chapter will appear in a book to be published by Springer on the history of mathematics education in modern Israel by Michael N. Fried and Orit Zaslavsky.

unclear the borders of the event are, not only as to when the event began and when it ended, but also more importantly what the event really was. In a much abbreviated timeline such as this one, where any thought of thoroughness must be banished, these problems are compounded by the necessity of choosing merely representative events. Still, a timeline of mathematics education in Israel, even a short one, can be a useful tool as a means to coordinate events discussed in more detailed accounts, such as those included in other chapters of this book, and as a springboard for deeper questioning.

Besides the problem with the very idea of a timeline, there is a further difficulty in the phrase "*in the State of Israel*". As can be seen in Chapter 35 by Leo Corry, the distinctive character of mathematics education in the State of Israel was, in many ways, already determined *before* the establishment of the State itself. Thus, besides the vocabulary adopted for teaching, many textbooks and approaches to mathematics teaching after 1948 continued to develop and change along a trajectory so completely grounded in earlier times that the War of Independence and induction of the State of Israel into the community of nations might appear as mere background events.

On the other hand, with the establishment of the modern State of Israel in 1948, education became a state affair. Questions of a national curriculum, equity, school standards, and so on became newly concrete and urgent. The Compulsory Education Law of 1949 (later modified in the State Education Law in 1953) was one of the very first pieces of legislation produced in the first Israeli Parliament. Mathematics education, in *this* regard, was subsumed under these more general educational and national issues.

For the presentation of our timeline below, these considerations have two implications. First, we cannot ignore developments before the creation of the modern State of Israel, even though our focus is on the post-1948 period. Second, many if not most of the events after 1948 are of a national character and shaped education in all disciplines in Israeli schools. Even where reforms were directed towards mathematics specifically, they were often embedded in developments having to do with mathematics *and* science education.

Finally, a word about the structure of this chapter. Each item in the timeline has a distinct year (sometimes with an adjoined letter, such as 1965a).

The commentary is indexed by that year. In this way, the commentary provides extra details only about each item in the timeline: it is not meant to be a continuous text.

2. Timeline

2.1. *1903–1934*

1903	The establishment of the Hebrew Teachers' Association
1905–1913	The founding of the first three Hebrew Schools: Herzliya Gymnasium 1905 (Tel Aviv), Jerusalem Gymnasium Rehavia 1909, Hebrew Reali School 1913 (Haifa)
1912	The cornerstone of the Technion in Haifa was laid
1918	The cornerstone of the Hebrew University of Jerusalem (HUJI) was laid
1928	The Establishment of the Mathematics Institute at the Hebrew University
c.1933	Introduction of the completion of secondary school examination, the "Bagrut," including the mathematics examination
1934	The founding of the Daniel Sieff Research Institute, precursor of the present Weizmann Institute of Science

Commentary 1903–1934

1903: The Hebrew Teachers' Association grew out of an earlier Hebrew teacher association in the 1890s, and it was dedicated to teaching in Hebrew. In 1928, the Hebrew Teachers' Association became the Teachers' Association (*Histadrut ha-Morim*) which remains a central teachers' union today.

1905–1913: As discussed in Corry's chapter, Hebrew was revitalized as a language for all affairs. "Hebrew Schools" were marked by an insistence not only to teach Hebrew but also to teach *in* Hebrew. It was in this context that Abraham (Baruch) Rosenstein at the Herzliya Gymnasium worked out the vocabulary and conventions for teaching mathematics in Hebrew, most of which are still in use.

1912: The Technion, which is one of Israel's foremost universities for science, mathematics, and technology, was then called the Technikum. The first students were enrolled in 1924.

1918: Among HUJI's board of governors were Albert Einstein, Sigmund Freud, and Martin Buber. The legitimacy of Hebrew school education was influenced by the international reputation of the Hebrew University (see below, 1933).

***c*.1933:** The British authorities in Palestine threatened not to respect Hebrew secondary school certificates without a completion examination according to British standards. Hebrew high schools under the Department of Education of the Jewish National Council in Palestine led by Joseph Luria (1871–1937) developed and administered their own examination, the "Bagrut", with the cooperation of the Hebrew University. The reputation of the Hebrew University sufficed to make the secondary school certificates recognized by universities outside of Palestine. It must be pointed out, however, that students could take the British exams as well, and many did.

The secondary school examinations are still critical for students wishing to pursue more advanced education. The central subject areas for the examinations have remained relatively stable: Bible (or Koran, or Druze heritage), Hebrew (or Arabic) grammar and composition, Mathematics, English, Civics, History, Literature, and at least one elective, including the various sciences. There have been, however, changes in how the examinations are administered as well as in the specific subtopics covered by them.

The name "Bagrut" means, literally, "maturity", and is related nominally to old German *Maturitätsprüfung* or the Austrian *Matura*, both completion examinations of a similar type to the Israeli "Bagrut".

1934: Although this institute was established as an institute for advanced studies in science and mathematics, the Weizmann Institute, which grew out of it, became a crucial center for science and mathematics education in later years (see below, 1968).

2.2. *1948–1956*

1948	May 14th: The State of Israel is officially established. The Declaration of Israeli Independence asserts, among other things, the universal right to education
1949	Compulsory Education Law
1953	State Education Law
1954	First journal for teachers of mathematics and physics (*Pages for the Teacher*)
1956	The first official curriculum for secondary schools (grades 9–12)

Commentary 1948–1956

1948: The section of the Declaration of Independence referring to education is very similar to article 26 of the United Nations Declaration of Human Rights published a few months afterwards. Living up to these principles would, however, continue to be a challenge for the country, a challenge not always met successfully.

1949: This was one of the first laws passed by the new Knesset which underlines the importance of education in the vision of the new state. The law stated that education should be free and compulsory for all children between the ages of 5 and 13 and for adolescents who had not completed their primary education (of which there were many from among the large population of refugees from the Arab countries and from post-war Europe). The law left in place three "trends" existing since the early 1920s — a general, socialist (associated with the Kibbutz and Moshav Movements), and religious trend — and even allowed a fourth ultra-orthodox trend. The politicization of the choice of trend, especially among new immigrants, quickly became political coercion. This led to a governmental investigation, the resignation of the Minister of Education, Zalman Shazar, and new elections. The fact that the government fell as result of the difficulties with the education law again brings into relief the centrality of issues connected with education; it also brings into relief the tremendous challenges the nation faced regarding massive immigration.

1953: The State Education Law (*chinuch mamlachti*) abolished the "trends" system and created two systems of State Education: one general and one religious. It also allowed for alternative systems which required the approval of the ministry of education; by that means, traces of the trend system persisted and still persist. The State Education Law asserted the full responsibility of the State for education for all its citizens; it set out the basic structures of educational policy-making and determined the role of the education ministry in teacher education.

1954: The first professional journal for mathematics and physics teachers appeared under the auspices of the Israel Mathematical Union. The journal was devoted to the pedagogy of mathematics, its history, curricular issues and exchange of teaching ideas.

1956: Before 1956, the 40 existing high schools established the curriculum according to the secondary school examination. Given the growing number of schools and students, the Ministry of Education published an official proposal for curricula in all subjects. The school mathematics curriculum consisted of three components: one for grades 9–10 and two for grades 11–12, one for students in the math–physics stream and one for those in the biology stream. An additional curriculum was published in 1957 by an *ad hoc* committee devoted to high school mathematics for students majoring in the humanities.

2.3. *1958–1965*

1958	Teachers' Organization (*Irgun HaMorim*) for secondary school teachers
1959	First trials of innovative secondary school mathematics curricula
1960a	First issue of the *Gilyonot Mathematica* (*Mathematics Pages — For Young Learners and Enthusiasts*)
1961–1962	Dushkin Committee — The creation of three-year teaching colleges
1963–1965	Prawer Committee — the establishment of middle school

Commentary 1958–1963

1958: The establishment of a teachers' union for *high* school teachers underlines an important division in educational thinking and policy between primary (then, grades 1–8) and high school (grades 9–12). The State Education Law referred essentially to primary school; teachers in high school were mostly university-educated and saw themselves almost as a different tribe. The teaching of mathematics was affected by this distinction, and there were great differences in how primary school teachers and high school teachers approached the subject. Moreover, acceptance to an academic high school was subject to a selection test, the "seker," which prevented many students from pursuing higher education and produced social divisions and a kind of educational segregation. This would come to a head with the Prawer Committee (see below, 1963).

1959: The trials of new curricula included aspects of calculus and other subjects not taught before. These efforts were led by the worldwide renowned algebraist Professor A. Shimshon Amitsur on behalf of the Ministry of Education.

1960a: This newsletter, published by the Israel Mathematical Union, contained articles and problems for students interested in mathematics. Its first co-editors were Abraham Ginzburg, Shmuel Avital, and Michael Maschler (who became central figures in mathematics education). The first issue had an introduction by Abraham Fraenkel (the set theorist and one of the founders of Hebrew University's mathematics department). Earlier newsletters for students included the *Mathematical Quarterly* (first issue, 1944) and the *Mathematics and Physics Pages* (first issue, 1942).

1961–1962: The Minister of Education, Abba Eban, set up this committee led by Alexander M. Dushkin of the Hebrew University to examine the preparation of primary school teachers. The committee sought ways to raise the level of primary teachers and narrow the gap between them and high school teachers. To this end, teacher colleges with a three-year program were proposed. The committee also maintained that mathematics should be obligatory, but that recommendation was not implemented.

1963–1965: The Minister of Education, Zalman Aranne (1899–1970), appointed Joshua Prawer (1917–1990) from the Hebrew University to head a committee to examine compulsory education and many of the problems in the entire structure of schooling. The Prawer Committee report, submitted in 1965, radically changed the structure of the Israeli education system from an 8 + 4 system (i.e., 8 years of free compulsory primary school, optionally followed by high school) to a 6 + 3 + 3 system (i.e., 6 years of primary school followed by a 3-year comprehensive middle school stage, grades 7–9, and 3 years of high school, grades 10–12). The leaving age for compulsory education was raised to 15. The recommendations of the committee were fiercely opposed by the teachers' union who saw the creation of middle school as a threat to the prestige of primary school principals and teachers. The recommendations were accepted in 1968. The curricular demands of the new "middle school" had a great influence on the design of the school mathematics curriculum, the production of new textbooks, and the training of mathematics teachers.

2.4. *1964–1970*

1964–1965	Amos de Shalit Committee — promoting the teaching of science (physics, chemistry, biology) and mathematics
1964	Youth division established at the Weizmann Institute
1965a	Technion Senate approved a degree program in science education, including mathematics and other disciplinary areas
1965b	Accelerated development of comprehensive high schools
c.1965c	Cuisenaire rods and similar mathematical manipulatives began to be introduced
1966	The Instructional Television Trust televised broadcasts of mathematics, biology, and English lessons
1967a	*The Israeli Center for Science & Technology Education* (MALAM) was established

(*Continued*)

(*Continued*)

1967b	Following the Prawer Committee's recommendation, the first middle school mathematics curriculum was published
1967c, 1970	Israel participates in international examinations in mathematics and reading with exceptionally good results
1968	Department of Science Teaching was formed at the Weizmann Institute
1970	Department of Science Teaching was formed at the Hebrew University

Commentary 1964–1970

1964–1965: The Minister of Education, Zalman Aranne, appointed the physicist Amos de Shalit from the Weizmann Institute as head of a committee to study the state of mathematics and science teaching in middle and high schools and to recommend ways of advancing it. Some of its specific tasks were to compare the Israeli curricula to other countries (e.g., the American SMSG and the British Nuffield program), review teaching practices, conduct curricula trials, develop textbooks and teaching aids. The committee's recommendations led to the development of new curricula in mathematics and sciences and to the establishment of the *Israeli Center for Science & Technology Education* in 1967.

1964: This was led by Dr. Moshe Rishpon to challenge talented students and bring them into contact with scientists.

1965b: Comprehensive high schools combined academic and vocational orientations and allowed a more diverse population of students. The date here is somewhat arbitrary: the promotion of such high schools occurred throughout the mid-1960s. By 1971, there were 20 comprehensive high schools in the country. It is still the case today that not all high schools in Israel are comprehensive.

1965c: The use of colored calculating rods, particularly the Cuisenaire rods for primary school mathematics was adopted in the mid-1960s and became quite popular beginning in the 1970s: the importance of these colored rods, among other concrete materials, for teaching arithmetic was stressed, for example, in a book published by the teachers' union on Methods of Teaching Elementary Mathematics (by Miriam Nussdorf-Blank, 1968). Manipulatives such as Cuisenaire rods and Geoboard had an enthusiastic following in Israel partly through the popularity of Piagetian cognitive psychology. Caleb Gattegno (1911–1998), a proponent of Cuisenaire rods, actively facilitated their use in Israel. The influence of Gattegno is indicative of the emphasis in Israeli mathematics education to be in tune with developments outside of the country.

1966: The broadcasts for 7th grade included arithmetic, number bases, properties of operations, fractions, proportion, and selected topics in geometry. Ninth grade programs included negative numbers, algebra, equations, functions, and word problems.

1967a: The center was established at the Hebrew University of Jerusalem and at the Weizmann Institute in order to develop textbooks and other materials for teaching mathematics, science, and technology.

1967b: A committee headed by A. Marcus, who was then the national superintendent of school mathematics, published the curriculum, whose basic outline remained intact until 2009.

1967c, 1970: Out of the 12 countries, Israel scored first in mathematics and in reading. However, it is still unclear how valid those results really were, and in the 1990s (see below), Israel performed badly in subsequent international examinations.

1968: The Department of Science Teaching was founded by the physicist Amos de Shalit. After de Shalit's untimely death the following year, at age 43, the department was taken over by Joseph Gillis, a prominent mathematician, working in both pure and applied mathematics and who worked at the famous Bletchley Park project to break the German Enigma code during World War II.

1970: The department was formed in cooperation with the School of Education.

2.5. *1971–1983*

1971a	A complete primary school mathematics curriculum
1971b	*Center for Educational Technology* (MATACH) established
1973	The "seker" examination was eliminated, opening high schooling to a greater number and variety of students
1976	The Unit for Science and Mathematics Teaching at the University of Tel Aviv was established
1978	The "units" system is introduced to the "Bagrut" allowing broader opportunities for completing secondary mathematics education
1979	Joseph Gillis initiates and coaches the Israeli team for the International Mathematics Olympiad
1983	The 7th International Conference of the International Study Group for the Psychology of Mathematics Education (PME) is held in Israel

Commentary 1971–1983

1971a: This curriculum was in effect until 1987.

1971b: In accordance with a government decision in 1970, MATACH was established by Yad Hanadiv (see 2011 below) to produce learning materials in mathematics and other subjects, particularly materials taking advantage of modern technologies. Soon after its establishment, Pearla Nesher became the scientific advisor of MATACH. She also produced in those years the series of books, *One, Two, and...Three*, which were used widely in Israeli primary schools until the 1990s. Nesher's involvement in MATACH was significant since, at the same time, she was an advisor for the ministry of education and, moreover, was at the center of a growing community researchers investigating primary school mathematics. Although the activities of MATACH were indeed initially focused on primary school students, its activities extended later to middle and high school students.

In this respect, it promoted or developed well-known technology-based programs such as *Geometric Supposer* and *Visual Math*, developed by Michal Yerushalmy.

1973: See the remarks on the "seker" in the commentary on 1958.

1976: Efraim Fischbein (1920–1998) was one of the founders of the unit at Tel Aviv University. He was influential internationally in advancing psychological perspectives of mathematics education, particularly those concerning the role of intuition in mathematics and science learning.

1978: The "unit system" was essentially a system of three levels: basic (three units), intermediate (four units), and advanced (five units). The advanced level was (and still is) required for some university programs, but not all: only the basic level is required to obtain a secondary school certificate. This unit system accorded with the 1975 recommendations of the Amitsur Commission.

1983: The significant international conference was held for the first time in Israel showed that work in mathematics education in Israel also was beginning to influence the international mathematics education community. Efraim Fischbein, mentioned above, was one of the founders and driving force of the PME in its early years.

2.6. *1986–2002*

1986a	Establishment of the *Israel National Center for Mathematics Education* (KESHER CHAM)
1986b	ALEH — Journal for high school mathematics teachers
1987	New primary school mathematics curriculum
1989	MISPAR CHAZAK — Journal for primary school mathematics teachers
1992	Harari Report "Tomorrow 98" on the state of science, mathematics and technology education in Israel

1995a	Unified high school mathematics curricula
1995b	Israel participates in the international TIMSS examinations for the first time
2000	Israel participates in the PISA examinations sponsored by the OECD for the first time
2002a	The creation of the national MEITZAV examinations
2002b	Report of the Ben-Zvi Committee

Commentary 1986–2002

1986a: This research and development center, founded by Nitsa Movshoviz-Hadar from the Technion, produced mathematics teaching materials disseminated through forums for teacher professional development and spurred research. It evolved into the National Teacher Centers. The center was called KESHER CHAM, a Hebrew acronym which stands for the "Advancement, Improvement, and Revitalization of Mathematics Education". It served as the National Center until 1977 when The Teachers Centers for elementary school mathematics teachers and its parallel for high school were established by the government and retained its R&D roles since then.

1986b: The journal *ALEH* was created to follow two previous teacher journals *Shvavim* and *Misparim*. Since its creation, 54 issues of *ALEH* have been published.

1987: The new curriculum, in effect until 2006, included the massive use of hand-held calculators from grade 1 and the introduction of probability. This curriculum established priorities among the topics to be taught.

1989: MISPAR CHAZAK (Strong Number) was originally published by the Beit Berl Academic College of Education. In 2000, its publication was taken over by the University of Haifa, and it was renamed MISPAR CHAZAK 2000.

1992: This report led by Haim Harari, then president of the Weizmann Institute, urged that new programs in science and mathematics be given high national priority and support. The wide implementation of the report produced a wealth of new learning materials in mathematics, courses for teacher leaders, and established regional (later abolished) and two national teacher centers, one for primary school and one for secondary school. The influence of the implementation of Harari report is visible today in many aspects of mathematics education (see http://education.academy.ac.il/Index4/Entry.aspx?nodeId=1110&entryId=20741).

1995a: A unified curriculum fused all the existing trial programs and was established as compulsory for all. Due to the changes in the *Secondary School exam*, this curriculum was implemented only in 2000.

1995b: These examinations, sponsored by the International Association for the Evaluation of Educational Achievement (IEA), take place every four years and are given to students in the 4th and 8th grades, though, in Israel, only 8th graders participate. As mentioned above (1967, 1970), Israel's performance was a grave disappointment.

2000: Israel's achievements on the PISA examinations (Programme for International Student Assessment) on reading, science, and mathematics were also disappointing and caused considerable soul-searching both in the media and among policy-makers. Among other things, this led to a pointed, but controversial, interest in the educational approaches of Singapore and Finland, owing to their impressive performance on international examinations; with that interest came also sharp criticism, by some, of educational approaches promoted by Israeli mathematics educators.

2002a: MEIZTAV (Hebrew acronym for "School Growth and Efficiency Measures") exams test students in the 5th and 8th grades in mathematics, science, Hebrew (Arabic), and English. It ought to be said that the MEIZTAV examinations were preceded by a similar national school assessment examination in the early 1990s under the direction of Joseph Bashi who served as Chief Scientist of the Ministry of Education from 1981 to 1993.

2002b: This committee reviewed learning and teaching materials, teacher education and professional development. A main outcome was establishing intensive courses to train professional mathematics teachers for primary schools. These courses lasted for four years and contributed a cohort of well-trained teachers specialized in mathematics.

2.7. *2003–2016*

2003	Use of Cuisenaire rods officially discontinued
2006	Establishment of the *National Authority for Measurement and Evaluation in Education* (RAMA)
2006	Introduction of the new primary school curriculum
2007	Haifa University is granted the opportunity to open two National Teacher Centers for mathematics according to a new more regulated system of tenders from the Ministry of Education
2011	Establishment of the Trump Family Foundation
2012	Gutfreund Report: Knowledge and Training of Secondary School Mathematics Teachers
2015	Strategic Program for the Promotion of Mathematics and Scientific Excellence
2016	The Hebrew University closed the Science Teaching Center and the Department of Science Education

Commentary 2003–2016

2003: This was a directive from the Ministry of Education.

2006: The *National Authority for Measurement and Evaluation in Education* (RAMA) was established by the ministry of education to monitor the process and outcomes of examinations — both national (e.g., MEITZAV) and international (e.g., TIMSS and PISA).

2006: This curriculum disallowed the use of hand-held calculators for grades 1–5 in favor of newer technologies and removed priorities of topics.

2007: Teacher Centers existed before 2007, as already noted in connection to the Harari Report (see 1992); however, they were established then under less rigorous regulations. Moreover, the MALAM (see 1967a) had a more direct role in their supervision than the Education Ministry.

2011: The philanthropic Trump Family Foundation (no connection to Donald J. Trump) was established for 10 years to improve the teaching of mathematics and science in Israeli high schools and to encourage more students to take mathematics and science at advanced levels. The Foundation acts in coordination with the Ministry of Education. The Trump Foundation is only one of a number of such philanthropic foundations having a strong influence in mathematics and science education all over the country. Other significant examples include the RASHI Foundation, YAD HANADIV of Rothschild Foundation, and KADIMA MADA (World ORT).

2012: The report, sponsored by the Initiative for Applied Education Research of Israel Academy of Sciences and Humanities, reflected the two-year work of a committee headed by Hanoch Gutfreund, former president of the Hebrew University. It brought together Israeli mathematicians, teachers of mathematics, and professional mathematics educators to examine the necessary knowledge for secondary mathematics teachers as well as the overall character of school mathematics teaching.

2015: This well-funded program was initiated by the Ministry of Education to address the declining number of students studying high school mathematics at the advanced level. Statistics have shown that the number of students studying at this level has since increased considerably. The program has also funded teacher training and retraining programs, hosted by several Israeli universities, to increase the number of teachers teaching in these advanced levels.

About the Authors

Michael N. Fried, PhD, chairs the Program for Science and Technology Education at Ben Gurion University. He is a mathematics educator and a historian of mathematics.

Hannah Perl was the director of the Science Division at the Ministry of Education Pedagogical Secretariat during the years 2011–2016. Previously, from 2003–2011 she served as the chief superintendent of School Mathematics for grades K-12. Until 2003 she was a mathematics teacher, a teachers' instructor and a member of the mathematics staff at the Science Teaching Center at the Hebrew University of Jerusalem. Hannah holds BSc and MSc degrees in mathematics and a PhD in science teaching, all from the Hebrew University.

Abraham Arcavi is a faculty member of the Department of Science Teaching at the Weizmann Institute of Science. Abraham's main interests are research and development in the areas of student and teacher learning.

Chapter 2

How Did a Crisis in Mathematics Education Lead to a Positive Reform?

Muhana Fares

Ministry of Education, Israel Government, Jerusalem, Israel

This chapter describes an educational reform aimed at increasing the number of students studying advanced level mathematics in high schools. The reform was initiated in view of the deteriorating number of students choosing to study mathematics at the advanced level: between 2006 and 2013, there was a decrease of 30%. This decrease reflected on the level of students who studied engineering and science in the universities. These phenomena could have a negative influence on academia, industry, security, and education. Thus, the Ministry of Education initiated a "Program for the Promotion of Mathematics and Scientific Excellence" which is discussed here.

1. Introduction

Mathematics is one of the core subjects taught in the education system in Israel, and successfully passing the secondary school examinations in mathematics is a prerequisite to obtaining a secondary school certificate. The importance of mathematics is based on it being a requirement for all core activities in modern society and the infrastructure for the study and development of science and technology. Mathematics education occupies the attention of policy-makers at all levels: academia, industry, security, and education. Following the strong criticism of the State Comptroller's Report and the accompanying reports (i.e., the Ben Zvi Report (2008) and the Harari Report (1992)), it became clear that change and reform were required to provide an answer to the need for more high school students graduating with higher-level mathematics. Rulings and decisions regarding "Who should learn what?" lead to organizational decisions at school and classroom levels.

In this chapter, I will present the "Program for the Promotion of Mathematics and Scientific Excellence" as a critical event that can deal with this crisis on different levels: policy, budget, and pedagogical.

Robinson and Aronica (2015) write that reforming any situation requires three concept levels: criticism of the status quo, a vision of what is desired, and a theory of change by which the transformation from one situation to the other can be accomplished. A central question is the essence of the vision. Is the vision about the "number" or the "percentage" of students learning mathematics?

2. The National Program for the Promotion of Mathematics and Scientific Excellence

The State of Israel has demonstrated significant achievements in research and development and in the number of successful start-up companies thanks to the work of its outstanding and gifted graduates. Over 90% of the undergraduates in engineering faculties in Israeli universities studied and completed mathematics at the advanced level in high school. However, ironically, between 2006 and 2013, there was a 30% decrease in the number of students aiming to study mathematics at the advanced level (see Fig. 1). Furthermore, the data showed a decrease in the number of students who excelled in this group. A decrease in the number of students wanting to study mathematics at the advanced level thus affects the human resources that can join high-tech industries, which will ultimately have a negative effect on Israel's economy.

A recent published study by the Taub Center (2015) found a positive correlation between studying mathematics at the advanced level and achievement in the job market and/or income level. Indeed, we are witness to the frustration of many students who did not have an opportunity to realize their potential and their desire to study in a progressive, intellectual environment. It is, therefore, clear that this reality has an impact both on the personal (fulfillment of academic and employment potential) and national (the future of security, economics, and academics in Israel) levels.

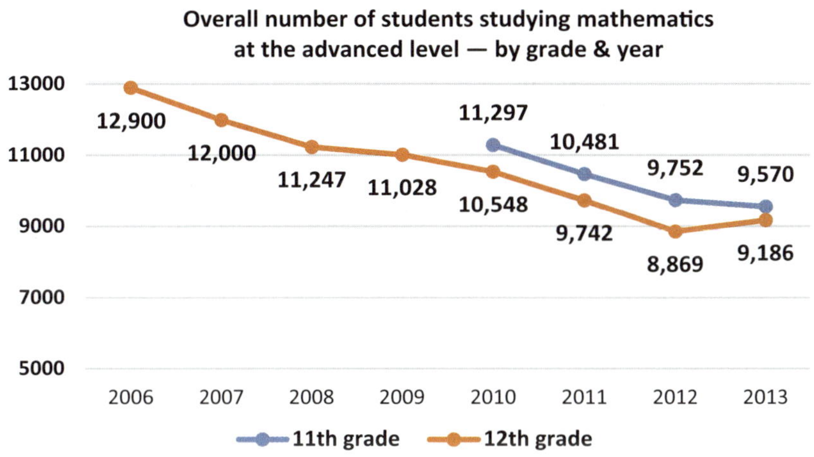

Fig. 1. Overall number of students studying advanced-level mathematics by grade and year (11th and 12th grades).

Changing this reality will hopefully have a positive effect on the social aspect.

3. The Reasons Behind the Problem

The decline in the number of students studying mathematics at the advanced level is a result of more than one factor. The factors, presented below, were gathered through discussions within the Ministry of Education and from information garnered in the framework of the "5 × 2 initiative" (2015) — a joint project of the Ministry of Education and the Business Coalition.

First, today ***students*** have feelings of low efficacy, prefer higher marks at lower levels of difficulty and with less effort, have experienced learning gaps since middle school, and exhibit a lack of perseverance. In addition, Israeli students are at the age when they are facing upcoming military service, and many believe that it is too early to invest time and energy for the purpose of higher-level learning that will be completed only in about six years. Second, a large portion of ***teachers*** lack mathematical knowledge and have feelings of low efficacy. There is a shortage of teachers able to teach the advanced level. Programs training primary and middle school teachers offer only a low level of mathematical education. There are a low number of graduates who pursue teaching due to issues of pay, professional status, and promotion opportunities. Third, many ***schools*** do not offer an advanced-level study program (only 566 out of 1100 schools taught mathematics at the advanced level in 2012). However, a major factor is the schools' orientation towards percent eligibility and there is a lack of awareness about the importance of learning at the advanced level and about the very real difference between intermediate and advanced levels. Fourth, many ***parents*** show a lack of awareness of the importance of advanced level studies. There are gaps (on the social level) of the extent of extra help provided. Parents tend to encourage children to aim for higher marks. Additionally, school evaluations are based only on the percentages of eligibility for secondary school certificate without any quality parameters. There is also a lack of consistency in the degree of difficulty of secondary school exams, a lack of sufficient resources to support the educational tracks, and a lack of differential policies adapted to different sectors and populations.

This challenging and multifaceted situation requires carefully designed and professionally performed reform.

4. Action Plan for Promoting Mathematics and Scientific Excellence

The action plan takes into account the constraints discussed in the previous section and is accompanied by a budget that addresses the goals and objectives of the program. Most importantly, this action plan sets out clear, precise annual objectives. As mentioned above, the plan's goals do not restrict itself to simply increasing the number of students in advanced-level programs; there are wider objectives. The program assumes that increasing the number of students in advanced-level mathematics programs would also increase the number of students studying chemistry and physics at high levels (Today, the number of students at an advanced level in each subject has remained constant — approximately 8000 students each.) As stated previously, this action plan provides measurable and reasonable annual goals (see Table 1 and Fig. 2).

In order to increase awareness and motivation, lectures, meetings, and tours are being offered to students, parents, teachers, and the community.

Table 1. Quantitative goals.

Population	Objective	In 2012	Goal for 2019
Students	Students with motivation, ability to withstand pressures and hardships, and having intellectual curiosity and commitment	8869	17,738
Teachers	Increase in the number of mathematics teachers qualified to teach the advanced level, an increase in training and professional development, strengthening the mathematical and pedagogical knowledge with some of the teachers	Approx. 1000	1800
Schools	Addition of advanced-level programs for mathematics and science in all schools in the country	566	800

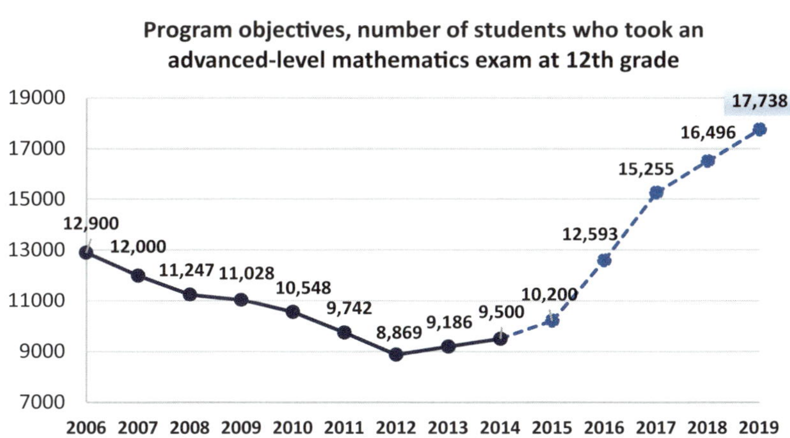

Fig. 2. Program objectives, number of students who took an advanced-level secondary school exam in mathematics by years.

Communities of teachers are being established with the help of leading educators. There is to be improved certification for active teachers, students are to be prepared for a degree in mathematics education, and engineers and graduates in the sciences are being encouraged to undergo teacher training (a teaching certification program for those holding an academic degree). B.Ed. graduates can undergo retraining to teach higher-level mathematics and science, and an "adopt a teacher" program has been initiated whereby a veteran teacher mentors a teacher just starting out with the advanced-level program. Roles are better defined, and mathematics department chairs in schools, teachers, teacher educators, and home-room teachers are offered professional development programs. Teaching assistants and support staff are recruited and trained.

Student support includes the following: 3000 additional hours of mathematics instruction to promote excellence in middle school per year, 3000 extra hours per year for individual instruction based on the number of students in 10th grade, and social involvement in mathematics. Supportive technological environments are also of central role in the advancement of the reform. This includes expansion of high school and virtual mentoring, experimenting with the use of Massive Open Online Courses (MOOCs), establishing a technology infrastructure for databases (teachers and students) on the school, district, and national levels, and increasing the amount of technological learning aids available (e.g., videos, practice sessions, demonstrations).

A steering committee was nominated by the Minister of Education which included representatives of the Ministry of Education, academia, the business sector, the non-profit sector, the military, and other stakeholders.

5. Intermediate Achievements

Although it is difficult to provide a definitive view of the program's success at this early stage, I think that its results so far are impressive and encouraging based on three factors: the change in public discourse, the change in attitudes and conceptions in the schools, and initial increases in the numbers of students and teachers. There is no doubt that the program has been well perceived by most of the media. Many articles have been written, most of which support the program and its goals. Some object to how mathematics is viewed quite differently from other subjects. The main change observed after the summer 2016 secondary school exams was a sharp rise in the number of students applying for advanced-level programs. The summer 2016 data show that 12,846 students took the advanced-level secondary school exam in mathematics.

6. Discussion and Conclusions

In my opinion, this program is an example of how a crisis situation at the national level — one with potentially serious consequences on different layers of society — can be used as a starting point to deal with the problem overall and change the crisis into an opportunity to promote and correct social issues. I believe this was due to several reasons: (1) correct analysis of the problem (many experts helped in analyzing the situation and determining the obstacles involved); (2) full support of Ministry of Education senior management; (3) tangible partnerships with outside agencies (e.g., the 5 × 2 initiative) that took part in the analysis and discourse, and took actual steps to help implement the program; (4) policy changes at administration level; (5) a significant budget increase; and (6) the program's holistic nature that included resources, teachers, a new policy, a change in the perception of teaching, and the creation of coalitions both within and outside the Ministry of Education.

I assume that the data will be in accordance with the defined objectives. I also assume that there will be some changes in the program and the various actions after gaining more experience. Some people fear that the program will lose its momentum due to political factors in the case of changes in personnel in the Ministry of Education. I, personally declare that I have no such concerns because the program has proven its importance regardless of one political viewpoint or another.

References

Ben Zvi, N. (2008). A report submitted to the Ministry of education (in Hebrew).

Harari, H. (1992). "Tomorrow 98": A report submitted to the Ministry of education (in Hebrew).

5 × 2 initiative (2015). About the initiative. Retrieved from: http://www.5p2.org.il/about-the-5x2-initiative/.

Israeli Ministry of Education (2014). *Teaching Mathematics* (in Hebrew). Retrieved from: http://www.mevaker.gov.il/he/Reports/Report_248/afbac200-9404-4280-8003-39e7d8d78953/224-ver-4.pdf.

Israeli Ministry of Education (2015). *Annual Reports*. Testing Division.
Lamm, Z. (2000). *Pressure and Resistance in Education: Articles and Conversations*, ed. Y. Harpaz. Jerusalem: Sifriat Poalim Press (in Hebrew).
Robinson, K. and Aronica, L. (2015). *Creative Schools: The Grassroots Revolution That's Transforming Education*. New York: Viking Press.
Taub Center (2015). Regarding the connection between number of units studied in mathematics and salary level. Retrieved from: Taub Center.

About the Author

Muhana Fares, Director of the National Programs Division in the Ministry of Education, has a BA in mathematics and statistics from the University of Haifa, and an MA in education systems management from the University of Haifa. In addition, he is a graduate of the Mandel Leadership Institute.

From 2009 to 2014, Fares served as the National Commissioner of Druze and Circassian education, and prior to that, he was Northern District Superintendent of Druze and Circassian education.

Fares founded and directed both the Yarka High School of Sciences and the community school in Hurfeish. Previously, he taught mathematics and computer science at several schools in the Galilee in Israel.

Chapter 3

A Start-Up Nation at Risk: Israel's Quest for Excellence

Eli Hurvitz

The Trump Foundation, Modi'in, Israel
ehurvitz@trump.org.il

Israel experienced a sharp decline in the number of its high achievers in mathematics in high schools between 2006 and 2012. The steep falloff placed the "start-up nation" at risk and was cause for alarm. This paper tells the story of how Israel awoke to this trouble and articulates the strategy that enabled it to reverse and restore its performance within five years.

1. Introduction

Human talent is Israel's most important resource. Israel takes pride in its scientists and entrepreneurs who have successfully built a small "start-up nation" and a striving economy in a very difficult region and against all odds. For a country under a continuous security threat and demanding military needs, staying the course with education, research and development is a remarkable achievement.

However, in 2012, this start-up nation was almost at risk. A sharp decline over less than a decade in the number of high school graduates majoring in mathematics and the sciences indicated an emerging problem. With less than 10% of students ranked "high achievers", the dangers of a drying pipeline of talent has turned the public eye towards this issue, which may shake the economy and affect the resilience of an already highly fragmented society.

This paper aims to tell the story of how Israel awoke to this growing challenge and its possible consequences. It will analyze the factors, which

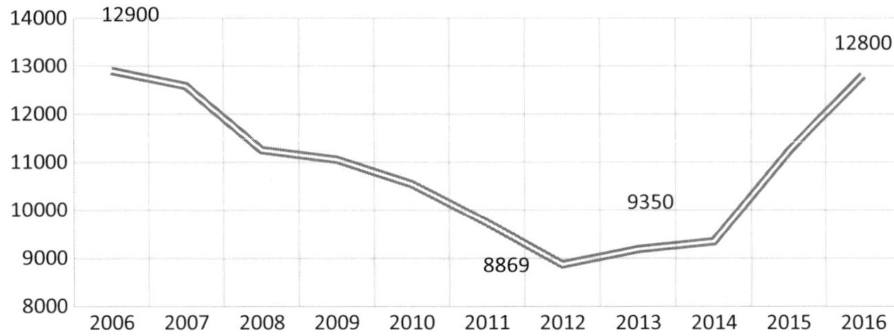

Fig. 1. Number of 12th grade students who passed the advanced-level secondary school exam in mathematics between 2006 and 2016 (Ministry of Education Data, 2017).

led to a negative trajectory and articulate the steps taken to reverse it. This unique endeavor was highly focused and aligned, leaning on a collaborative effort of government, philanthropy, academia, industry, schools and teachers, who decided to act on this issue together.

Five years into the effort, the decline has changed course and in 2017, more than 16,000 12th grade students are preparing for the advanced-level secondary school exam in mathematics. For Israelis, who are accustomed to mostly bad news, such a success is not easily digested. Local critics argue that the jury is still out on whether this positive transformation will endure over time and questions have been raised regarding its long term effects (Fig. 1).

2. The Decline (2006–2012)

Initially, there were many who dismissed the indicators presented in national and international tests and objected to the pessimistic forecast for Israel's educational performance in mathematics. They argued that mandatory army service, research universities and the vibrant start-up scene serve as centers of excellence and hubs for innovation that compensate for poor schooling.

However, as data accumulated and the media rung the alarm bells, mindsets began to shift. As a sense of urgency emerged, the challenge was to bring as many concerned stakeholders as possible to identify, understand and agree upon the root of this decline.

A collaborative effort was launched, convening round tables, instigating research and harnessing expertise, together, starting with an attempt to make sense of what was happening.

Three factors soon became clear. It became evident that high school students are not majoring in mathematics as before because standards were raised without notice nor preparation, veteran teachers were retiring in big numbers leaving a vacuum behind them, and the incentive eco-system for excelling students became ineffective (Hurvitz and Alon, 2015; Raz, 2016).

2.1. *Curriculum shift*

Graduation with a secondary school certificate in Israel requires taking a final examination in mathematics. In high school, there are three main levels of study: basic, intermediate, and advanced. Mathematics teachers in school are responsible for distributing students into these tracks, based on previous performance, assessment of abilities and student preference.

Due to "math wars" and red tape, the curriculum for these tracks, however, has not been formally amended in decades. As a result and as a matter of practice, the final exam became the policy driver of what is supposed to be taught and textbooks are reverse engineered to comply with the exam. Since the secondary school exam is not calibrated and its level of difficulty varies, the Ministry of Education found it convenient to use on the exam as a catalyst for introducing its changing policies.

Changes in the exams over the years reveal a significant evolution. Specifically since 2004, as modularity was canceled and the advanced-level exam started demanding more thinking and articulation and less technique. It emphasized transferability between mathematical representations, reasoning and applicability. Policy-makers explained these changes were correlated with cutting edge research and aligned with developments around the world (Arnovitz and Visman, 2014).

Expectations of teachers were elevated accordingly. Although the quota of teaching hours remained constant, teachers were expected to abandon teaching the curriculum in a linear manner and teach mathematics concepts in a spiral fashion instead. This shift, for which teachers and students were not prepared, has led to confusion and difficulty. In 2013, following an especially difficult final exam, a public protest led the Ministry of Education to slightly improve grades, while maintaining the high standards.

The intended outcome of this policy was the creation of a distinct gap between the intermediate and advanced level tracks. One may argue that raising the standard was indeed the right thing to do, if it had only been executed properly, starting with a formal school curriculum change. However, as a consequence of this policy, many students found the new hurdle impossible to overcome and decided to make do with a lower track.

2.2. *Teaching gap*

There are excellent mathematics teachers in Israeli schools. However, in the 2000s, a substantial number of them were approaching retirement age, including many who immigrated to Israel from the former Soviet Union in the 1990s. This severe shortage was reflected in closing study tracks in many schools, particularly the advanced-level classes. In 2010, of the 1129 schools in Israel in which students took secondary school exams, only 520 offered an advanced-level track in mathematics.

Adding to this was a highly selective, almost fixed mindset teaching approach, which was widely practiced. The high standards of the curriculum, which were not aligned with the number of teaching hours, led many teachers to teach at an accelerated pace, focusing on the material and giving less individual attention to struggling students. The inevitable dropdown of students to lower levels in fact provided some relief, as it resulted in smaller classes of only the very best students.

Many advanced-level students found the challenge overwhelming. The typecast of an Israeli advanced-level student is one that was sometimes raised to achieve effortless success. In middle school they excelled in almost every subject-matter, but suddenly they needed to invest time and confront difficulties and they did not have the grit for it. Nothing in their previous school experience had prepared them for this, and they were shocked to receive lower grades than usual. Female students dropped down epidemically, while the boys blamed others for it and then soon they too joined the intermediate level track.

As a result, the potential for excellence was leaking all across the country — in central Israel and in its periphery, among Jews and Arabs, religious and secular, girls and boys. Many students, who were able, but sometimes not resilient enough, were in need of a highly capable clinical teaching. But by this point, the teaching cadre was losing members on a daily basis, leaving a very difficult situation behind.

2.3. *Negative incentives*

For more than a generation, the eligibility for a secondary school certificate has been the end goal for high school education in Israel. This certificate is in most cases a prerequisite for higher education, which is considered an important key to well-paid employment. In 1990 however, only 30% of Israeli youth were eligible for a secondary school certificate and just 23% pursued an academic degree.

In order to expand access, the government decided to open regional colleges and close vocational training tracks. Local authorities and schools were measured against the rate of eligibility, with league charts published in the national press. The outcome of this policy has been a higher participation rate in education, reaching a world standard of approximately 50% in both certification eligibility and higher education (Trajtenberg, 2014).

Nevertheless, one of the unintended consequences of this success was the sacrificing of excellence. If a student was having difficulties in the advanced-level track, the school would have tried to avoid the risk of ineligibility, and would recommend them to drop down to a lower track. Students and parents believed that it is better to excel in the intermediate level than to struggle in the advanced.

They were not wrong in this widely-accepted assumption, because the universities were practically encouraging it. The registration incentives that universities provided to advanced-level graduates had eroded along the years. Parents and students were using online calculators provided by the universities to evaluate their prospects for acceptance to a favored department. They found out that the difference between intermediate and advanced levels had become almost insignificant.

The problem was that they did not take into consideration that times have changed. Since the labor market did not grow at the same pace as the number of graduates from higher education, employers became more selective. They started hiring university graduates from only the top rated departments, where acceptance criteria include advanced-level mathematics. In practice, secondary school certification eligibility was no longer sufficient to pave the road to future success; however Millennials, who were not aware of the changing circumstances, felt deceived (Kimhi and Horovitz, 2015).

3. The Awakening (2012–2017)

In order to turn the wheel and reverse the negative trajectory, philanthropy, government, academia, industry and a wide coalition of players developed a shared threefold strategy. Its first component was the teaching capacity which required long-term and deep investment in professional infrastructure. The second was the convening of stakeholders to forge agreement, alignment and collaboration. The third was the launch of a new government policy which energized and organized the joint endeavor and provided the necessary resources.

3.1. *Teaching capacity*

The number of university graduates of mathematics departments in Israel is much lower than the need for mathematics teachers in schools. In any case, upon graduation the majority of these young adults are interested in pursuing an academic or a high tech career. Therefore the pipeline for a new generation of teachers would have to come from a different route.

In 2012, following a popular social protest that galvanized the public, a new channel emerged comprised of career changers. At the age of forty, many high tech employees began to express their desire to give back to society and make an impact. They reported that they now cared more than before about work-life balance issues and sometimes wished to return to what they really wanted to do in life.

A number of teacher training programs were opened for hundreds of candidates, including school-based residencies, hybrids for maintaining academic and teaching careers, tailored M.Ed. programs, a special track in the Teach First Israel movement, and more. A designated placement agency tried to help several of them to find a teaching position and each new teacher was planned to receive two years of induction assistance, which includes individual and group coaching.

Concurrently, with philanthropic and government assistance, veteran teachers across the country took first steps to organize their professional development in learning communities and peer-led instructional coaching. With guidance of lead teachers and support from academic institutions, many interested teachers meet routinely to discuss evidence of student work and their own teaching practice in order to improve instruction. In some cases, diagnostic assignments, simulations and classroom-based videos are used to facilitate this clinical discussion.

3.2. *Cross sector collaboration*

Substantial resources have been invested in mathematics and science education in Israel over the years, by government, philanthropy, academia, industry and the Israeli Defense Forces. They fund multiple programs operated by non-profits, universities and colleges, businesses, local government and schools. These programs touch on various angles, including classroom and afterschool teaching, teacher training, motivation, parents' education, and more.

Instead of criticizing the negative national trajectory, these separate players decided to act jointly. A collaborative Collective Impact effort, titled "Five times Two" (hereafter abbr. 5p2), brought them together to study the causes of the decline, to landscape and map the field, and then to define shared vision, mission, goals and measures (Hurvitz, 2015).

3.3. *Government policy*

As soon as the professional capacity of teachers and the cross sector coalition of stakeholders were both in place, the government called all involved parties to join a new "National Program". The program defined the goal of doubling the number of advanced-level graduates within five years, reaching 18% of the matriculating cohort.

The program comprises several elements, including an increase in teaching hours to schools, a TV campaign calling on parents and students to choose the advanced-level track, and significantly raising the university registration bonus for advanced-level graduates. Members of the steering committee of "5p2" serve as the public board of the program.

This unique strategy is now at the stage of providing both first aid and steroids as well as taking steps to building professional and institutional muscles. It integrates a focused and measurable government policy, with convening and coordination among relevant players. However, its long term achievement will be determined in the coming years based on its success to creating solid professional foundations and entrusting teachers to further developing their clinical teaching practice.

References

Arnovitz, G. and Visman, I. (2014). *Analysis of Changes Made in the Physics and Mathematics Five Unit Matriculation Tests Between 1990 and 2014*, The Trump Foundation.

Hurvitz, E. (2015). Three moments of introspection: The story of the "5p2" initiative — The viewpoint of philanthropy, in *Collective Impact in Israel*, JDC-Israel, pp. 77–82.

Hurvitz, I. and Alon, S. (2015). The "5p2" initiative to expand excellence in science education — Insights after one year, in *Collective Impact in Israel*, JDC-Israel, pp. 84–97.

Kimhi, A. and Horovitz, A. (2015). *Do the Math*: *The Connection Between Math Matriculation Units of Study and Salary Levels*, Taub Center.

Raz, T. (2016). *National Program on Excellence in Mathematics*: *Evaluation Report*, National Authority for Measurement and Assessment in Education.

Trajtenberg, M. (2014). *The Israel Higher Education System*, Council for Higher Education.

About the Author

Eli Hurvitz is the executive director of the Trump Foundation, a philanthropic foundation (not related to the President of the USA) dedicated to mathematics and science education in Israel. He is a member of the Israel National Board of Education.

Chapter 4

Supervision of Mathematics Teaching by the Ministry of Education

Hannah Perl[*,‡], **Dorit Neria**[*,§], **Ruth Segal**[#,¶] **and Niza Sion**[*,‖]

[*]*Ministry of Education, Israel*
[#]*Kesher-Cham, Technion — Israel Institute of Technology, Israel*
[‡]*hannah.perl1948@gmail.com*
[§]*doritne@education.gov.il*
[¶]*rutisegal@gmail.com*
[‖]*sionniza@gmail.com*

Current major roles of the supervision of mathematics teaching are described. These include developing and implementing the national curriculum and national assessment; supporting mathematics teachers and improving instruction via mathematics instructors; teachers' professional development; working with the advisory committee; supervision and approval of mathematics textbooks and other teaching materials.

1. The Structure of the Israeli Education System

Education in Israel is compulsory from preschool (ages 3–5) until the end of high school at grade 12. The school system comprises three parts: 6 years of primary school (grades 1–6) followed by 3 years of middle school (grades 7–9) and 3 years of high school (grades 10–12).

Primary and middle school teachers are employed directly by the Ministry of Education. High school teachers are employed by municipalities and the private sector but are mostly funded by the Ministry of Education. Schools in the Israeli system are divided not only according to whether they are religiously oriented or not but also whether they are Hebrew speaking

Table 1. The Israeli education system — distribution of sectors.[a]

		Non-Jewish			
	Jewish	Arab	Bedouin	Druze	Circassian
Percent of Students	73.9%	18.7%	5.4%	2%	0.03%

[a]Data from: http://main.knesset.gov.il/Activity/Info/MMMSummaries19/Education_3.pdf.

Table 2. The Israeli education system.

	Primary school	Middle school	High school
Number of students	972,000	281,000	411,000
Number of schools	1800	800	1820
Number of classes	36,000	10,600	17,060
Number of teachers	64,000	27,000	

(secular, religious, and ultra-Orthodox) or Arabic speaking (Muslim, Christian, Bedouin, Druze, and Circassian). Table 1 shows the distribution of schools according to language, and Table 2 provides the distribution of the number of students, schools, classes, and teachers, generally.

2. The Structure of Mathematics Supervision in Israel

The supervision of mathematics teaching in the Ministry of Education is the responsibility of two chief superintendents — one in charge of pre-primary and primary schools (K-6) and the other in charge of middle and high schools (grades 7–12). Also included in the mathematics supervision staff is a mathematics curriculum specialist and mathematics instructors.

The role of the chief superintendents is to lead and oversee all aspects of mathematics education in the school system. Their broad responsibilities extend to curriculum development and its implementation, learning environments, teachers' mathematical and pedagogical development, assessment and approval of learning materials. The chief superintendents also collaborate with other departments in the Ministry of Education in matters relating to mathematics education but are under the jurisdiction of those other departments, for example, the departments of primary schools, secondary schools, teacher's education, etc.

The mathematics supervision staff belongs to the science division in the pedagogical secretariat of the Ministry of Education. In addition to the two chief superintendents there are two special sectorial superintendents of school mathematics: one for the Arabic-speaking schools and the other for the ultra-Orthodox schools. Although these superintendents belong to their respective sectors they are professionally directed by the chief superintendents and are an integral part of the mathematics supervision team.

The superintendents are accompanied by a professional advisory committee consisting mainly of representatives from academic institutions — mathematicians and mathematics education researchers, mathematics educators, teachers, school principals and the supervision staff. The committee advises the Ministry of Education and the superintendents' vision in policy issues, curriculum, and other central topics. The presence of representatives from universities and especially from the schools themselves gives an important input to decision making processing.

The supervision of mathematics funds two academic teachers' centers: one for primary school teachers and one for secondary school teachers (middle and high schools). These teaching centers provide the academic foundation for the continuous professional development of leading teachers, preparation of innovative teaching materials, editing the teachers' journals and special projects in collaboration with the chief superintendents.

In this short paper, we will focus the role of supervision in the following aspects of Israeli mathematics education: curriculum development and implementation; professional development of mathematics instructors; approval and direction of new teaching material; national and international assessment.

2.1. *Curriculum*

Teaching in all schools in Israel, K-12, follows a national curriculum. The curriculum is developed by special subcommittees of the advisory committee, appointed by the director of the pedagogical secretariat in the Ministry of Education. There are different committees for the different age groups: pre K, 1–6, 7–9 and 10–12. The committee members include mathematicians, mathematics education researchers specializing in the relevant age group, mathematics educators, leading teachers, and curriculum experts. In order for a revision of the curriculum, a new curriculum to be implemented has to be approved by the advisory committee, the relevant chief superintendent, the director of the science division and signed by the director of the pedagogical secretariat. The whole process of requesting a new or a revised curriculum and developing it until its approval is led by the chief superintendent, the mathematics curriculum expert, and the head of the advisory committee.

The national curriculum takes into account general principles such as vertically spiral progression along the years so that links from one grade to the next enable students to progress in their mathematical education, deepen their understanding and expand it. A central concern in designing the curriculum is mathematical rigor. Rigor in this context is not a matter of making mathematics harder or by introducing advanced topics at earlier grades, as is commonly assumed. Rather, it refers to conceptual understanding, procedural skills, fluency, reasoning and application (Alberti, 2012). Affective aspects are also taken into consideration, and the curriculum includes instructions of how to make mathematics lessons more exiting and relevant for the students.

The current primary school mathematics curriculum was published in 2006 and was gradually implemented. It comprises three major themes: arithmetic (whole numbers, simple fractions and decimal numbers), geometry and data interpretation and representation (measurements are included in all relevant topics). Every student in grades 1–6 must study at least 125 hours of mathematics every school year. At this stage, there is one curriculum for all students, with no ability grouping, with the exception of special needs students. Mathematically talented students receive extra enrichment in their schools with guidance from the mathematics supervision staff.

The new middle school mathematics curriculum (grades 7–9) was approved in 2010. It comprises three major topics covered in spiral fashion: number and quantity, algebra and geometry (transition from intuitive geometry to deductive formal Euclidian geometry). Probability and statistics are integrated in all relevant topics. Every student in grades 7–9 must study at least 150 school hours of mathematics each year. There is one core curriculum for all students and an expanded program for STEM bound students. The expansions are both in contents and in skills. Starting from the 8th grade, students can be directed into classes of different ability levels.

The high school mathematics curriculum is divided into three levels: basic, intermediate, and advanced. These differ from one another other in content (topics and depth), level of skills required and the number of school hours expected to be taught.

The basic level curriculum is designed for students who are not likely to major in mathematics, science, engineering or subjects requiring high level mathematics (in 2016, approximately 62% of the students were tested in this level). It is the minimal requirement for secondary school certification. This curriculum is aimed at developing mathematical literacy and knowledge of basic mathematical concepts and skills — mathematics and mathematical abilities that every high school graduate must have in the modern world.

The intermediate level is focused on mathematics needed for the social sciences (such as a special emphasis on statistics) and sciences which need an intermediate level of mathematics (in 2016, approximately 23% of the students were tested in this level).

The advanced level is the highest level available and is intended for students who are interested in advanced mathematics and in subjects that rely on high level mathematics. The mathematics contents at this level include advanced and abstract topics, and are aimed at developing a broader view of mathematics. The skills required are high level thinking and high performing skills. (In 2016, approximately 15% of the students were tested in this level).

The differences between the levels notwithstanding, the curriculum is designed to be continuous and coherent across K-12: it aims at gradually developing students' mathematical knowledge, skills and high level thinking by moving from basic concrete levels to more abstract thinking. For example, the concept of multiplication is developed step by step from primary to high school. In primary school, students study multiplication of natural numbers and progress to multiplication of fractions and decimal numbers. In middle school, they learn how to multiply negative numbers and algebraic expressions. Finally, in high school, they learn to multiply functions, vectors, and complex numbers. This progression is followed not only on a procedural plane but also on a conceptual plane — a revision — from multiplication as repeated addition to the more abstract construct which involves multiplying different kinds of objects.

Currently a new curriculum is being developed that, among other things, will include a fourth level intended for exceptional talented students. Before being approved, however, pilot studies will have to be carried out. Also during the development of the curriculum, drafts will be made public on the internet and comments from mathematics educators, teachers and other interested people will be taken into consideration.

The case of the new curriculum just described, illustrates how designing a new contemporary curriculum is an extremely complex and long process which presents many challenges. The committees have to consider all the varied aspects of mathematics teaching in the country, such as ministerial educational policies, different ability levels of students, teachers' mathematical and pedagogical knowledge, social ramifications, academic institutions' requirements, preparation of teaching materials, etc. The subcommittee also takes into consideration curricula from other countries and new developments in mathematics and mathematics education research.

2.2. *Mathematics teacher instructors*

Mathematics teacher instructors are an important part of the supervision team. Their role is to serve as a link between the chief superintendents and the mathematics teachers in the districts and in the schools. All mathematics teacher instructors are highly educated, experienced teachers, which have leadership qualities. Their role as teachers as well as instructors gives them a unique perspective of how the Ministry's policy is actually applied. Their opinion is highly valued in the processes of implementation of mathematics education policy and curriculum, defined by the Ministry of Education and the supervision.

Teacher instructors assist in implementing national goals and aims, teaching reforms, special projects, and other activities that promote mathematics education of all schools in Israel.

There are three main categories of teacher instructors: national, district, and school instructors. The number of instruction days at each category is allocated by the Ministry of Education and the administration in the districts, in accordance with the aims to be achieved and national goals and projects.

The national instructors work directly with the two chief superintendents and assist them in all their tasks and decision making processes regarding policy issues. They are the link between the supervision staff at the Ministry headquarters and the district mathematics instructors.

The district instructors are responsible for implementing the national mathematic policies and goals in their districts. They work collaboratively with the mathematics supervision staff in the Ministry's headquarters and with the district administration, in adapting the national goals to the needs and specifications of the districts. One of their major tasks is to lead the school instructors in their district.

The school instructors are guided by the district instructors to work directly with the mathematics teachers at schools, on a one-to-one basis, to improve the teaching and learning of mathematics in the school. They meet with the school mathematics team, usually once a month or more if needed. School instructors update the teachers with relevant information and assist the teaching staff with all mathematics teaching issues that arise in the school, observe mathematics lessons and give feedback to the teachers. The instructors do not have administrative authority (e.g. they cannot hire or fire teachers, nor can they make official recommendations in management issues), nevertheless, their opinion and input are highly respected. They are an important force affecting teaching practices in the classroom.

In all three categories there are sectorial mathematics instructors for the Arab speaking population and the ultra-Orthodox community. In the middle and high schools, the private sector provides their own instructors and supervisors for their schools. They are not appointed by the Ministry of Education, but work cooperatively with the Ministry's staff.

2.3. *Supervision and approval of mathematics textbooks and learning materials*

Textbooks are designed to reflect policy. They represent an interpretation of policy in terms of concrete actions of teaching and learning. A curriculum can be implemented only if it is accompanied with high-quality materials. Even more so, textbooks are the print resources most consistently used by teachers and their students in classrooms. Serving such a central role, textbooks can dictate the content of mathematics instruction in schools as well as the instructional approach of the textbook material (Reys *et al.*, 2003).

Because of the significant influence textbooks have on teaching, all textbooks and national learning materials must be approved by the Ministry of Education before they are allowed to be used in the classroom. In the pedagogical secretariat there is a special section responsible for approving textbooks and learning materials. This section works in full collaboration with the mathematics supervision staff. Learning material is more than printed text, so today all textbooks must also have a digital version for students, and a teachers' manual.

All mathematics textbooks (printed and digital) must adhere to the standards set by the Ministry of Education and the mathematics supervision, both mathematically and pedagogically. Writers must include all mathematical content and pedagogical directives defined by the national school curricula. Mathematics textbooks are evaluated by at least two experts: a mathematician and a mathematics educator. The approval process usually takes a few rounds of comments and corrections. A textbook is approved only after all evaluators approve the final version and the last approval must be given by the relevant chief superintendent.

2.4. *National and international assessment*

National assessment of mathematics in Israel consists principally in an exam called the *MEITZAV* given in the 5th and 8th grades, and the *Bagrut*, the national secondary school examination, generally given in the last two years of high school. Israel also participates in two international assessment programs: TIMSS at 8th grade and PISA.

2.4.1. *The MEITZAV*

The *MEITZAV* (Hebrew acronym for "School Growth and Efficiency Measures") is a system that includes student achievement tests as well as questionnaires designed to glean information about the school climate and pedagogical environment (administered to principals, teachers and students). The purpose of the *MEITZAV* at the school level is to provide school principals and teaching staff with data useful to help realize student potential, improving the pedagogical climate and enhancing instruction in school.

At the system level, the *MEITZAV* is intended to provide a snapshot of how well Israel's students have mastered the curricula of core subjects, including mathematics (Beller, 2013).

As already mentioned, mathematics tests are administered to students at two grade levels: grades five and eight. The achievement tests are designed in accordance with the mathematics curricula and are intended to examine the extent to which primary and middle school students meet the expected level required of them by the curricula (Beller, 2013). The *MEITZAV* tests are developed and administered by The National Authority for Measurement and Evaluation in Education (RAMA). The mathematics tests are written in collaboration with supervisors of mathematics and their staff.

Schools are tested once every three years in external tests (external *MEITZAV*). In the years in which the school is not tested externally schools may choose to take the same test internally (internal *MEITZAV).* In this case the schools evaluate it themselves and are not obliged to report the results to any external entity. Internal tests are accompanied by pedagogical guidance material for teachers. The school can use internal *MEITZAV* grades as they see fit for improvement and internal learning and as part of the annual assessment provided students. The process of teachers' grading internal *MEITZAV* tests enhances their professional development as it exposes them to professional indicators/rubrics that define expectations from students and enables them to learn from students' answers about their knowledge and comprehension levels (Beller, 2013).

2.4.2. *Matriculation exams*

The Israeli secondary school examinations are called "*Bagrut*" which literally mean "maturity". These examinations are critical for students who want to pursue further advanced education. In order to receive certification students must pass a series of tests in required subjects such as Bible or Koran, Hebrew, English, Civics, History, Literature, Mathematics — and a number of elective subjects, including the sciences and technology. The exams can be taken at a variety of levels but at least one elective has to be taken at the highest level. These tests can be taken at the end of the eleventh and twelfth grades (with several exceptions). Mathematics examinations are compulsory but can be taken at any one of three levels corresponding to the basic, intermediate and advanced level curricula, discussed above. At each level there are two separate written examinations which can be completed at different dates. The tests are national and are issued at specific dates, three times a year (January, June and July), under the supervision of the testing department in the pedagogical administration of the Ministry of Education. The tests are developed by special committees of teachers, mathematicians, mathematics educators and assessment specialists. The questionnaires must reflect the curriculum and its educational directives and must be approved by the chief superintendent of school mathematics. For evaluation, all tests are transported to a special assessment center where they are scanned and sent digitally to specially trained teachers for grading. The evaluation procedure, including training the teachers in assessing the tests, is led by the chief superintendent and the supervision team. The final grade the student receives is the average of the test grade and the school teacher's evaluation of the students work in class. Universities and other institutions of higher education give bonus entrance credits to students who complete the advanced level mathematics exam. This examination is also a necessity for students who want to study engineering, computer sciences and physics at prestigious universities.

2.4.3. *International assessments*

Israel participates in two international assessment programs: The PISA study, and TIMSS.

International studies make possible the comparison of student achievements across many countries in several subjects as well as the study of other educational issues. Furthermore, results of these studies enable comparison between different sectors and different population groups within each participating country (Beller, 2013).

The results of international surveys and assessments are carefully studied by the supervision team and the Ministry's policy makers, if necessary changes are proposed in educational policy, curricula and pedagogy. The 2003 findings of both international studies underscored the fact that the Israeli middle and high school curricula had to be updated, adding more mathematical literacy and high level thinking at all levels. It was also understood that in order for students to effectively deal with the new challenging tasks, teaching practices must change and the quality of teaching must improve. The Ministry's policy makers, the chief superintendent together with the instructors of all three categories led many changes and are still amidst the process. The results can already be seen. Israel's PISA and TIMSS achievement scores have improved over the years. In the present 2015 research cycle, there is no significant change in Israel's achievement scores as compared to those of the previous round of PISA research conducted in 2012 and 2011 TIMSS.

3. Regarding Future Policy

There are many challenges the supervision of mathematics has to face in the coming years. One of them regards the importance of teachers' knowledge. Many researchers have stressed relationship between teachers' knowledge and their students' achievements (Coe *et al.*, 2014). In mathematics education, it has been found that teachers' content knowledge, beliefs about mathematics teaching and learning, and knowledge of pedagogies and mathematics have positive influences on the effectiveness of teachers' pedagogies (Blömeke and Delaney, 2014). Yet, in Israel, as in many other countries (Jensen *et al.*, 2016), the training of primary school teachers includes minimal time developing expertise in mathematics. Recently, a committee of mathematicians and mathematics educators has debated on this issue and recommended changing the nature of training, and emphasizing the role of professional mathematics teachers in primary schools. The implementation of the recommendations will be challenging and will influence all primary school teachers training programs.

The dispersion of mathematics scores in Israel is one of the largest among the countries and economic entities participating in the international studies, and the largest among OECD countries. The main achievement gaps appear when comparing students from the two language sectors (Hebrew and Arabic speakers), and when comparing students from different economic, social, and cultural backgrounds. The percentage of top-performing students in all three areas of literacy was similar to the OECD average, while the percentage of low-performing students in all three disciplines was 1.5 times greater than the OECD average. These findings are being addressed at the highest level of the Ministry of Education and concrete efforts are already being made to reduce these gaps.

Two other major challenges are shared by other countries. During the last decade the number of students passing an advanced-level secondary school examination in mathematics has been steadily declining and, consequently, the number of STEM students in the universities, mostly in engineering, has also declined. In mathematics, the Ministry of Education has launched a national action plan to increase the number of students studying advanced level mathematics. As part of this plan, the Ministry has provided schools with extra teaching hours, innovative teaching materials, virtual mathematics classes, communities of teachers, school instructors and special mentors for the teachers. The numbers are already beginning to increase, but the improvement must continue further and be sustained afterwards.

A major handicap in this connection is the lack of well enough qualified mathematics teachers able to teach advanced level mathematics in high school. The number of teachers receiving teaching certificates from universities is inadequate at present. For this reason, special teacher training programs have been initiated, including teaching certification programs for those holding an academic degree, and sending less qualified teachers back to the universities to complete their mathematical and pedagogical education. Despite these efforts, the Ministry recognizes that for the long term a more structured strategy will need to be designed.

References

Alberti, S. (2012). Making the shifts, *Educational Leadership*, **70**(4), 24–27.

Beller, M. (2013). *Assessment in the Service of Learning: Theory and Practice*, Ramat Gan, Israel: RAMA.

Blömeke, S. and Delaney, S. (2014). Assessment of teacher knowledge across countries: A review of the state of research, in *International Perspectives on Teacher Knowledge, Beliefs and Opportunities to Learn*, Netherlands: Springer, pp. 541–585.

Coe, R., Aloisi, C., Higgins, S. and Major, L. E. (2014). What makes great teaching? *Review of the Underpinning Research*, Retrieved from: http://dro.dur.ac.uk/13747/.

Jensen, B., Roberts-Hull, K., Magee, J. and Ginnivan, L. (2016). *Not So Elementary: Primary School Teacher Quality in Top-Performing Systems*, Washington DC: National Center on Education and the Economy.

Reys, R., Reys, B., Lapan, R., Holliday, G. and Wasman, D. (2003). Assessing the impact of "standards"-based middle grades mathematics curriculum materials on student achievement, *Journal for Research in Mathematics Education*, pp. 74–95.

About the Authors

Hannah Perl was the director of the Science Division at the Ministry of Education Pedagogical Secretariat during the years 2011–2016. Previously, from 2003–2011 she served as the chief superintendent of School Mathematics for grades K-12. Until 2003 she was a mathematics teacher, a teachers' instructor and a member of the mathematics staff at the Science Teaching Center at the Hebrew University of Jerusalem. Hannah holds BSc and MSc degrees in mathematics and a PhD in science teaching, all from the Hebrew University.

Dorit Neria is the chief superintendent of School Mathematics in Preschool and Primary Schools (K-6) in the Ministry of Education. She also teaches at the Department of Science and Technology Education at Ben-Gurion University. Dorit has been a mathematics teacher since 1989. In her PhD she researched cognitive aspects of mathematically talented students.

Ruth Segal received her PhD in mathematics education at Technion in 2013. She is a lecturer at Shaanan Religious Academic College of Education and at Oranim Academic College of Education. She taught high school mathematics for 26 years. She integrates technology in mathematics instruction and research. She also served as the national teachers' instructor for high schools (grades 10–12) in Israel. Presently she also works at Project RAMZOR, Kesher-Cham R&D Center, Technion — Israel Institute of Technology.

Niza Sion is the national teachers' instructor for all middle schools (grades 7–9) in Israel since 2000. Together with the chief superintendent she leads all aspects of mathematics education at that level: curriculum development, assessment, and teachers' professional development and guides other middle school instructors. Niza has been a mathematics teacher since 1986 and has master's degree in education specializing in developing national programs.

Chapter 5

Mathematical Education in Israeli Religious High Schools

Thierry (Noah) Dana-Picard[*,§]
and Sara Hershkovitz[#,¶]

[*]*Jerusalem College of Technology (JCT), Jerusalem, Israel*

[#]*Center for Educational Technology (CET), Tel Aviv, Israel*

[§]*ndp@jct.ac.il*

[¶]*sarah@cet.ac.il*

The religious population in Israel is heterogeneous; consequently, there are several kinds of high schools catering to this population. We give a schematic description of the population and of the schools, then present trends at the different levels of study. We relate briefly to outstanding and gifted students, using yeshiva high school for gifted students at JCT as an example.

1. Introduction

Despite popular opinion, the religious population in Israel is not homogeneous. Thus, the variety of educational institutions catering to this population is extremely varied.

There are two main distinctions:

(1) Level of religious observance: the first distinction is according to a socio-religious criterion. Broadly we can identify the following main categories: (a) Traditionalist, (b) Modern-Orthodox, (c) Orthodox, and (d) ultra-Orthodox (called Haredi in Hebrew). By Orthodox, we mean people who live according to the Jewish religious laws. There are various levels of observance, actually we could represent the different levels not as a discrete set as we did, but as a continuum, which would make our

Table 1. Initial classification of religious high schools.

	Traditional	Modern-Orthodox	Orthodox	Ultra-Orthodox
Co-ed	X	X	—	—
Boys only	X	X	X	X
Girls only	X	X	X	X

study impossible. These levels imply also specific conceptions regarding education.

These categories are rather coarse, they contain subcategories, but they enable us to describe features of mathematical education in the corresponding institutions.

(2) Co-education:

  (a) Is the institution co-ed?
  (b) If not, which segment of the population does the institution cater to? Boys only or girls only?

We summarize the answers in Table 1.

As the categories are not well-defined (in the mathematical sense), we will not translate the table into percentages. Nevertheless, note that for the ultra-Orthodox population, co-education is inexistent, but for what we call the traditional all three possibilities exist.

For every category of the religious population there may exist different kinds of high-schools:

(1) Talmudic high schools for ultra-Orthodox boys. No secular studies are taught beyond primary school. Thus, their mathematical education is extremely minimal and they are not candidates to secondary school certification.
(2) High schools for ultra-Orthodox girls. Here the students learn mathematics up to a certain level, depending on sociological reasons. In the past, the vast majority of these girls took the basic level secondary school exam in mathematics. Today a greater number of girls study advanced-level mathematics. Mathematics education is organized at three levels (basic, intermediate, and advanced (see Appendix)).

(3) Yeshiva high school (boys) and Ulpana (girls) for Orthodox students. Half of the time is devoted to religious studies and half to secular studies. The pace of mathematical education, is similar to general high schools, but again here we see in some of these schools (particularly in the boys schools) learning towards the secondary school exam in a concentrated format.
(4) Comprehensive religious high schools. These schools parallel general high schools in Israel with adaptations in the curriculum to meet the requirements of modern-Orthodox and traditional populations. Here also there exist subcategories regarding mathematical education.

The schools have some freedom to organize their perennial program, according to specific educative goals. In this chapter, we relate to the traditional and modern-Orthodox population. The ultra-Orthodox population is studied in Chapter 6 in this book.

The different levels for mathematical education in high schools (basic, intermediate, and advanced) are described in other chapters of this book; we do not elaborate on them here. As secular schools are under the authority of the Ministry of Education, so are religious schools. The majority of religious high schools belong to a national chain, the predominate chains are Amit, Bnei Akiva, and Noam-Tzvia.

2. General Trends in Mathematics Learning in Religious Schools

In this section, we relate to high-schools for the traditional, and modern-Orthodox populations. In these schools, mathematics is taught according to the national school mathematics curricula. Nevertheless, note that a certain unknown percentage of religious students studies in non-religious institutions.

We rely on data originating from the Ministry of Education.

The percentage of religious students learning basic-level mathematics was stable along the years, showing a slow increase until 3 years ago. In parallel their achievements improved (from an average grade of 78.5 in 2011 to 79.7 in 2014). At the same time, the percentage of students learning at the intermediate-level showed a low in 2013, but in 2014 restored its 2011

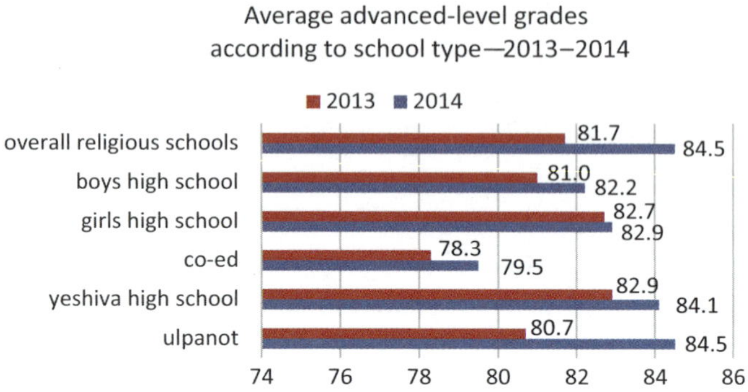

Fig. 1. Data dispatching according to school type.

level; over this period the average grade improved significantly, from 81.3 in 2011 to 83.2 in 2014.

In parallel, the percentage of religious students learning at the advanced-level decreased from 10.3% in 2011 to 9.5% in 2014. The average grade increased from 83.6 in 2011 to 84.0 in 2014, showing perhaps an improved motivation to learn mathematics, which will be useful at the entrance to academic studies.

Efforts have been made nationwide, to draw more students to learn mathematics at the highest level. These efforts have been fruitful for the religious population. Not only did more students study at this level, but their achievements improved as well. Figure 1 shows refined data, dispatched according to the different kinds of religious schools. The increase in the girls' achievements is noticeable. It may reveal a new approach to scientific studies, which had formerly a low appeal.

3. Present and Future Developments

3.1. *Institutional culture*

In most yeshiva high schools and Ulpanot (plural for Ulpana), mathematics education follows the same pathway as in secular schools. The percentages of students studying the advanced-level curriculum may be somehow lower than the general population. This is due generally to the fact that these students also study in an advanced curriculum other subjects unique for the religious population. Time and energy are not infinite, so choices have to be made, either by the student or by the institution. This is an original part of what Artigue (2002) calls *institutional culture.*

3.2. *Development of specific materials adapted to the population*

A few projects exist to develop specific textbooks for the Orthodox and ultra-Orthodox populations. This is true already at the primary level, where the first notions on numbers and arithmetic are taught together with the notion of Gematria: in traditional reading of biblical texts, a numerical value is given to the letters according of their ordering, then a numerical value is given to words and sentences, by addition of the values of the letters. It is very important to note that textbooks and the various learning materials need to be adapted for the religious community. The problems presented and the visuals need to be appropriate for the student population being served. For example, an exercise beginning with "X students went to the cinema, Y students went to the beach and Z students went to both the beach and the cinema. . ." may not be acceptable for a great part of the religious population who do not go to movies. Adaptation to the general culture of the population is based on three principles:

(1) Take the sociological context and its cultural part into account. For example, the synagogue is a central place in life.
(2) Use the names currently met in this population, mostly biblical names.
(3) Use graphics suitable for the population. For example, the boys presented on the pictures wear a yarmulka and the girls wear long skirt.

New trends in the ultra-Orthodox and in the modern-Orthodox population appeared in the last years. Students who learnt only basic mathematics in high school wish to enter academy towards degrees which request a higher level in mathematics, such as engineering curricula. Their lacunas in mathematical knowledge are filled while learning in ad hoc programs such as pre-academic studies, existing in universities and many colleges.

In these programs, an effort is made to adapt the way the curriculum is taught to the cultural environment of the students. Dana-Picard and Hershkovitz (2017) describe the relevance of studying mathematical aspects of pieces of Jewish Art (e.g., architecture and Judaica) to the national school mathematics curriculum, building bridges for Orthodox students between mathematics and their own world. By that way, the transition from Talmudic studies to mathematics is made easier; after all, numerous mathematical tools are needed to understand the Talmud. It must be noted that in many occurrences, the Sages of the Talmud had to build mathematical tools which did not exist at that time, but elaborating on this remark is beyond the scope of this chapter.

Today, educative programs may not be aimed at learning mathematics for their intrinsic interest, even not as a monument which has to be visited (Chevallard, 2015), but for their practical value to obtain a diploma and then join the work force and make a living.

3.3. *The usage of technology*

Today technology is ubiquitous; therefore students must be acquainted with it very early, for developing new knowledge and applied skills, which are part of the general mathematical culture. Internet resources are used on a low scale, as large parts of the religious population is reluctant to have this at home, but in school and beyond Computer Algebra Systems and Dynamic Geometry Systems are more widely accepted and used. Following Artigue (2002), we expect from software and computational tools to be pedagogical instruments. The tools should help to avoid practices too much orientated towards pure lecturing and procedural work, and rather push towards more profound understanding and acquisition of new mathematical skills.

Regarding a population learning Jewish traditional studies in the same institution in a more or less integrated framework, the mutual influence of both fields may enhance more profound understanding, the Talmud providing practical situations and the technology-based mathematics providing tools for understanding Talmud. Research only begins in this domain.

Using Artigue's taxonomy, working with technology should not be aimed at achieving only *pragmatic value* (procedural technical work), but also *epistemic value* (mathematical understanding). The new technology skills are part of the new mathematical knowledge acquired by the students.

As an example, we mention the *Torah u-Mada* (Torah and Science) program for gifted students established over 18 years ago at JCT, a yeshiva high school. A similar program for girls, *Talia*, has been opened 8 years later, with modifications to fit religious girls' education. Together with their high school program, students may learn academic courses towards a B.Sc. in various applied fields. A description is given by Dana-Picard and Komar (2008).

The academic courses are delivered either in a special class (e.g., Calculus and Linear Algebra) or together with regular JCT students (e.g., Differential Equations). The courses are delivered in technology-rich environment. The teams teaching mathematics and Jewish studies have non-empty intersection.

This year a new program has been opened for gifted middle school students. The mathematics program is taught in a technology-rich environment, using in particular the curriculum developed by the Center of Educational Technology (see Chapter 39 in this book).

It seems that opening joint programs of high school studies together with academic studies is becoming popular and will be ubiquitous in the next years. These programs are generally built with a university or a college located in the same area. Some of them work with the Open University. The *Torah u-Mada* yeshiva is a kind of exception as it is located on the campus of the Jerusalem College of Technology. Of course, the mathematics courses are the core of these programs.

4. Final Observations

Overall, the traditional and modern-Orthodox populations are studying similarly to the secular population. However, the number of students, especially boys, studying mathematics at the advanced-level is low. As most high school graduates do not carry on immediately to higher education, due to military service (or national civil service), a trend exists to study mathematics at a low level, and if needed after military service they return and complete the needed level according to the choice of university degree they are applying for. This is noticeable in the religious boys' milieu as for them the elapsed time in military duty is usually longer.

Nowadays, the different components of the religious population show very quick changes. The relation to educational needs seems to proceed towards a greater involvement in mathematics, science, and technology.

References

Artigue, M. (2002). Learning mathematics in a CAS environment: The genesis of a reflection about instrumentation and the dialectics between technical and conceptual work, *International Journal of Computers for Mathematical Learning*, **7**, 245–274.

Chevallard, Y. (2015). Teaching mathematics in tomorrow's society: A case for oncoming counter paradigm, in S. J. Cho (ed.), *The Proceedings of the 12th International Congress on Mathematical Education*, Dordrecht: Springer, pp. 173–187.

Dana-Picard, Th. and Komar, M. (2008). Academic studies for gifted high school students, in R. Leikin (ed.), *Proceedings of the 5th International Conference on Creativity in Mathematics and the Education of Gifted Students*, The Center of Educational Technology (Tel Aviv). Available: http://cmeg-5.edu.haifa.ac.il/Part%206%20-%20PP-Cmeg5%20proceedings.pdf

Dana-Picard, Th. and Hershkovitz, S. (2017). Glimpses at mathematics and Jewish art, to appear *in The Symmetry Journal*.

About the Authors

Thierry (Noah) Dana-Picard has two PhDs, one from the University of Nice (France) and one from Bar-Ilan University (Israel). For many years, he has been involved in teacher training. He is now a professor of mathematics at Jerusalem College of Technology, head of chair for Education, Mathematics and Judaism at JCT. He is President Emeritus at JCT.

Sara Hershkovitz, PhD, is head of Mathematics Department, Center for Educational Technology. She is a professor at Shaanan Religious Academic College of Education, Israel.

Chapter 6

Excellence in Mathematics in the Ultra-Orthodox Community: Fantasy or Reality?*

Reuven Gal[#,§], Yehuda Morgenstern[#,¶] and Yael Elimelech[‡,‖]

[#]Samuel Neamann Institute for National Policy Research, Israel
[‡]Manhi' — Jerusalem Municipality Educational Authority, Israel
[§]reuvgal@sni.technion.ac.il
[¶]yehuda.morgenstern@sni.technion.ac.il
[‖]yaelim20@gmail.com

In this chapter, we describe the current ultra-Orthodox (*Haredi*) education system, with special emphasis on mathematics studies. We then illuminate some of the recent transformations in this regard, both within the Yeshivot of the Haredi men, as well among the Haredi women's education system, predominantly the Beit-Ya'akov schools. We conclude by forecasting, cautiously, an imminent shift towards excellence in mathematics among various sections in the U-O society.

1. Introduction

In early 2011, the ultra-Orthodox (or *Haredi*) population in Israel[a] consisted of 833,500 people, which represented 10.8% of the total Israeli population (7.7 million), or 13.6% of the Jewish population in Israel (6.13 million). Of these, 63.4% were under 20 years of age (Ben Moshe, 2011).

According to the "Israel 2028" report, in 2028 the ultra-Orthodox will account for more than a fifth of the Jewish population in Israel. The

---

*This chapter is derived from a field study conducted at Samuel Neamann Institute and solicited by the Trump Foundation, Israel. See Gal, Morgenstern and Elimelech (2016).

[a]There are many (predominantly — four) methods for defining and estimating the size of the ultra-Orthodox Jewish population in Israel. See Friedman *et al.* (2011).

percentage of Haredi pupils in first grade will then amount to 40% (Brodet, 2008).

The Haredi education system is based on consecrating excellence in studying the *Halakha* (the collective body of Jewish religious laws). In fact, contrary to other education systems in Israel, the Haredi education system promotes a one study track, backed by one ideology — Torah studies are at the heart of excellence. This system structure stems from two main reasons: first, the *mitzvah* (commandment) of Torah studies, compulsory for every religious Jew, but mostly applied in the Haredi society; second, keeping the young generation inside the Haredi society. Consequently, the pursuit of excellence is directed solely toward the studies of the Holy studies; the rest — foreign languages, science, and history — have been deemed secondary, unnecessary, and in some places even forbidden.

However, in the last decade, the Haredi society is undergoing a major transformation. The technology and communication revolution have shaken the walls of the Haredi society and deeply influenced it. In addition, economical changes within the Haredi society and various government policies (e.g., reinforcing job-seeking and academic studies) have had their impact as well. Consequently, more and more Haredi men and women are now integrated into the workforce, higher education is no longer a taboo, and systems of studying and training are developed independently.

2. Current Tendencies in Haredi Education Systems

The common categorization in the research literature related to the Haredi society comprises three groups: *Misnagedim* (Lithuanian Jews), *Hasidic* Jews and *Sephardic* Jews. While in general this is the most accurate segmentation, for the purpose of this chapter we discuss a different categorization: (1) women; (2) men — mainstream (or conservative); and (3) men — other streams.

The number of students in the Haredi education system (from primary school to high school) in 2012 was 271,703 male and female Haredi students, about 45 percent more than in 2003 (188,081) (see Malach, Hoshen and Cahaner, 2016).

In the 2014–2015 school year, there were 4319 senior year Haredi students who took the secondary school exams in mathematics (3205 women; 1114 men). Figure 1 presents the percentage distribution of students taking

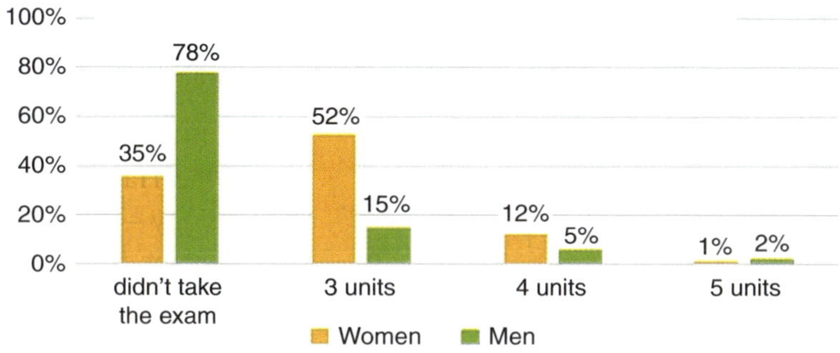

Fig. 1. Percentage distribution of students in the secondary school exam in 2014, by exam level.

Source: Data received from **Nerit Katz**, Ministry of Education.

the exam in 2014, by exam level. No more than two percent (22 boys and 32 girls) took the advanced-level exam, and few took the intermediate-level exam as well. Over one-third (35%) of the female Haredi students and the majority of the male students (78%) did not take the secondary school exams in mathematics at all. Interestingly, the rate of excellent Haredi students (who scored at least 85% in the exams) was about 50 percent — a rate similar to the average in the general population.

3. Education System for Men: Current Situation

3.1. *Talmud Torahs*

Talmud Torahs (*Cheders*, or *Heders*) are primary schools (grades 1–8) for boys in the Haredi education system. The main purpose of these institutions is to prepare the children to Haredi life, since education is the main tool for the preservation of legacy and tradition in the Haredi society.

The minimal secular studies that are provided are grammar, reading, writing, basic English, and elementary arithmetic. Arithmetic studies include the four operations, fractions, and percentages. However, even this "mathematical base" is shaky, since it does not develop fluency in the multiplication table, division, percentages, and decimal fractions. In some cases, arithmetic studies can be extended and improved, depending on teachers' will and capabilities (Gal, Shahaf and Morgenstern, 2012).

3.2. *Teachers*

The issue of teachers for secular studies is very problematic. At present, many are *Yeshiva* students with general training in teaching, yet some teachers have no training at all. Some were born abroad, in English-speaking countries, or had their earlier education in the general system.

The situation improved to some extent recently, after new educational programs for Haredi teachers were launched, with full training in mathematics and natural sciences. In 2013–2014, fourteen such training programs were available for Haredi teachers, which certified 830 Haredi female teachers and only 50 Haredi male teachers (Malach, Cahaner and Regev, 2016).

3.3. *Mainstream yeshivot*

Yeshiva ketana (*small yeshiva*) is the high school institutional leg of the Haredi education system, for post-*Bar Mitzva*-age students (ages 13 through 17.5, on average). Thereafter, they learn in a *yeshiva gedola* (*large yeshiva*), which usually involves a dormitory, until they get married. Then, they live with their wives and families, while continuing their studies in a *Kollel* (*gathering*).

Upon arrival to a *small yeshiva*, the young boy dedicates his time exclusively to Torah studies. Mainstream *yeshivot* rarely effectively integrate secular studies — especially mathematics — with holy studies. This is due to an overarching objection among all segments of the Haredi society. Some believe that the intensive nature of religious studies facilitates students' mental capacities and improves their learning-curves in general, thus enabling a faster attainment of mathematics at a later point in time. This hypothesis, however, has not been empirically validated yet.

3.4. *Integrated yeshivot*

Alongside the traditional mainstream *yeshivot*, there are other Haredi institutions which integrate science-related subjects. Such institutions include high-school *yeshivot* (which typically are not exclusively Haredi), Modern Orthodox *yeshivot*, *yeshivot* for Anglo-Saxons, and vocational *yeshivot*. Contrary to mainstream *yeshivot*, these *yeshivot* conduct secular studies, and some even prepare students for the secondary school exams. However, few such *yeshivot* exist, because Haredi leaders oppose increasing their number.

4. Education System for Women: Current Situation

Haredi girls graduate from primary school after seven years and start the 8th grade in a school system which is equivalent to middle and high school. After graduation, at eighteen, most continue to study for two more years in a seminar (grades 13–14), predominantly in order to obtain a teaching certification. Officially, only these final two years are structured as a teaching-training seminar. However, the entire system, starting from the 8th grade, is called "seminar". Most of these institutions belong to the well-known chain named *Beit Yaakov* (*House of Jacob*). A parallel education network, dedicated for the Sephardic sector, is called "*Ma'ayan HaChinuch HaTorani*" (*The Fountain of Religious Education*).

The total number of Haredi female students has increased from 90,425 in 2003–2004 to 132,670 in 2012–2013. Of those, the number of students in secondary education (i.e., middle and high schools) increased from 37,000 to 55,000, respectively (Malach, Hoshen and Cahaner, 2016).

Secular studies take place in grades 9–12 and include mathematics, English, language studies, history, literature, natural sciences, and geography. However, the large majority do not take the official matriculation exams, recognized by the Ministry of Education. Instead, each year about 4500 Haredi girls take the Szold Certificate Examinations (also called "the external exams", and named after their creator, the Henrietta Szold Institute), according to an arrangement existing for many years.[b]

Figure 2 presents the percentages and the absolute numbers of Haredi girls, in 12th grade, who took the basic, intermediate, and advanced-level secondary school exams in mathematics, between 2009 and 2014. As can be seen, the total number of girls who took the mathematics exam (across all three levels) has slightly increased over the years; however percentage-wise, there was no significant change since 2009.

Following the 2012 ruling by the Ministry of Education to officially recognize the Szold Certificate as equivalent to the secondary school certificate,

[b] In 2012, the Ministry of Education recognized the Szold Certificate as equivalent to the secondary school certificate.

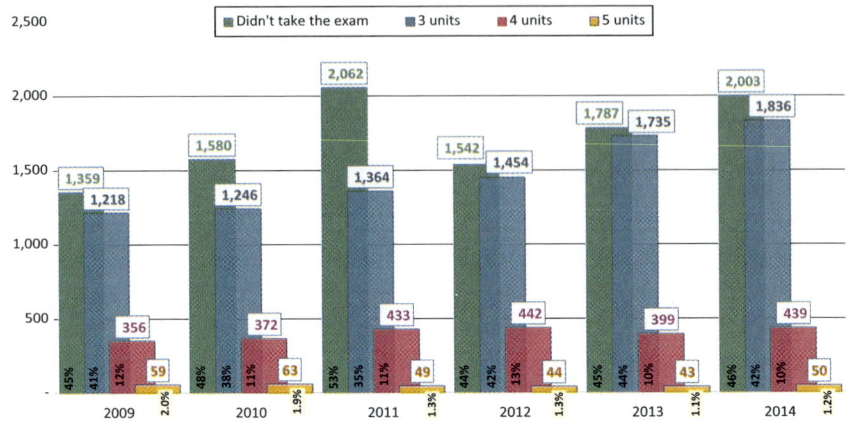

Fig. 2. Number and percentage of girls who took an exam in mathematics in Haredi education system, 2009–2014.

Source: Data received from **Nerit Katz**, Ministry of Education.

many Haredi high schools for girls encouraged their students to take the secondary school exams. However, only a very small percentage of these students took the advanced-level exam in mathematics. Notwithstanding, our field study (Gal, Morgenstern and Elimelech, 2016) shows that these numbers will significantly grow in the next couple of years, for two reasons: Several high schools only recently began teaching mathematics, and their students haven't began taking the exams yet; and a significant increase of certified female teachers will lead to a growth in the number of classes preparing female Haredi students for the advanced-level secondary school exam in mathematics.

5. Conclusion

The Haredi community in Israel is undergoing a significant transition. Particularly, its education system is transforming quite quickly. Assuming that in the next decade, greater numbers of Haredi students will be reaching high school age, we may cautiously predict increasing numbers of students studying mathematics and taking exams. While for Haredi men this will occur through the growing numbers of alternative *Yeshivot*, Haredi women will reach this level through their main education system — the Seminars. The ethos of excellence, treasured for generations through Torah studies, may also become a leverage for excellence in mathematics and sciences for Haredi students who care to live by *Torah im Derech Eretz*.[c]

References

Ben Moshe, E. (2011). Change in Israeli population structure and composition by cultural-religious sector in the next twenty years — and its implications on the labor market. Jerusalem: Ministry of Industry, Trade and Labour, Department of Research and Economy (in Hebrew).

Brodet, D. (ed.) (2008). *Israel 2028: Vision and Strategy for Economy and Society in a Global World*. Haifa: Samuel Neamann Institute, Technion (in Hebrew).

Friedman, I., Shaul-Menna, N., Fogel, N., Romanov, D., Amedi, D., Feldman, M., Shayek, R., Shifris, G. and Portnoi, H. (2011). *Measurement and Estimates of the Population of Ultra-Orthodox Jews*. Jerusalem: Central Bureau of Statistics (in Hebrew).

Gal, R., Morgenstern, Y. and Elimelech, Y. (2016). *Field Study on the Potential for Excellence in Mathematics and Physics within the Ultra-Orthodox Education System in Israel*. Haifa: Samuel Neamann Institute for National Policy Research, Technion (in Hebrew).

Gal, R., Shahaf, M. and Morgenstern, Y. (2012). *Report on EMCS Studies (English, Mathematics, and Computer Science)*. Haifa: Samuel Neamann Institute for National Policy Research, Technion (in Hebrew).

Malach, G., Cahaner, L. and Regev, A. (2016). *The Five-Year Plan of CHE-PBC for the Haredi Population for the Years* 2012–2016 — *Evaluation Study and Recommendations for the Next Five Years* (2017–2021). Jerusalem: Council for Higher Education, Planning and Budgeting Committee (in Hebrew).

Malach, G., Hoshen, M. and Cahaner, L. (2016). *The Yearbook of Ultra-Orthodox Society in Israel* 2016. Jerusalem: Israeli Institute of Democracy, Jerusalem Institute of Israel Studies (in Hebrew).

About the Authors

Reuven Gal is a senior research fellow at Samuel Neamann Institute and head of the Ultra-Orthodox Integration Project.

[c]*Torah with "the way of the land*" is a philosophy within Orthodox Judaism which advocates a synergy between traditionally observant Judaism and the modern world.

Yehuda Morgenstern is a member of the Gur Hassidic community, a graduate of the Technion (BSc in Industrial and Management Engineering) and a research fellow at the Samuel Neamann Institute. He is the co-founder and director of a system dedicated to integrating Haredi students into universities and coaching them throughout their studies.

Yael Elimelech, an Ultra-Orthodox Jerusalemite, received her BA in Business Management and an MBA in Organizational Consulting, both from Kiryat Ono College. She is currently the head of pedagogical development of ultra-Orthodox schools at "*Manhi*" — Jerusalem Municipality Educational Authority.

Chapter 7

Mathematics Education in the Arabic-Speaking Sectors in Israel

Shaker A. Rasslan* **and Amal Sharif-Rasslan#**

The Arab Academic College of Education, Haifa, Israel

**shaker1953@gmail.com*

#amalras@arabcol.ac.il; amal.rasslan@gmail.com

This chapter describes mathematics education for Arabic speakers (about 21% of the total population) in Israel. Mathematics education among Arabic speakers is derived from, and directly related to, the mathematics education of the majority of Hebrew speakers. We discuss mathematics education outcomes (MEITZAV, secondary school exams, PISA, and TIMMS) and examine the new mathematics curricula and mathematics teacher training (in-service and pre-service) in primary school, middle schools, and high school. The overall picture is not encouraging given the disparity between Hebrew and Arabic speakers. However, since 2013 there has been a positive change in terms of the M.Ed. in mathematics education curriculum in Arab-sector teacher training institutions and an increase in the number of students enrolled in the advanced-level track in mathematics.

1. Introduction

The population of the State of Israel stood at 8630.9 thousands in 2016 (CBS, 2017, Table B/1), and at 8463.4 thousands in 2015, which broke down into 74.84% Jews and 20.77% Arabs, and 4.38% "others". The distribution by religion corresponds to 74.84% Jews, 17.54% Muslims, 1.96% Christians of whom 1.54% are Arab Christians, 1.62% Druze, and 3.97% which are not classified by religion (CBS, 2016, Chapter 2, Table 2.1, Table 2.13).

There are no significant differences in terms of *gender* sectors, and the rate is about one-to-one (CBS, 2016, Chapter 2, Table 2.3). The greatest differences between sectors lies in socio-economic status, since the Arab sector in Israel is the lowest, with more than half of all families and more than 50% percent of all children living below the poverty line (The Institute of Insurance Report, 2016). The number of schools, classes and students in primary education (2015/16) were 3326, 30,565 and 740,414 in the Hebrew-speaking school system and 621, 9827 and 245,800 in the Arabic-speaking school system. In secondary education (middle and high school) there were 1567 schools, 20,503 classes and 515,353 students in the Hebrew-speaking system and 433 schools, 7180 classes and 191,580 students in the Arabic-speaking system.

1.1. *Students*

Secondary school graduation exams are a significant threshold in the learning process in Israel, which increases the chance to integrate in higher education. In 2015, the percentage of examinees eligible to receive certification was 73.8% (77.8% in Hebrew education against 61.4% in Arab education). In *Hebrew education*, this percentage broke down into ultra-Orthodox 33.9%, religious state education 80.2% and state education 81.8%. In *Arab education*, it broke down into Christians 75.6%, Druze 73.9% and Muslims 58.6%. A perennial look between the years 2008–2015 yields that there was an increase in the percentages of examinees eligible to receive certification especially in Arab education. By *gender*, in 2015 the percentages of *female* examinees eligible to receive certification were higher than *males* in *Hebrew* and *Arab education* as well (80%–75.2% and 70.8%–49.9% respectively) (CBS, 2016, Chapter 4, p. 67). However, in the last few years, a remarkable change in scholastic achievement in Arab-sector secondary schools has occurred. The percentage of Arabic-speaking students in the 12th grade who passed the secondary school exams in intermediate and advanced-level subjects (*science* and *technology* paths) has increased dramatically (CBS, 2016, Chapter 8, Table 8.30), as well in Hebrew education.

1.2. *Teachers' qualification and gender issues*

Teachers' qualification (*including math teachers*) emerged as a key driver of education in the 2015/16 school year: 92%, 97%, and 91.3% of the Arab-sector primary and secondary school teachers respectively have academic grade of wages (CBS, 2016, Chapter 8, Table 8.7, 8.18). In addition, 18.4%, 27.6%, and 30.5% of all Arab-sector primary and secondary school teachers have a Master's degree or higher. In this respect, there is a significant difference between the Arab sector and the Hebrew sector (26.0%, 44.7%, and 42.7% in the Jewish school system respectively). In many countries, the percentage of female teachers is increasing rapidly and to some extent teaching positions are exclusively held by women. This process is taking place in developed and developing societies. In this sense, the Arab population in Israel should be considered as a developing society. It is worth noting that the percentages of female mathematics teachers in primary and secondary schools increased from 50%, 32.3%, and 25.9% respectively in 1990 to 70%, 69%, and 54.4% respectively in 2015–2016 (CBS, 1991, 2016).

2. Mathematics Education among Arabic-Speaking Sectors

2.1. *Mathematics curricula*

In Israel, new primary school mathematics curricula for all sectors have gradually been implemented since 2006, and for middle school since 2010. Since 1989 there is no mathematics curriculum for high school, and in the 2007/8 school year the Ministry of Education published a revised Table of Specifications for secondary school exam in mathematics, which specifies the subjects required in the basic, intermediate, and advanced levels. This prompted the State Comptroller to write in 2014 that: "*As of September of* 2013, *no mathematics curricula for high schools have been prepared... Teachers face challenges as regards teaching material and mathematics students find it difficult to learn*" (p. 959). In 2010, the professional committee approved a "*new high school mathematics curriculum*" for *all sectors*, as a direct continuation of the middle school mathematics curriculum. This curriculum is in advanced stages of final approval. In 2014–2015, the Ministry of Education set up "*the national plan for meaningful learning in mathematics and science*" to encourage *excellence*. On 2015, this plan started to work as a part of other plan called "*mathematics first*" for the advanced-level track, which was expanded later to double the number of advanced-level mathematics students (RAMA, 2017, Report No. 2, p. 32).

2.2. *Evaluation of the new mathematics curricula*

2.2.1. *New primary school mathematics curricula*

The Weizmann Institute of Science evaluation report (2006) examined six different new mathematics instructional materials (IM) (textbooks, teachers' guides, etc.) for primary schools and found that these IM address few high-level thinking activities. As a result, some IM were dropped and others were either updated or rewritten. Worse, in the **Hebrew education**, especially in the **religious state education**, it was found that the percentage of schools who taught mathematics fewer than the *standard hours* (6 hours a week) was higher compared to other sectors. However, in **Arab education**, the appropriate percentage was very small (1%–2% in the Arab sector and 0% in the Druze and Bedouin sectors). The State Comptroller's Report (2014, p. 960) determined that: "*standard hours had not been implemented in all schools in* 2012 *and* 2013; *dozens of schools did not teach the required standard hours for first grade to third grade*".

2.2.2. *New middle school mathematics curricula*

The evaluation reports (RAMA, 2012–2013) on the implementation of the new mathematics curricula found that teachers were often unfamiliar with the educational material and that heavy schedules made it difficult to teach all the material required by the new programs. In addition, not enough hours were allocated to the Exhausting Students' Mathematical Potential (*Mitzui ve-Metzuyanut*) project, in the **religious, Druze and Bedouin sectors**.

2.2.3. *The New Mathematics Test Program in 5-unit track*

The Henrietta Szold Institute (2016) examined mathematics achievements in the advanced-level track between 2010–2014 by sector and by gender. The results showed *a decrease* in the percentages of participants in all sectors (Jews, Arabs, Druze and Bedouin) up to 2013 with the percentages in the

Bedouin sector significantly lower than the other sectors. In 2014 there was *a slight increase* in all but the **Arab sector** where the percentage remained the same. The percentage was higher in the **Jewish sector** (10.8%) compared to 7.1%, 9.4% and 3.3% in the **Arab, Druze, and Bedouin sectors**, respectively. In the Jewish sector there were more males than females candidates (57%–43%) but more females than males in the Arabic-speaking sectors. The male–female percentages were 41%–59%, 38%–62% and 27%–73% in the **Arab, Druze**, and **Bedouin sectors**, respectively.

2.2.4. *The Israeli "National Program for Meaningful Learning in Mathematics and Science"*

The worrying and alarming data above (Section 2.2.3) directly affect technology education, the hi-tech sector, the Ministry of Defense, academic institutions, the Ministry of Education and the mathematics education community in Israel. As a result, in 2014/2015 the Ministry of Education set up the "National Program for Meaningful Learning", in 98 selected schools included schools from the Arab-speaking sectors. The objectives of this program are as follows (RAMA, 2016): (1) Increase the number of students studying advanced-level mathematics by 15%, and raise this percentage by 100% within five years. (2) Increase the number of students eligible to receive certification in the sciences and technology by 100% within 5 years. (3) Realize this potential in different sectors. (4) Reduce the gender gaps. According to a very recent study (Yedioth Ahronoth, 2017) there has been a dramatic increase in the proportions of Arab speakers students who study advanced-level mathematics: 34.5% in the Arab sector, 44% in the Druze sector, and 130% in the Bedouin sector. Where it was written: "*Closing the gap with the Jewish sector is still a long way away, but the budding revolution provides hope that it will soon be felt in the labor market in coming years*" (Yedioth Ahronoth, daily newspaper, 30/01/2017, p. 34). By contrast, the use of such *politically correct statements*, draw rosy conclusions too early as regards the impact of the new curricula.

2.2.5. *New mathematics curricula in Arabic-speaking schools*

The new mathematics curricula, program and IM introduced to the Jewish sector were translated into Arabic and are now used in Arabic-speaking schools (except East Jerusalem, where the mathematics curriculum is under the responsibility of the Palestinian authority). In primary education, some of the six IM (Section 2.2.1) have been implemented. Two other IM were written originally in Arabic.

- In middle schools, the two IM used in Jewish schools were translated into Arabic and are currently used in Arab schools.
- In the Arab high schools, part of the IM in the Jewish high schools has been adopted. Only one IM has been partly translated into Arabic.
- The shortage of mathematics textbooks in Arabic for the high school level only aggravates the difficulties mentioned above for both teachers and students.

2.3. *Mathematics teachers' training*

Studies in both Israel and internationally have argued that teachers' ability to teach well is the most significant factor contributing to the success of the school system. It is worthy to mention that the percentages of mathematics teachers with professional training is *higher* in **Arab education** than in **Hebrew education** (*primary* schools: 51%–20%, *middle* schools: 75%–54% and *high* schools: 79%–62% respectively) (CBS, 2017, Press Advertising 15/03/2017).

The continuing education policies promoted by the Ministry of Education after the reforms are based on the studies showing that professional development (PD) programs can have a significant impact on teachers' knowledge and skills and hence student achievement.

From 2009 to 2012 the State Comptroller's Report indicated "(i) *At all levels of education, more than half* (55%) *of all mathematics teachers with more than two years of experience did not attend any PD courses on issues relating to this profession*; (ii) *The highest percentage of attendance was for high school teachers* (61%)." (The State Comptroller Report, 2014, p. 990). In the **Druze** sector, the percentage of teachers, who did not participate in PD, was relatively low, compared to **other sectors**. These findings are chilling in light of the decline on *national and international mathematics achievement tests*.

Table 1. Achievements of Israeli students on MEITZAV exams by sector and grade.

5th grade	8th grade
• No real changes were observed in achievements in 2015 compared to 2013 • In 2015, compared to 2013, Hebrew-speaking students' achievements increased by 5 points and Arabic-speaking students increased by 6 points • There was still a **gap** between the Hebrew speakers and the Arab-speakers sectors: Hebrew speakers were ahead by 42 points in 2015	• In the 2015 academic year there was an increase of 13 points in all schools compared to 2013 • Hebrew-speaking students' achievements in 2015 increased by 15 points and Arabic speakers increased by 9 points compared to 2013 • There was still a **gap** between Hebrew speakers and the Arab speakers in 2013/15. The Jewish sector scored 50 points higher than the Arab sector

2.4. *National and international mathematics achievements tests*

The School Growth and Efficiency Measures (MEITZAV) exam, PISA (Program for International Student Assessment), and TIMSS (Trends in International Mathematics and Science Study) are discussed in detail below (RAMA, 2015, pp. 19–21, 38–40, 117).

2.4.1. *MITSAV mathematics tests for 5th and 8th grades*

Table 1 summarizes the MEITZAV results in mathematics for 5th and 8th graders, *separately*.

2.4.2. *PISA mathematics tests for 9th–10th graders*

Table 2 summarizes the PISA results for Israel domestically and internationally. High- and low-achievers in PISA results are presented as well.

2.4.3. *TIMSS mathematics tests for 4th–8th graders*

Table 3 summarizes the TIMSS results (2015) for 8th graders; 4th graders did not participate the exams.

Table 2. PISA (2015) — internationally and domestically.

Internationally	Domestically
• Israel's average score in mathematics was 470 points; hence 20 points lower than the OECD average (490 points) • Israel ranked 39th out of 70 countries and entities in general; and out of the 35 OECD countries, Israel ranked 30th	• There was still a **gap** between Hebrew speakers and Arabic speakers of about 104 points
High- and low-achieving students percentages	
• The percentages of high-achievers in Israel were still below the figures in the OECD (9%–13% in 2012 and 9%–11% in 2015) • The percentages of low-achievers in mathematics literacy was invariant, and for the corresponding rates in OECD countries (34%–23% in 2012 and 32%–23% in 2015) • **Gender**: Hebrew speaking male students scored better than females, as well as in the OECD countries	• There were still **gaps** between high-achievers in the two sectors with considerably more in the Jewish sector (12%–0% in 2012 and 12%–1% in 2015) • There were still **gaps** between low-achievers in the two sectors, with fewer among Hebrew speakers (24%–67% in 2012 and 22%–64% in 2015) • **Gender**: Female Arab students scored better than male students

3. Higher Mathematics Education Among Arab Speakers

3.1. *Mathematics teachers' training institutions*

Today, there are five academic teachers' training institutions (Table 4) for Arabic speakers (Arabs, Druze, and Bedouin) (Ministry of Education, 2016). These institutions are located all over the country (North, Central region, and South). In addition, a small fraction of Arab students is enrolled in teacher training at universities and teacher training institutes for Hebrew speakers (due to constraints on admission requirements, convenience, beliefs). Note that Al-Qasemi college offers only high school track in mathematics, and Kaye-Bedouin offers mathematics specialty only in primary school track.

Table 3. TIMSS 2015 — internationally and domestically — 8th graders.

Internationally	Domestically
• The average total score in mathematics in Israel was 511 points. This average was significantly higher than the average overall scores of the 39 countries that participated in the study cycle which was 481 points. Accordingly, Israel was ranked 16th • 13% of all Israeli students scored at the level of "high-achievers", which is the 8th rank	• The **gap** between Hebrew speakers and Arabic speakers was about 73 points
High- and low-achieving students	
• The percentages of high-achievers in Israel are higher than the rates in all TIMSS countries (12%–3% in 2011 and 13%–5% in 2015) • The percentages of low-achievers in mathematics are below the corresponding percentages in all TIMSS countries (13%–25% but equal at 16% in 2015) • **Gender**: Hebrew-speaking male students scored better than females in the TIMSS countries as a whole	• The **gaps** between high-achievers in both sectors were larger (15%–5% in 2011 and 9%–6% in 2015) • The **gaps** between low-achievers were much smaller for the Jewish sector (7%–28% in 2011 and 9%–31% in 2015) • **Gender**: Female Arab students scored better than male students

Table 4. Mathematics teachers' training institutions for arabic speakers by level of education and by class range, study format and degree.

	Primary			Secondary school		
	Class range	Study format	Degree	Class range	Study format	Degree
	K1–K6	Double major	B.Ed.	K7–K10	Double major	B.Ed.
Arab college	√	√	√	√	√	√
Al-Qasemi				√	√	√
Bet-berl	√	√	√	√	√	√
Kaye-Bedouin	√	√	√			
Sakhnin	√	√	√	√	√	√

3.2. *The M.Ed. curriculum in mathematics education*

In the last five years (2012–2017), The Council for Higher Education (MALAG) has officially recognized "The Program for M.Ed. in Mathematics Education" in two institutions (The Academic Arab Colleges of Haifa and Al-Qasemi) which have improved the quality of learning and teaching of mathematics.

3.3. *Mathematics teaching certificate*

A recent study (Matlob, 2015) showed that about 45.8% of Arab academics who learned hi-tech subjects (sciences, engineering) turn to education to earn a living. Many flock to teaching certification programs for those holding an academic degree in institutions for Arabic speakers to improve their employment opportunities. It should be noted that a small number have taught in other institutions and universities for the same reason.

3.4. *Mathematics teaching staff*

The mathematics lecturers' staff involved in mathematics teachers' training courses plays a major role in shaping the professional image of the teacher. Data collected by the authors show that 88% (50 out of 57) hold a Ph.D. or M.A./M.Sc. degrees in Pure Mathematics or Mathematics Education. Of these, 72% (36 out of 50) have Ph.D. in Pure Mathematics or Mathematics Education. All the staff in the M.Ed. program has a Ph.D. in Pure Mathematics or in Mathematics Education. Many combine research and teaching, and take an active part in local and international conferences.

4. Conclusion

Mathematics education among Arab speakers in Israel is derived and directly related to the mathematics education of Hebrew speaking. This is a special and comprehensive document, which examined the products of mathematics education among Arab speakers: achievements in secondary school examination and MEITZAV, PIZA, and TIMSS tests, school mathematics curricula and instructional materials, mathematics teachers (in-service and pre-service) training and their professional development.

The overall picture is unambiguous and not encouraging given the disparity between Hebrew and Arabic speakers sectors. The gaps between the two sectors are deep.

The *macro-policy* led by those responsible for education and in particular, for mathematics education is important, but a *micro-policy* regarding other weak and challenged sectors should be implemented as well. Special attention must be paid to the Bedouin sector and students enrolled in basic and intermediate-level mathematics. These students find it difficult to learn mathematics, whether they are Jews or Arabs, boys or girls, or high or low on familial SES. Final remark: it should be emphasized that this chapter is a descriptive and not a research one.

References

CBS, Central Bureau of Statistics (1991). pp. 612, 615–616.

CBS, Central Bureau of Statistics (2016). Chapter 2 (Table 2.1, 2.3, 2.13), Chapter 8. (Tables 8.3, 8.7, 8.8, 8.18, 8.19, 8.30, 8.63) pp. 1–2.

Henrietta Szold Institute (2016). Teaching advanced–level mathematics among Arab society, pp. 9–11 (in Hebrew).

Matlob (2015). Education and employment, Arab academic in Israel.

Ministry of Education (2014). Data of entitled to certificates among matriculation examinees, (in Hebrew).

Ministry of Education (2016). *Division A*, Teachers Training Institutions, pp. 1–15.

National Insurance Institute (NII) (2016). Poverty and social gaps, pp. 21, 30 (in Hebrew).

Prime Minister Office (2013). The authority of economic development of the Arab, Druze and Circassian sectors, p. 15 (in Hebrew).

National Authority for Measurement and Evaluation (RAMA) (2012). A program to advance achievements in seventh grade in mathematics, mother tongue and science and technology in the 2012 school year (in Hebrew).

National Authority for Measurement and Evaluation (RAMA) (2013). A program to advance achievements in eighth grade in mathematics, mother tongue and science and technology in the 2013 school year (in Hebrew).

National Authority for Measurement and Evaluation in Education (RAMA) (2015). Achievements tests, pp. 19–21, 38–40, 117 (in Hebrew).

National Authority for Measurement and Evaluation in Education (RAMA) (2016). Evaluation of the program "Mathematics First" in the academic year 2015 (in Hebrew).

State Comptroller's Report (2014). Mathematics education, 64C, pp. 957, 982, 990.

Weizmann Institute of Science (2006). Evaluation six mathematics curricula's. p. 76.

Yedioth Ahronoth (30/01/2017) Henrietta Szold Institute and Trump Foundation: Teaching advanced-level mathematics among Arab society, p. 34 (in Hebrew).

About the Authors

Shaker A. Rasslan earned his MSc and PhD in mathematics education from the Technion and the Hebrew University of Jerusalem, respectively. His post-doctoral research in Mathematics Education was at The Mathematics Education Research Center at Warwick University, UK. He is a reviewer of research reports for the PME. His main research interests include advanced mathematical thinking, cognitive development, concept image and creativity.

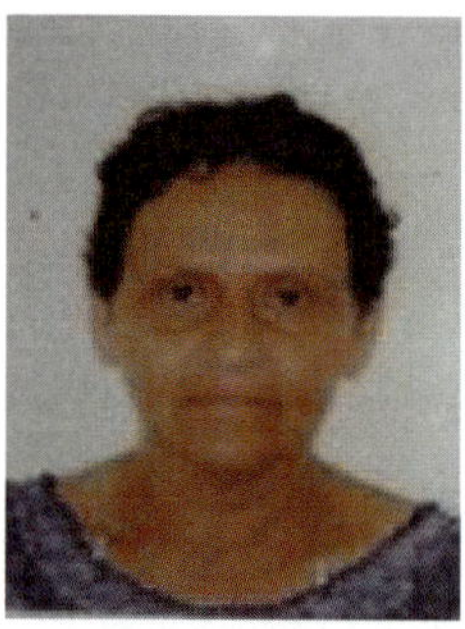

Amal Sharif-Rasslan earned her BSc, MSc and PhD degrees in mathematics and mathematics education from the Technion. She is the head of the Mathematics Department and the MEd program for mathematics education at the Arab Academic College of Education in Haifa, Israel.

Her main research topics cover mathematics, mathematics education, and in particular advanced mathematical thinking, and applied mathematics in cardiology. She was the recipient of the William Harvey Award in 2015.

Part II

Issues and Innovations Related to Mathematics Education at Preschool and Primary School (Grades K-6) in Israel

Chapter 8

New Developments and Trends in Preschool Mathematics Education in Israel

Ornit Spektor-Levy* and Taly Shechter#

Da-Gan Center, The School of Education,
Bar-Ilan University, Ramat-Gan, Israel
**ornit.spektor-levy@biu.ac.il*
#talyshechter@gmail.com

This is an overview of preschool mathematics education in Israel through a description of the triangular relationship between: (a) theory and academic research, (b) policy and teacher training, and (c) the practice of teaching and learning mathematics. Underlying the preschool mathematics curriculum and professional development programs is the accumulated academic knowledge of early mathematical thinking and pedagogical approaches, adapted to the age and cognitive capabilities of preschoolers (3–6 years of age). Two examples of authentic, extended mathematical practices of preschool teachers and their preschoolers are presented.

1. Introduction

Children's informal experiences with quantitative situations (e.g., choosing the bigger ball, sharing a cake, counting stairs) begin at a very early age. In most countries, such experiences are formalized in preschool educational frameworks, in accordance with governmental educational policies and mathematics curricula. Early childhood mathematics curricula generally have three overarching, universal goals: (a) to lay the foundations for the children's mathematical thinking; (b) to encourage their curiosity and inclination to engage in mathematics so as to implement mathematical practices in other domains; and (c) to make mathematics enjoyable through play-based experiences (English and Mulligan, 2013).

2. Early Mathematical Thinking

2.1. *What do we know from research?*

As new insights emerge on children's mathematical thinking, it is now recognized that young children can learn more than was previously expected (English and Mulligan, 2013; NRC, 2009). Preschoolers spontaneously comprehend basic principles of counting, develop efficient methods of calculation and possess effective strategies for addition and subtraction (Clements, Copple, and Hyson, 2002). Children as young as 5 years old exhibit interest in mathematical activities during every-day play (Seo and Ginsburg, 2004). These natural behaviors and cognitive processes are the foundation of mathematical education in preschool (3–6 years of age).

Preschool activities that require quantifying manipulations, representations, and explanations constitute valued goals in mathematics education (National Mathematics Advisory Panel, 2008), are essential components of mathematical knowledge, and play an influential role in promoting future mathematics learning (Klein *et al.*, 2008; Moore and Geary, 2016). Preschool children's knowledge of mathematics predicts their later school success throughout primary (Mervis, 2017) and even high school (National Mathematics Advisory Panel, 2008; NRC, 2009). Therefore, children must learn to mathematize their informal experiences by abstracting, representing, and elaborating them mathematically. If they do not, they miss the opportunity to learn the language of mathematics in all its multifaceted forms (Clements and Sarama, 2011).

2.2. *Pedagogy: contextual, active and authentic mathematics education adapted to age and development*

Daily activities in preschool often include mathematical opportunities that can be highlighted to set the stage for the exploration and discussion of key mathematical concepts, or the application of mathematical ideas and skills in new contexts. However, everyday activities are not sufficient, and should not alone constitute the total mathematics experience for young children (Greenes, Ginsburg, and Balfanz, 2004). Educators should plan and provide systematic and structured mathematical experience in playful and enjoyable ways (Clements and Sarama, 2011). Clements (2001) believes that "high quality teaching in mathematics is about challenge and joy, not imposition and pressure. Good early childhood mathematics is broader and deeper than mere practice in counting and adding" (p. 270).

Preschool life offers ample opportunities to engage children in a range of mathematical activities, such as counting the children while taking attendance, setting the table while matching the number of plates to the number of children, running a preschool "grocery store", etc. Such a holistic environment exploits the preschoolers' high level of motivation to learn and develop in a self-directed, self-motivated manner at a stage when children first develop their beliefs, habits, and feelings about mathematics.

Strategies of metacognitive instruction were found to improve mathematical thinking by encouraging children to think out loud and explain the process through which they reached the solution, thus training them to steer and control problem solving processes, raise assumptions, use different strategies, and above all, ask themselves cognitive questions (Mevarech, 1995). Teachers should model these processes for their students and implement peer teaching: through the process of explaining, children naturally execute metacognitive processes by clarifying the process of solving the problem, substantiating it, leading and controlling it (Whitebread *et al.*, 2007).

3. Ministry of Education Programs and Actions to Enhance Mathematics Education in Preschool

3.1. *The national preschool mathematics curriculum*

The first national preschool mathematics curriculum was introduced as a framework program on 1995. It was a short document that emphasized the need to engage in mathematics in early years and specified the mathematical capabilities that young children should acquire before entering primary school. It was only in 2008 that the first national preschool mathematics curriculum was introduced. It was the first time that the Ministry of Education had announced mathematics as obligatory in preschool. The curriculum provides topic guidelines, and a set of suggested processes and principles for mathematics instruction for young children (see Appendix) that reflect the accumulated knowledge described in Section 2 above. The main objectives of the curriculum are: to lay the foundations for the development of children's mathematical thinking; to stimulate children's curiosity and desire to engage in mathematics so as to develop observation, investigation, and inference skills; to foster a fondness for engaging in mathematics. The content-focused objectives are to ensure that children graduate preschool with knowledge and understanding of the concept of numbers, shapes and objects in the environment, and to equip them with mathematical tools to aid problem-solving in everyday life.

The above objectives are to be met while maintaining the unique characteristics of preschool. The preschool teacher should serve as a role model, using accurate mathematical language and demonstrating explicit mathematical thinking. The preschool environment should be designed to be mathematically challenging and provide children with opportunities to observe, examine, and investigate various mathematical phenomena. It also takes into consideration that preschoolers are at an age when they need conceptual aids to understand various mathematical situations (i.e., objects that help understand abstract concepts).

The preschool mathematics curriculum consists of three main subjects: number sense and number concept, spatial perception, and quantitative concepts in everyday life (for a more detailed description of each subject, concepts, and suggested instructional methods, see Appendix). Problem solving is integrated into each of the topics.

For each age group (3–4, 4–5, 5–6 years), the curriculum specifies the objectives and topics and includes pedagogical guidelines intended to clarify and define the subjects, to offer recommendations related to children's development, means of assessing the activities, and practical examples and ideas for learning activities.

3.2. *The National Teachers Center for STEM education in Preschool — Da-Gan Center: Professional development programs in preschool mathematics*

The National Teachers Center for STEM (Science, Technology, Engineering, Mathematics) in Preschool has developed intensive professional development courses for in-service preschool teachers. These courses foster mathematical knowledge and leadership, leading to more proficient teachers in preschool mathematical education. Participants delve deeply into the preschool mathematics curriculum while undergoing training in various instructional practices. They develop authentic, concrete, constructivist mathematical activities dealing with relevant contexts to children's everyday life, accompanied by appropriate conceptual, visual, and graphic aids to facilitate the preschoolers' understanding. Teachers learn to mediate instruction to meet the mathematical abilities of each child, to document the children's understanding and products, and to encourage teacher–child and child–child mathematical discourse while designing a physical learning environment that enhances mathematical thinking and practices. Participants are required to present evidences of implementation and preschoolers' progress. Other activities in the center that support mathematics education include conferences, annual magazine, development of teaching materials, and running an educational website.

4. Mathematics in Action: Initiatives and Activities in Preschool Classrooms

Three important initiatives of preschool mathematics education in Israel are described in this book: *Origametria — Paper Folding for Teaching Geometry in Preschool and Primary School*, by Oberman (see Chapter 9); *Educating the Eye*: *The Agam Program for Visual Thinking*, by Hershkovitz, Markovitz, Rosenfeld, Ilani, and Eylon (see Chapter 10); Professional Development for Preschool Teachers: The CAMTE Framework and Repeating Patterns, by Tirosh, Barkai, Levenson and Tsamir (see Chapter 10). Another initiative was developed by Teubal and Guberman (2014), who study non-verbal graphic texts for children of preschool age, such as children's graphical representations of numerical values, or their use of maps as a tool to represent spatial relationships.

Nevertheless, there is not one program that covers all subjects of the preschool mathematics curriculum. Most mathematics engagement activities in preschool classrooms are independent initiatives of the preschool teachers. Therefore, next, we will describe two exemplary cases of authentic, integrative, longitudinal projects, planned by preschool teachers and aimed at developing children's mathematical thinking and understanding in an authentic context.

At Preschool A, the teacher believes that engaging in mathematics is an integral part of the preschoolers' everyday life, and the availability of mathematical stimuli in the classroom space allows them to engage in the subject naturally during play — whether spontaneously or in teacher-mediated, structured activities. The teacher designed a Mini Sports Center, with spaces located both inside and outside the classroom. In the Sports Center, the preschoolers develop motor skills and spatial orientation. Through track and field games they develop counting, addition, and subtraction skills by calculating their scores. In the Medical Clinic area, socio-dramatic play is combined with learning mathematics: the children can take a number and await their turn. The receptionist calls them into the clinic by number. They practice identifying numbers during an eye test. They pay for medicine that they buy at the pharmacy with play money, etc.

At Preschool B, the children set up an actual restaurant in the classroom (Fig. 1). First, the children considered the concept of a restaurant and what running a restaurant entails, then they chose the name of the restaurant after holding a poll, voting, counting the votes, and illustrating the results in a

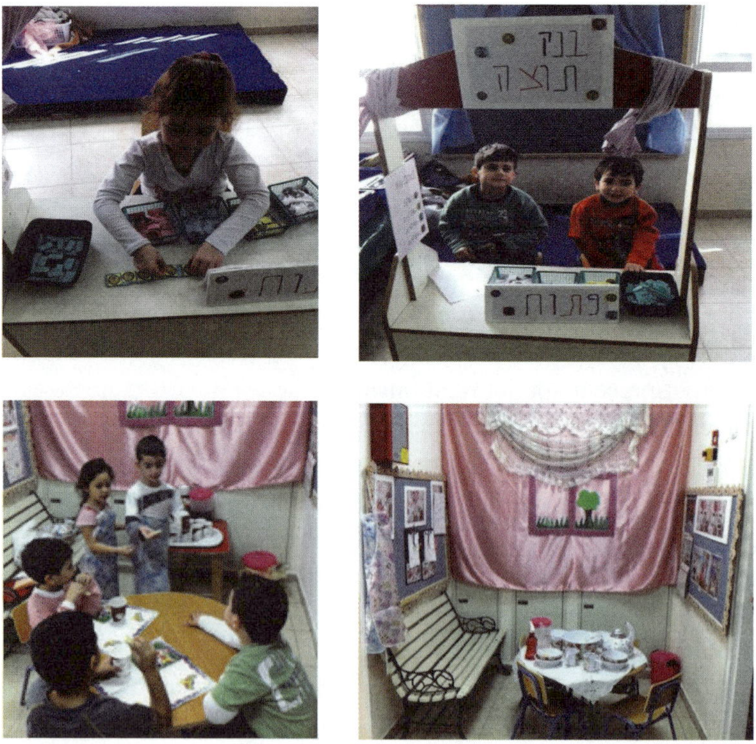

Fig. 1. Counting play coins in the bank. Paying at the restaurant.

diagram. They set up a bank and a cash register for the restaurant using play coins they made themselves. They planned the dishes on the menu, with a focus on healthy food. They designed the menu posters, and used mathematical representations to display the prices. Next, they divided the roles amongst themselves and decided who would be the server, the cashier, the host, and the guests. The food was prepared on the spot and served at the tables. Engaging in purchase and sale during play evoked opportunities for solving various mathematical problem solving, and ways of documenting and presenting the processes of reaching them.

In summary, we have provided a short summary of what is known from research about early mathematical thinking and the pedagogical approaches adapted to the age and cognitive capabilities of preschoolers. The Israeli preschool mathematics curriculum and national professional development programs are based on this accumulated knowledge. Teachers are expected to provide children with a learning environment that promotes mathematical thinking, discourse, and practices, as demonstrated by two examples of authentic practices.

References

Clements, D. H. (2001). Mathematics in the preschool, *Teaching Children Mathematics*, **7**(5), 270.

Clements, D. H., Copple, C. and Hyson, M. (2002). Early childhood mathematics: Promoting good beginnings. A joint position statement of the National Association for the Education of Young Children (NAEYC) and the National Council of Teachers of Mathematics (NCTM).

Clements, D. H. and Sarama, J. (2011). Early childhood mathematics intervention, *Science*, **333**(6045), 968–970.

English, L. D. and Mulligan, J. T. (eds.). (2013). *Reconceptualizing Early Mathematics Learning*, Springer Science & Business Media.

Greenes, C., Ginsburg, H. P. and Balfanz, R. (2004). Big math for little kids, *Early Childhood Research Quarterly*, **19**(1), 159–166.

Klein, A., Starkey, P., Clements, D., Sarama, J. and Iyer, R. (2008). Effects of a prekindergarten mathematics intervention: A randomized experiment, *Journal of Research on Educational Effectiveness*, **1**(3), 155–178.

Mervis, J. (2017). No easy answers, *Science*, **355** (6325), 568–571.

Mevarech, Z. R. (1995). Metacognition, general ability, and mathematical understanding, *Early Education and Development*, **6**(2), 155–168.

Moore, A. M. and Geary, D. C. (2016). Kindergartners' fluent processing of symbolic numerical magnitude is predicted by their cardinal knowledge and implicit understanding of arithmetic 2 years earlier, *Journal of Experimental Child Psychology*, **150**, 31–47.

National Mathematics Advisory Panel (2008). Foundations for Success: The Final Report of the National Mathematics Advisory Panel (U.S. Department of Education, Office of Planning, Evaluation and Policy Development, Washington, DC).

National Research Council (NRC) (2009). Mathematics learning in early childhood: Paths toward excellence and equity, ed. Committee on Early Childhood Mathematics, in C. T. Cross, T. A. Woods and H. Schweingruber (eds.), Center for Education, Division of Behavioral and Social Sciences and Education. National Academies Press. [NCJ].

Seo, K. H. and Ginsburg, H. P. (2004). What is developmentally appropriate in early childhood mathematics education? Lessons from new research, in: D. H. Clements, J. Sarama and A. M. DiBiase (eds.), *Engaging Young Children in Mathematics: Standards for Early Childhood Mathematics Education*, pp. 91–104. Routledge.

Teubal, E. and Guberman, A. (2014). *Graphic Texts: Literacy Enhancing Tools in Early Childhood*, Springer.

Whitebread, D., Bingham, S., Grau, V., Pino Pasternak, D. and Sangster, C. (2007). Development of metacognition and self-regulated learning in young children: Role of collaborative and peer-assisted learning, *Journal of Cognitive Education and Psychology*, **6**(3), 433–455.

About the Authors

Ornit Spektor-Levy is a faculty member at the Science Education Center, School of Education, Bar-Ilan University. She received her PhD from the Weizmann Institute for Science. She is the director of the Israeli National Teacher Center for STEM in preschool. Her educational research focuses on STEM education in early childhood, scientific communication, development of scientific curiosity and literacy, and professional development of science teachers.

Taly Shechter is a PhD student at the School of Education, Bar-Ilan University. She is deputy director of the Israeli National Teacher Center for STEM in preschool and pedagogical coordinator of preschool teachers' professional development programs. Her study and expertise focus on the development of engineering and mathematical thinking among preschool children.

Chapter 9

Origametria — Paper Folding for Teaching Geometry in Preschool and Primary School

John Oberman

Shaanan Religious Academic College of Education,
Kiryat Shmuel, Haifa, Israel
shaanan@shaanan.org

New opportunities are being researched and examined for teaching geometry in preschool (ages 4–6) and primary school (grades 1–6). Israeli preschool and primary school mathematics curricula have been historically viewed as preparation for the formal teaching of geometry in high school. Our understanding that spatial awareness and geometrical thinking develop at a very early age has led us to build a dynamic learning environment based on the techniques of paper folding, together with digital technologies using animations. The chapter will describe the elements of the Origametria program and its implementation in kindergartens and schools in Israel. Additionally, it will describe the changing attitudes of teachers with respect to teaching geometry, together with the increased motivation of pupils to use their intuitive spatial reasoning for fostering geometric thinking to reach higher levels of abstraction.

1. Introduction

The name *Origametria* is a combination of the words *origami* and *geometry*. The term was created by the Israeli Origami Center (IOC) to describe its innovative program to teach curriculum geometry through origami. After several years of observation, the program was formally approved by the Israeli Ministry of Education in 2010. As of the 2016–2017 school year, the program has received formal approval for use from preschool to grade 6. In Origametria, each lesson involves a simple origami model. During the folding process the teacher leads a discussion on a relevant geometry topic such as symmetry, angles, or area, depending on the grade level.

The program was developed by Miri Golan, Director of the IOC, in close collaboration with mathematicians and specialists in the field of mathematics education. The program was designed to enhance self-esteem and a sense of accomplishment, while developing principles of basic geometry and learning skills such as motor skills, spatial perception, logical and sequential thinking, hand-eye coordination, focusing and concentration, aesthetics, and three-dimensional perception. Since 2011 the IOC has developed a program named "Computerized Origametria", in which paper folding is shown with animations. In-service training for K-6 educators in the years 2015–2017 has taken place in more than 26 teacher centers in Israel. Several fields of research, both in Israel and abroad, have enabled the developers to gain further insight into possibilities of fostering and developing geometric thinking (Boakes, 2009; Golan, Klemer and Oberman, 2014).

2. Theoretical Background

Considerable research has identified the difficulties that learners have in grasping geometric concepts and the resulting misconceptions about geometric concepts, shapes, and figures (Clements and Sarama, 2000). Many of the difficulties arise as the result of learning geometry in static situations. Other difficulties arise by presenting new concepts when the cognitive development of the learner has not yet reached the necessary stage to absorb these concepts with the existing knowledge.

The issue of misconceptions and errors in geometry learning by young learners is more prevalent than has been shown by research. Many teachers are themselves unsure about many geometrical concepts and as a result transfer their misconceptions to the learners (Cunningham and Roberts, 2010).

With Origametria, two paths are able to support the teaching and learning of geometry: One path uses paper folding to strengthen the natural intuition that young children have about geometrical concepts (Copley, 2000); the other path leads to supporting teachers by using a computer-based program. Our understanding that spatial awareness and geometrical thinking develop at a very early age has led us to develop a dynamic learning environment based on the techniques of paper folding, together with digital technologies using animations. This allows teachers with minimal expertise to use paper folding to expand their own understanding of these concepts

Fig. 1. An example of a folding sequence (presented as an animation) and related discussion questions concerning recognition of polygons and their properties.

and use these actions to minimize the development of misconceptions (see Fig. 1. "**Folding instructions**: In each non-colored triangle fold one of the shorter sides towards the longest side of this triangle according to the dotted line. **Discussion questions**: Which polygon has been formed after these folds? What can you say about the diagonals of this new polygon? Explain").

The computer-based program gives clear instructions for every step of the folding sequence, supported with animations. In addition, there are detailed questions for group discussion with the students. The steps allow students to check their work and then attend to the relevant geometric concepts in the questions.

3. Origametria in Primary School

3.1. *Lesson structure*

First, a specific topic of geometry must be selected for teaching. The choice of topic depends on the requirements of the primary school mathematics curriculum for each grade. Each year the topics taught are based on spiral instruction. Once a topic is selected, an origami model is found which focuses on that topic in its folding sequence. For example, a model that consists of many different triangles, or specifically isosceles triangles, or polygons, or bisections. While the students fold the model, the teacher stops after each new step to check whether new instances of the chosen topic could be identified, and whether they are the same or different from previous instances. The process of identification can be characterized by the following four ideas:

- *Geometric insight*: understanding the topic using folds.
- *Exploration*: the continuous search throughout the folding process.
- *Properties and context*: studying the topic in various ways by using the folded paper or while folding.
- *Lesson summary*: At the end of the lesson we check whether each student has acquired the geometric knowledge taught in that particular class. After every lesson the pupils leave with a folded model, which serves as a motivational tool for future learning.

3.2. *The method of teaching*

Here are six of the principles underlying the Origametria program:

(1) *The style of teaching*: We avoid negative statements during our classes and do not criticize students' folding. Instead of "this is incorrect", we say "this is beautiful; now, open this and fold...". This ensures that students that experience difficulties can fold like all other students.
(2) *We never check the accuracy of folding*: We stress that *accuracy* is an important yet relative term and changes from student to student. When students ask their teacher whether a fold is accurate, the teacher returns the question to the students, asking whether they think it is the most accurate fold they can make. This obliges the student to inspect their work and to fold as accurately as they can.
(3) *Our teachers never touch a student's model*: Instead, teachers may show a step using their own paper again and again. This gives students a stronger sense of ownership and accomplishment as their work is entirely their own, folded without any assistance.
(4) *The choice of models*: We carefully select the models and improve their folding sequences so that the teachers can explain the folding procedure once and the students should be able to follow easily.
(5) *Positive reinforcement*: After each step, the teacher examines students' folds and gives positive encouragement.

(6) *The model is never named while being folded*: During the folding process, the completed model is not revealed. This has the double effect of focusing students' attention on the geometry of the current step and also of expanding students' imaginations. The teacher concludes the lesson with a surprise reveal (Golan, 2011; Golan and Jackson, 2009).

3.3. *Example showing the development of geometrical thinking and knowledge of concepts*

The following example demonstrates the construction of four geometrical concepts: composition and decomposition of shapes, parallelism, diagonals, and line symmetry.

Example. Learning the properties of polygons by using the composition and decomposition of shapes.

The students made various polygons by paper folding. Using these polygons, new shapes were constructed. This activity allowed the students to deepen their understandings of the properties of polygons which were used in the composition and decomposition of the polygons (Fig. 2).

4. The Kindergarten Origametria Program

The Kindergarten Origametria Program (KOP) trains preschool teachers to use origami as a learning tool for the primary school geometry curriculum.

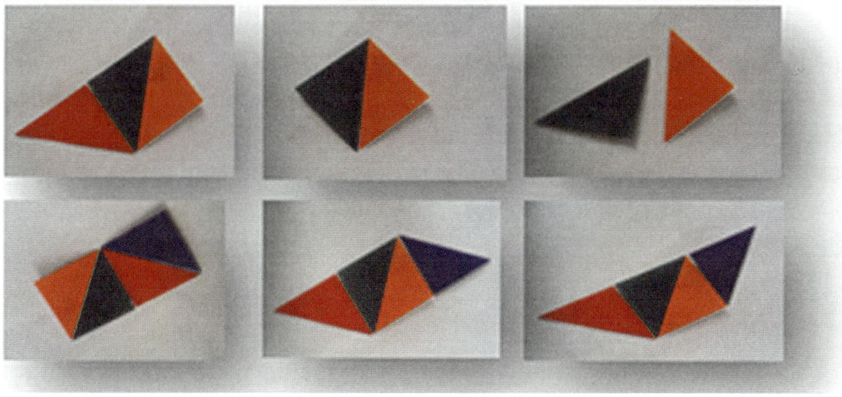

Fig. 2. Different compositions of folded triangles.

The program relies on the imagination of preschool children (ages 4–6) as the focus of the activities, involving them in a series of creative folding activities that acknowledge their limited motor skills. The KOP is a tool for the preschool teacher to conduct activities in a creative environment, building a strong mathematical base for students. The emphasis on imaginative play and experiment as a means to learning, though common in many preschool activities, is an unconventional and unusual approach to using origami as a learning tool. In a recent presentation Miri Golan, an experienced teacher, a world renowned origami artist, and the creator of the KOP program, stated that each course runs for a total of 32 hours, over eight sessions, spread over a period of several months. Each course accepts 20 preschool teachers (PTs). The KOP course comprises three parts which are described in the following subsection.

4.1. *Learning topics of geometry based on the KOP (10 hours)*

The PTs gain basic geometric knowledge, terminology, and definitions relevant for the KOP. Each topic is introduced ahead of its introduction to the children in order to promote self-efficacy among PTs. This knowledge is gained through paper folding sessions, led by an instructor. Learning is thus folding-based, not lecture based. During the folding process, PTs describe and analyze the geometry of the paper.

4.2. *The method of using the KOP activity in preschools (12 hours)*

The PTs are introduced to methods for conducting KOP activities. Between course sessions, each PT introduces what they have recently learned as an activity in their own kindergarten and reports back to the class on those experiences. They are encouraged to take photographs and to make short movies to better illustrate their experiences. Feedback is given by the instructor and other PTs.

4.3. *Additional geometric information (10 hours)*

The PTs are taught basic three-dimensional geometry (i.e., cubes, cuboids, and pyramids) and the relationships between various triangles and quadrilaterals.

References

Boakes, N. J. (2009). Origami instruction in the middle school mathematics classroom: Its impact on spatial visualization and geometry knowledge of students, *RMLE Online*, **32**(7), 1–12.

Clements, D. H. and Sarama, J. (2000). Young children's ideas about geometric shapes, *Teaching Children Mathematics*, **6**(8), 482–488.

Copley, J. V. (2000). *Geometry and Spatial Sense in the Early Childhood Curriculum*, Washington, DC: National Association for the Education of Young Children.

Cunningham, R. F. and Roberts, A. (2010). Reducing the mismatch of geometry concept definitions and concept images held by pre-service teachers, *Issues in the Undergraduate Mathematics Preparation of School Teachers, Vol.* 1.

Golan, M. (2011). Origametria and the van Hiele theory of teaching geometry, in P. Wang-Iverson, R. J. Lang and M. Yim (eds.), *Origami 5: Fifth International Meeting of Origami Science, Mathematics, and Education*, Boca Raton, FL: CRC Press, pp. 141–150.

Golan, M. and Jackson, P. (2009). Origametria: A program to teach geometry and to develop learning skills using the art of origami, in R. J. Lang (ed.), *Origami 4: Fourth International Meeting of Origami Science, Mathematics, and Education*, Natick, MA: A K Peters, Ltd, pp. 459–469.

Golan, M., Klemer, A. and Oberman, J. (2014). Developing and fostering geometric thinking: A qualitative research program for teaching quadrilaterals using Origametria, *Origami6: 6th International Meeting on Origami in Science, Mathematics and Education*, Tokyo, Japan.

About the Author

John Oberman has been involved in Israel in the teaching of mathematics at all levels and in teacher training since 1965. He has been a member of every curriculum committee and a member of the Professional Committee of Mathematics. He is currently a lecturer in mathematics at the Shaanan Religious Academic College of Education, Haifa.

Chapter 10

Educating the Eye: The Agam Program for Visual Thinking

Rina Hershkowitz[*,‡], **Zvia Markovits**[#,§], **Sherman Rosenfeld**[*,¶], **Lea Ilani**[*,||] **and Bat-Sheva Eylon**[*,**]

[*]*Weizmann Institute of Science, Rehovot 76100, Israel*
[#]*Kibbutzim Academic College of Education, Tel Aviv, Israel*
[‡]*rina.hershkowitz@weizmann.ac.il*
[§]*zviam@oranim.ac.il*
[¶]*shermrosenfeld@gmail.com*
[||]*leailani1016@gmail.com*
[**]*bat-sheva.eylon@weizmann.ac.il*

"All of us are visual illiterates. The education we give to our children causes them to be visually illiterate as well. Therefore, the main goal of the Agam Program is to educate the eye"

Yaacov Agam (1984)

Although visual thinking contributes to the development of basic skills in many domains, this subject is neglected in preschool and primary school. The Agam Program is a structured curriculum of 36 units that teaches visual skills, visual vocabulary and visual thinking to young children. Research shows that this approach has many positive effects for young children from diverse backgrounds, in areas such as mathematical thinking, school readiness, and general intelligence.

1. Visual Thinking and the Agam Program

There is a growing recognition of the importance to integrate visual thinking into the education system in general and into early education in particular. Research indicates that the early development of visual thinking contributes to the development of basic skills in geometry and mathematical thinking

(Clements and Sarama, 2000; Denton and West, 2002) and also to other fields, such as writing, mapping skills and school readiness (Clements and Sarama, 1999; Eylon and Rosenfeld, 1990).

Visual thinking is a complex phenomenon that includes aspects such as visual perception, visual communication, dynamic visual thinking, spatial intelligence, spatial thinking, spatial sense and spatial perception. Research supports the claim that visualization is a central meta-cognitive skill and important mode of thinking, especially in mathematics (Alsina and Nelsen, 2006), science (Gilbert, 2005) and creative work (McKim, 1980). According to Gardner (1983), spatial intelligence — which emerges from spatial thinking — includes the capacities to perceive and recognize instances of the same element, transform and manipulate these perceptions, make mental images and transform them, and produce a graphic likeness of spatial information. Research suggests that these basic spatial thinking skills should be taught in early childhood. In their review on spatial thinking in young children, Gersmehl and Gersmehl (2007) conclude that: (1) brain structures for spatial reasoning are fully functional at a very early age, (2) training can enhance spatial thinking skills, and (3) practice in early grades is an important, perhaps even essential, part of the foundation for later learning in this area. The NRC report (NRC, 2006) recommends that visual thinking instruction should be infused across and throughout the curriculum from K-12. In practice, this recommendation is not followed in most countries.

The central goal of the Agam Program is to develop the visual thinking of young children, as a means to improve their over-all cognitive and emotional development. It was inspired, initiated and written by the artist Yaacov Agam (Agam, 1984). The program has been refined, implemented and tested by a team of researchers and educators in the Science Teaching Department at the Weizmann Institute of Science, since 1984. In 1996, Yaacov Agam won the UNESCO's Comenius award, which honors innovation in education. The award recognized both the originality of the program as well as the accompanying educational research.

2. The Structure of the Agam Program

2.1. *The program units and their integration*

The Agam Program has 36 units, each of which deals with a different visual concept or a combination of concepts. The approach integrates the

Fig. 1. Some manipulatives used in the Agam Program.

acquisition of each of these concepts via the same skills, i.e., identification, memorization, reproduction (when the object is present), reproduction from memory (when the object is not present) and creativity. The program is accompanied by a wide variety of manipulatives (e.g., plastic geometric shapes, memory cards and transparencies. See Fig. 1).

The program progressively and logically integrates the visual concepts of the units, starting with elements of a "visual alphabet" (in the first units), which can be used later to describe higher-order elements, such as "words", "sentences" or "phenomena" in mathematics, science, and other domains (in the other units). The first units introduce students to main geometric figures, directions, colors and size relationships. Each visual concept is first introduced by itself and then shown in combination with other visual concepts. For example, after the introduction of ***Circles*** and ***Squares*** as "visual letters" in the first two units, the third unit, ***Patterns*** uses these "letters" to create "visual words" or "visual sentences". Patterns are understood as visual periodic series whose elements are circles and squares of different sizes, colors and positions, with changing intervals between them. In this unit, all activities deal with linear (one dimensional) patterns, but the children's creativity is unbounded (see Fig. 2).

2.2. *Clusters in the Agam Program*

The Agam Program can also be understood as clusters of visual concepts. This framework is based on a content analysis of the units as well as

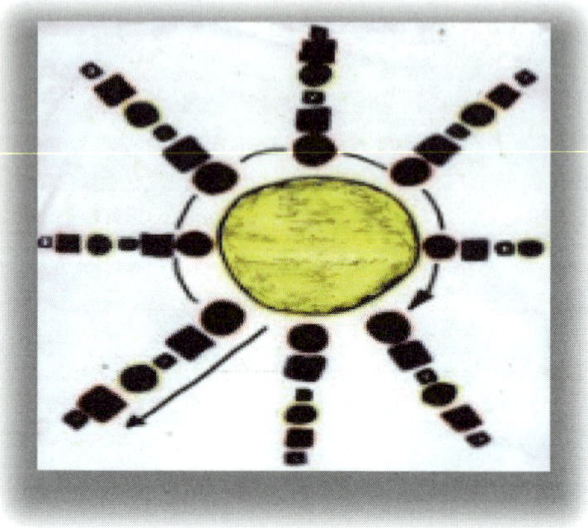

Fig. 2. The "sun" pattern, created by a 5-year old child. Note that each of the rays is a linear pattern and the sequence of all rays is an additional pattern in two dimensions.

empirical experience with implementing and testing the units in classrooms. Based on this analysis we reorganized the 36 units into the following seven clusters: Main Geometrical Figures, Directions, Colors, Size Relationships & Estimation, Advanced Concepts, Visual Skills, and Dimensions. These clusters are now used as a lens through which the teachers can view the program. Below we describe one cluster.

2.3. *The dimensions cluster*

The 4 units of this cluster (one for each dimension) are built as a logical sequence. A **dot** lacks any dimension. When it moves in different directions, it describes different one-dimensional lines. A **one-dimensional line**, when it moves in different directions, describes a two-dimensional surface. A **two-dimensional surface**, when it moves in different directions, describes **three-dimensional geometric objects in space.** In order to describe the movement of three-dimensional geometric objects through space, a **fourth dimension — time** — needs to be added. These four dimensions enable a full description of the physical reality.

It is interesting to note that Yaacov Agam's approach to time and movement has been a central focus of his work, as a pioneer of kinetic art. In his view, people tend to resist the passing of time, so they try to freeze it (e.g., with "still life" paintings) instead of giving time its proper place in our culture and perception of the world. Agam believes that through "educating the eye", young children can learn to (a) develop an understanding and comprehension of the four dimensions, and (b) live with nature in a dynamic and creative way, instead of developing fossilized and outdated cognitive processes.

3. Educational Research of the Agam Program

3.1. *Students in the Agam Program*

Educational research with 25 matched pairs of experimental and comparison preschools conducted in 1985–7 showed that the children in the experimental group significantly outperformed similar children in the comparison group, on tests measuring visual concepts, spatial skills and transfer effects. The children in the experimental group also demonstrated statistically significant improvements on IQ (general intelligence) and mathematical readiness. These effects were found equally for boys and girls, as well as for students from privileged and underprivileged backgrounds (Eylon and Rosenfeld, 1990; Markovits and Hershkowitz, 1997; Razel and Eylon, 1986, 1990). The program also made a dramatic effect on the educational achievement of many children who were less verbal than average (Eylon and Rosenfeld, 1990; Razel and Eylon, 1990). The research conducted in 2006–7 with two groups of 310 children (experimental and comparison) showed that students in the Agam Program significantly outscored comparison students in tasks involving "everyday visual thinking" (e.g., identifying a lost sun umbrella at the beach) and showed that the program contributes to the acquisition of mathematical and scientific concepts (Eylon *et al.*, 2014).

An example of a test item used in both research studies (1985–7 and 2006–7) is shown in Fig. 3. In this item, students are required to focus their attention on relevant visual information and to ignore non-relevant visual information. While there was no significant difference between the two groups in the pretest, statistically significant differences were found between the groups, favoring the experimental group after each of the two years.

3.2. *Teachers in the Agam Program*

A study with 25 preschool teachers, new to the Agam Program, investigated their prior knowledge and beliefs regarding visual cognition and its

Fig. 3. Students identify circles and squares in a complex environment.

enhancement (Markovits, Rosenfeld and Eylon, 2006). The results indicate that teachers often do not expose their students to visual thinking. For example, in order to graphically reproduce squares, the teachers were given a dotted paper equally distanced (11 dots × 16 dots) and were asked to draw as many different squares as they can. This item is part of the "Square" Unit (# 2) in the Agam Program. Most of the teachers drew typical squares of different sizes (squares formed by two horizontal and vertical lines, in relation to the position of the given paper), such as □, suggesting that the teachers were less aware of squares in different positions, such as ◇, and that probably they did not expose the children to squares in these different positions.

3.3. *Points of view and contribution of the Agam Program*

Two studies, one conducted during 2008 with 54 preschool teachers and the second conducted in 2013 with 24 preschool teachers, aimed to gather information about the opinions of teachers involved in the Agam Program regarding implementation, influence, the program's contribution, and other issues (Eylon *et al.*, 2014).

For example, in the 2013 questionnaire, in their response to an open-ended item the teachers gave arguments for expanding the Agam Program: 75% of the teachers said that it develops visual abilities, visual language and the ability to see and pay attention to details; 63% of the teachers said that the program develops new concepts and contributes to other learning domains in preschool; 33% of the teachers said that the children love the program, it creates a good experience for them and provides a variety in teachers' work; 29% of the teachers mentioned that the program fits different types of children with special needs and language difficulties; 17% said that the program develops creativity and 13% that it promotes intelligence and learning strategies.

4. Conclusions

The development of visual thinking is almost neglected in preschool and primary school. It seems that there is a naïve assumption that somehow students will acquire visual thinking abilities which they will be able to use latter on, when such abilities will be needed. Our work with the Agam Program challenges this assumption and presents strong evidence that visual cognition can be developed through systematic learning, leading to student advances in a wide range of academic domains. Visual cognition is also important in mathematics, as evidenced by the many positive responses of mathematics educators to publications which describe the program and its achievements (see references below).

Acknowledgments

The authors would like to thank Yetti Solomon for her statistical analysis of the research data, and other members of the Agam Program staff: Micha Razel and Anat Ben-Zvi.

References

Agam, Y. (1984). Seeing is learning — or should be, *Kidma: Israeli Journal of Development*, **8**(2), 24–31 (in Hebrew).

Alsina, C. and Nelsen, R. B. (2006). *Math Made Visual: Creating Images for Understanding Mathematics*, Washington, DC: Mathematical Association of America.

Clements, D. H. and Sarama, J. (1999). *Preliminary Report of Building Blocks Foundations for Mathematical Thinking, Pre-kindergarten to Grade 2: Research-Based Materials Development*, NSF grant No. ESI-9730804. Buffalo, NY: State University of New York at Buffalo.

Clements, D. H. and Sarama, J. (2000). Young children's ideas about geometric shapes, *Teaching Children Mathematics*, **6**, 482–488.

Denton, K. and West, J. (2002). *Children's Reading and Mathematics Achievement in Kindergarten and First Grade*, Washington, DC: National Center for Education.

Eylon, B. and Rosenfeld, S. (1990). *The Agam Project: Cultivating Visual Cognition in Young Children*, Technical Report. Rehovot, Israel: Department of Science Teaching, Weizmann Institute of Science.

Eylon, B., Hershkowitz, R., Ilana, L., Markovits, Z. and Rosenfeld, S. (2014). *Educating the Eye: The Agam Program*, Research Report. Rehovot, Israel: Department of Science Teaching, Weizmann Institute of Science.

Gardner, H. (1983). *Frames of Mind: The Theory of Multiple Intelligences*, New York: Basic Books.

Gersmehl, P. and Gersmehl, C. (2007). Spatial thinking by young children: Neurologic evidence for early development and "educability", *Journal of Geography*, **106**(5), 181–191.

Gilbert, J. K. (2005). *Visualization in Science Education*, Dordrecht: Springer Publishers.

Markovits, Z. and Hershkowitz, R. (1997). Relative and absolute thinking in visual estimation processes, *Educational Studies in Mathematics*, **32**(1), 29–47.

Markovits, Z., Rosenfeld, S. and Eylon, B. (2006). Visual cognition: Content knowledge and beliefs of preschool teachers, in K. Novotna, H. Moraova, M. Kratka and N. Stehlikova (eds.), *Proceedings of the 30th PME Conference*, Vol. 4, Prague, Czech Republic, pp. 145–152.

McKim, R. H. (1980). *Experiences in Visual Thinking*, Boston: PWS Engineering Publishers.

National Research Council (NRC) (2006). *Learning to Think Spatially*, Washington, DC: National Academies Press.

Razel, M. and Eylon, B. (1986). Developing visual language skills: The Agam Program, *Journal of Visual and Verbal Languages*, **6**(1), 49–54.

Razel, M. and Eylon, B. (1990). Development of visual cognition: Transfer effects of the Agam Program, *Journal of Applied Development Psychology*, **11**, 459–485. Retrieved from: http://www.sciencedirect.com/science/article/pii/019339739090021B.

About the Authors

Rina Hershkowitz is a consultant to the Science Teaching Department at the Weizmann Institute of Science. Rina is also a member of the Agam Program staff and its academic consultant since its inception.

Zvia Markovits is a staff member of the Science Faculty at the Kibbutzim Academic College of Education, Technology and the Arts. Zvia is a member of the Agam Program staff.

Sherman Rosenfeld, a member of the Agam Program staff since its inception, currently is an educational consultant.

Lea Ilani is a member of the Agam Program staff in the Science Teaching Department at the Weizmann Institute of Science.

Bat-Sheva Eylon, a staff member of the Science Teaching Department at the Weizmann Institute of Science, has headed the Agam Program since its inception.

Chapter 11

Professional Development for Preschool Teachers: The CAMTE Framework and Repeating Patterns

Dina Tirosh*, Pessia Tsamir#, Esther Levenson‡ and Ruthi Barkai§

Department of Mathematics, Science and Technology, School of Education, Tel Aviv University, Tel Aviv, Israel

**Dina@post.tau.ac.il*

#Pessia@post.tau.ac.il

‡Levensone@gmail.com

§ruthi11@netvision.net.il

In Israel, there is a mandatory preschool mathematics curriculum. Yet, most preschool teachers have little preparation for teaching mathematics to young children. In this chapter, we present the Cognitive Affective Mathematics Teacher Education (CAMTE) framework that was developed at Tel Aviv University and used to investigate different elements of preschool teachers' knowledge and self-efficacy for teaching mathematics. The framework was also used to plan and enact professional development for preschool teachers. The chapter offers a theoretical rationale for the framework and then demonstrates how the framework was used within the context of teaching repeating patterns.

1. Introduction

Concern for preschool mathematics education may be seen in the rise of national curricula in various countries which now make specific recommendations for including mathematics as part of the preschool program. In Israel, the National Preschool Mathematics Curriculum (2010, Hereafter abbr. INPMC, see Appendix) is mandatory and contains specific guidelines and aims for children from 3–6 years. With new standards come new demands for teachers and the necessity for providing teachers with the tools to meet those demands. Yet, in Israel, as in many countries, attention to mathematics teacher education is mostly given at the primary and secondary levels (Kaiser, 2002). Many prospective preschool teachers in Israel attend only two mathematics education courses during their four-year education degree. Thus, providing ongoing professional development focused on mathematics preschool education is imperative.

This chapter describes the Cognitive Affective Mathematics Teacher Education (CAMTE) framework for professional development that takes into account preschool teachers' knowledge as well as their related self-efficacy beliefs. Preschool, in this chapter, will relate to children ages 4–6, one and two years prior to first grade. The chapter begins by presenting the general framework and the rationale behind the framework. It continues by demonstrating how the framework was used specifically within the context of teaching patterning. The INPMC (Israeli Ministry of Education, 2010) encourages teachers to engage children with patterning activities such as identifying, drawing, and continuing repeating patterns, as well as using mathematical language to describe these patterns. At the preschool level, educators have specifically noted that exploring repeating patterns may promote children's appreciation of underlying structures, emphasizing the minimal unit of repeat (e.g., Starkey, Klein and Wakeley, 2004). This chapter will present some results of our research into preschool teachers' knowledge and self-efficacy for teaching patterning, as well as an excerpt from our program.

2. The CAMTE Framework

In framing the mathematical knowledge preschool teachers need for teaching, we draw on Shulman (1986) who identified subject-matter knowledge (SMK) and pedagogical content knowledge (PCK). In addition, we found it useful to differentiate between two components of teachers' SMK: being able to produce solutions, strategies, and explanations and being able to evaluate given solutions, strategies, and explanations. Regarding PCK, we draw on the work of Ball, Thames, and Phelps (2008) and differentiate between knowledge of students (e.g., their conceptions and misconceptions) and knowledge of tasks (e.g., tasks that promote mathematical knowledge and tasks that assess knowledge).

The CAMTE framework also draws on Bandura's (1986) social cognitive theory, which takes into consideration personal beliefs and self-perception when explaining human behavior. Bandura defined self-efficacy as "people's judgments of their capabilities to organize and execute a course of action required to attain designated types of performances" (1986, p. 391). Hackett and Betz (1989) defined mathematics self-efficacy as "a situational or problem-specific assessment of an individual's confidence in her or his ability to successfully perform or accomplish a particular [mathematics] task or problem" (p. 262). The CAMTE framework takes into consideration teachers' mathematics self-efficacy as well as their pedagogical-mathematics self-efficacy, i.e., their self-efficacy related to the pedagogy of teaching mathematics. Table 1 illustrates the CAMTE framework within the context of teaching repeating patterns.

In this chapter, we focus on Cells 1 and 5. We report on a study which investigated preschool teachers' knowledge and self-efficacy for producing solutions to pattern tasks. We then illustrate how the results of that study were used in planning and implementing our professional development program.

Table 1. Applying the CAMTE framework to patterning knowledge and self-efficacy.

	Subject-matter		Pedagogical-content	
	Solving	Evaluating	Students	Tasks
Knowledge	Cell 1: Defining, drawing, extending repeating patterns	Cell 2: Identifying examples of repeating patterns	Cell 3: Knowledge of students' ways of drawing, extending, and comparing repeating patterns	Cell 4: Designing and choosing tasks for promoting and evaluating children's knowledge of repeating patterns
Self-efficacy	Cell 5: Mathematics self-efficacy related to defining, drawing, extending repeating patterns	Cell 6: Mathematics self-efficacy related to evaluating solutions	Cell 7: Pedagogical mathematics self-efficacy related to children's conceptions	Cell 8: Pedagogical mathematics self-efficacy related to designing and evaluating tasks

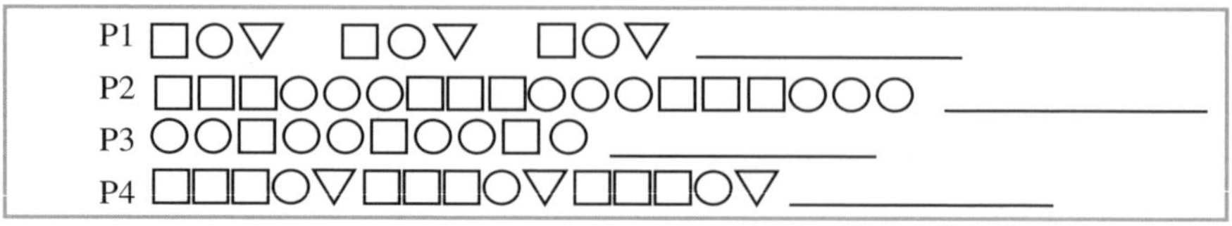

Fig. 1. Continue the pattern.

3. Investigating Knowledge and Self-efficacy

Several elements of knowledge and self-efficacy are related to Cells 1 and 5. This section focuses on teachers' knowledge and self-efficacy for continuing repeating patterns (see Tirosh *et al.*, 2015 for a more extended discussion of other related aspects). A group of 25 teachers were asked to rate their agreement (on a scale of 1–4, 4 being the strongest agreement) to the following statement: I am able to continue a repeating pattern. Following that statement, teachers were shown four patterns (see Fig. 1) and were requested to draw a continuation of each pattern.

Results indicated that teachers had high regard for their ability to extend repeating patterns ($M = 3.85$, $SD = 0.36$). Their self-efficacy was mostly in agreement with their knowledge. All continuations that teachers drew for the first two patterns and for the last pattern were acceptable ways of continuing the pattern.[a] Figure 2 summarizes the different ways each of the patterns ($P1$, $P2$, ...) was continued ($C1$, $C2$, ...) and the number of teachers who drew each continuation (listed in parentheses). For each pattern, the first continuation presented is the minimal set of items that completes the pattern with a complete unit of repeat. Starred (*) continuations indicate continuations which result in a pattern ending with an incomplete unit of repeat. Note that in the third pattern, one common continuation is repeating all of the elements that are presented in the pattern (10 elements).

Approximately 90% of the continuations ended the pattern in a complete unit of repeat, even when the length of the unit of repeat was relatively long (as in P2 and P4). In addition, for Patterns 1, 2, and 4, most teachers added just enough elements to end the pattern with a complete unit. Pattern 3 was the only pattern presented on the questionnaire that did not end with

[a]We acknowledge that sequences may be generated in an infinite number of ways.

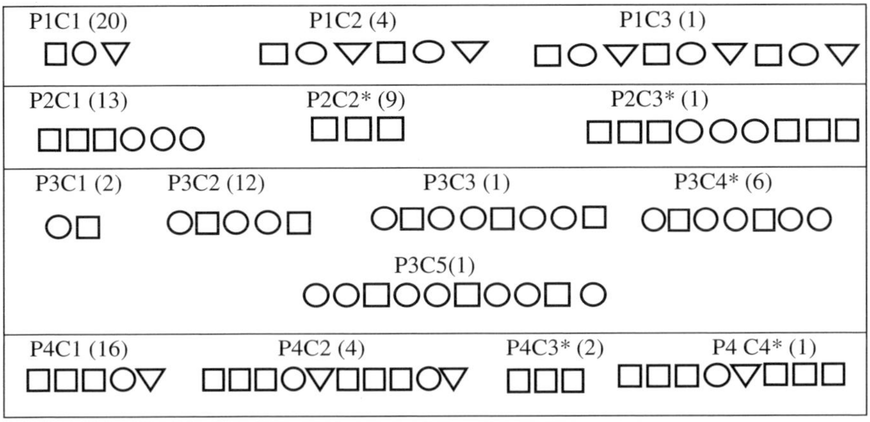

Fig. 2. Frequency of teachers' continuations of the patterns.

a complete unit. To end that pattern with a complete unit, it was possible to add less than a complete unit. Only two teachers did this.

The above results revealed a strong tendency to end patterns with a complete unit of repeat. Yet, repeating patterns, such as repeating decimals, do not always present themselves by ending in a complete unit. This finding highlighted an aspect of teacher's knowledge in need of more attention, namely, that ending in a complete unit is not a critical attribute of continuations of repeated patterns. In the following section, we describe part of our professional development which focused on this issue.

3.1. *Setting*

All preschool teachers who participated in our program had a B.Ed. degree in education and practiced in municipal preschools. The professional development program consisted of 21 hours. The teachers met over a period of about four months in the local professional development center in their area. All lessons and tasks were planned by the authors of this chapter; the fourth author was the actual facilitator (F). Taking into account teachers' obvious interest in planning and teaching lessons (Watson and Sullivan, 2008), we designed patterning tasks that teachers could implement with children, but could also be used to engage teachers with the mathematics involved in patterning, and promote their knowledge of patterning tasks and children's ways of solving those tasks.

Present the child with one pattern at a time. Prepare 2 or 3 separate containers, each containing cut-outs of triangles, squares, or circles. When presenting the first pattern, place before the child two containers, one with blue squares and one with red triangles. For each pattern ask: What comes next? Repeat the question two more times.

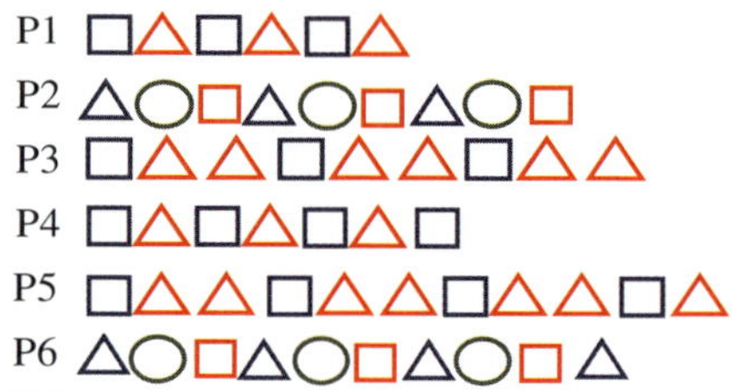

Fig. 3. A task for children and for teachers.

3.2. *Program segment*

During the second session of the program, the following activity was presented to the teachers (see Fig. 3). From the perspective of the minimal unit of repeat, the patterns presented have essentially three different pattern structures: ab, abc, and abb. The first three patterns end with a complete unit of repeat; the last three do not. This was planned in light of the results discussed in the previous section.

The task above stimulated a discussion among the teachers. After reviewing the first three patterns and their different structures, the teachers examined the fourth pattern (P4):

1	F:	What is the difference between P1 and P4?
2	Maya:	It (P4) has another cycle.
3	F:	What do you mean by another cycle? I still have three repetitions of the minimal unit of repeat.
4	Sophie:	ABA
5	F:	Ok. The first ended with a complete unit and this one doesn't. It has the first element of the next cycle... Some children choose which element comes next in the pattern by going back to the first element and adding that one. If the pattern ends in a complete unit, then that strategy works. But if they use that strategy here, what will they place?

6 Sherry: A square.
7 F: But what is generally accepted as coming next, is a triangle. That (placing a square) happens because most of the pattern tasks presented to children have patterns that end with a complete unit of repeat.

The minimal unit of repeat for both P1 and P4 is AB, or more specifically, □△. P1 has three cycles of the minimal unit of repeat while P4 has $3\frac{1}{2}$ cycles. The facilitator begins by promoting teachers' mathematical knowledge of patterns (Cell 1 of the CAMTE framework). This knowledge includes using precise mathematical terminology such as "minimal unit of repeat". In Line 5, the facilitator describes one frequent strategy children use when asked to find the next element of a pattern. This promotes teachers' knowledge of students' ways of solving problems (Cell 3). The facilitator ends (Line 7) by explaining to the teachers how important it is to engage children with various patterns, including those which do not end in a complete unit of repeat (Cell 4).

4. Some Final Remarks

Obviously, there is more to teaching mathematics in preschool than the topic of repeating patterns. We have also used the CAMTE framework to both investigate and promote preschool teachers' knowledge and self-efficacy for teaching number concepts and geometry (e.g., Tsamir *et al.*, 2015). Those studies showed that knowledge and self-efficacy are not only topic specific, but also task specific. As such, there is a need to continue investigating teachers' knowledge and self-efficacy for teaching different areas of preschool mathematics in order to provide the professional development so needed.

Acknowledgment

This study was supported by The Israel Science Foundation (ISF grant No. 1270/14).

References

Ball, D. L., Thames, M. and Phelps, G. (2008). Content knowledge for teaching, *Journal of Teacher Education*, **59**(5), 389–407.

Bandura, A. (1986). *Social Foundations of Thought and Action: A Social Cognitive Theory*, Englewood Cliffs, NJ: Prentice Hall.

Hackett, G. and Betz, N. (1989). An exploration of the mathematics self-efficacy/mathematics performance correspondence, *Journal for Research in Mathematics Education*, **20**(3), 261–273.

Israeli Ministry of Education (2010). *The Israel National Preschool Mathematics Curriculum (INPMC)*. Jerusalem, Israel: Author [in Hebrew].

Kaiser, G. (2002). Educational philosophies and their influence on mathematics education — An ethnographic study in English and German mathematics classrooms, *ZDM — International Journal on Mathematics Education*, **34**(6), 241–257.

Shulman, L. S. (1986). Those who understand: Knowledge growth in teaching, *Educational Researcher*, **15**(2), 4–14.

Starkey, P., Klein, A. and Wakeley, A. (2004). Enhancing young children's mathematical knowledge through a pre-kindergarten mathematics intervention, *Early Childhood Research Quarterly*, **19**(1), 99–120.

Tirosh, D., Tsamir, P., Levenson, E., Barkai, R. and Tabach, M. (2015). Preschool teachers' self-efficacy and knowledge for defining, drawing, and continuing repeating patterns, *Presented at the 21st MAVI Conference in Milan, Italy*.

Tsamir, P., Tirosh, D., Levenson, E., Barkai, R. and Tabach, M. (2015). Early-years teachers' concept images and concept definitions: Triangles, circles, and cylinders, *ZDM — International Journal on Mathematics Education*, **47**(3), 497–509.

Watson, A. and Sullivan, P. (2008). Teachers learning about tasks and lessons, in D. Tirosh & T. Wood (eds.), *The International Handbook of Mathematics Teacher Education*, Rotterdam: Sense Publishers, Vol. 2, pp. 109–134.

About the Authors

Dina Tirosh is a full professor of mathematics education in the Department of Mathematics, Science and Technology Education at Tel Aviv University.

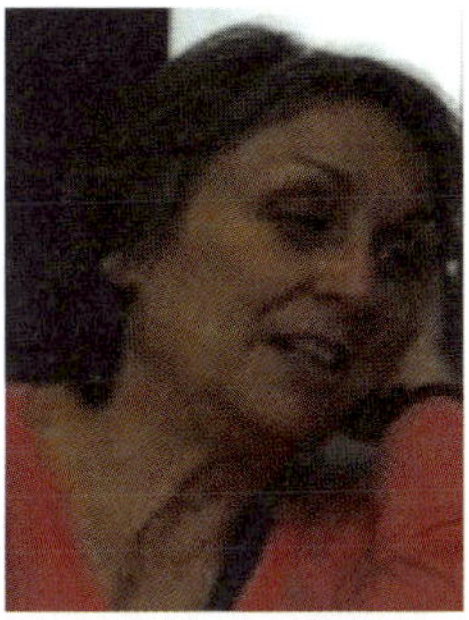

Pessia Tsamir is a full professor of mathematics education in the Department of Mathematics, Science and Technology Education at Tel Aviv University.

Esther Levenson is a researcher and teacher educator in the Department of Mathematics, Science and Technology Education at Tel Aviv University, and a lecturer at the Kibbutzim Academic College of Education.

Ruthi Barkai is a senior lecturer and a researcher in the Department of Mathematics, Science and Technology Education at Tel Aviv University, and a lecturer at the Kibbutzim Academic College of Education.

Chapter 12

Time to Know — A Socio-constructivist Initiative to Integrate Computers in the Teaching and Learning of Primary Mathematics

Dovi Weiss*,‡ and Tali Wallach#,§

**Kibbutzim Academic College of Education, Tel Aviv, Israel*

#Achva Academic College of Education, Arugott, Israel

‡dovi.weiss@smkb.ac.il

§tali.wallach@gmail.com

This chapter describes the Time to Know (T2K) Initiative. T2K created a digital teaching platform and interactive core curriculum for one-to-one classrooms. The platform was designed to empower the teacher in conducting different instructional practices based on the socio-constructivist approach. The T2K platform will be presented and discussed by going through a typical lesson.

1. Introduction

Time to Know (T2K) was established in 2005 with the sole purpose of harnessing technology to improve the traditional education system in Israel. A total of 150 experts were recruited to establish an innovation lab that addressed the following issues:

- Single way of teaching helps no one.
- Lecturing format offers a low level of engagement.
- Current teaching provides a low level of formative assessment.
- Classrooms lack 21st century skills preparation.

After extensive techno-pedagogical research, field testing, and R&D development, the first Digital Teaching Platform (DTP) was invented. The DTP was the first system designed for teaching in a one-to-one classroom; as opposed to an LMS (Learning Management System), designed specifically for e-learning. The solution was implemented in a selected number of Israeli schools.

With significant improvement in pedagogical performance, roll out was extended to hundreds of schools and thousands of students in Israel and abroad.

The guiding principles for creating the T2K solution were as follows:

- Holistic solution that addresses teaching, learning, and assessment.
- Teacher-centric approach that empowers the teacher and does not replace her.
- Socio-constructivist approach (Palincsar, 2005) for conceptual learning (not only practicing) and for promoting 21st century skills (Griffin, McGaw and Care, 2012).
- Formative assessment including real-time monitoring (Black and Wiliam, 2006).
- Digital core curriculum (not supplemental) that is orchestrated by the teacher with the assistance of the digital teaching platform.

2. T2K Environment, What is it All About?

Let us now describe how the T2K environment is applied to the classroom by taking a peek at a 4th grade learning process and describing one typical mathematics lesson. The teacher is Rachel. Her plan for this lesson is to facilitate her students to recognize that 1/2 can be represented as 2/4, 3/6, etc., and that these are actually "different names" for the same fraction. She plans this lesson as a first step toward generalization, where students will be able to suggest equivalent fractions by themselves.

2.1. *Lesson introduction*: *Is* 3/6 *equivalent to* 1/2?

Rachel starts her lesson by showing a short animated film. The students are exposed to a scenario where two well-known characters, Menny and Cheryl, are traveling through a city. They reach Febo's chocolate store, and see a flashing billboard advertising chocolate bars that are half dark chocolate and half white chocolate (Fig. 1).

Menny looks at the sign and says: "This is false advertising! The sign says that it's 1/2 dark chocolate but the flashing lights show 3/6."

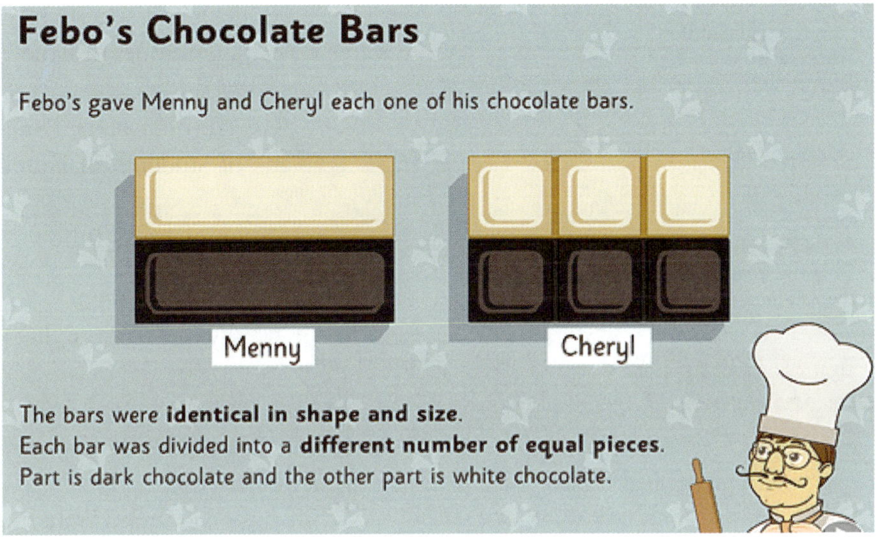

Fig. 1. The flashing billboard.

Cheryl says: "The advertisement is correct."

The scenario ends with the question "Who is right, Menny or Cheryl?"

The film lets the teacher raise an engaging dilemma in a realistic context and to follow-up by leading a class discourse about who's right. The students are motivated to present their views and to explain their reasoning (critical thinking and communication). The teacher does not rush to judge the student's explanations or to give the correct answer. She keeps it as an open question and suggests that the students use their computer to find a solution.

2.2. *Lesson exploration*: *Is* 3/6 *equivalent to* 1/2?

In the exploration stage, students work with computer support. They can work individually, in pairs, or groups, depending on the teacher's choice and the student's preference. In our case study, most students worked in pairs with each assigned to a computer device. On every computer, there appears an interactive mathematical application (Fig. 2). The students explore "hands on". They move the cursor to the different parts of the "chocolate bar", showing information in the checking area, to explore

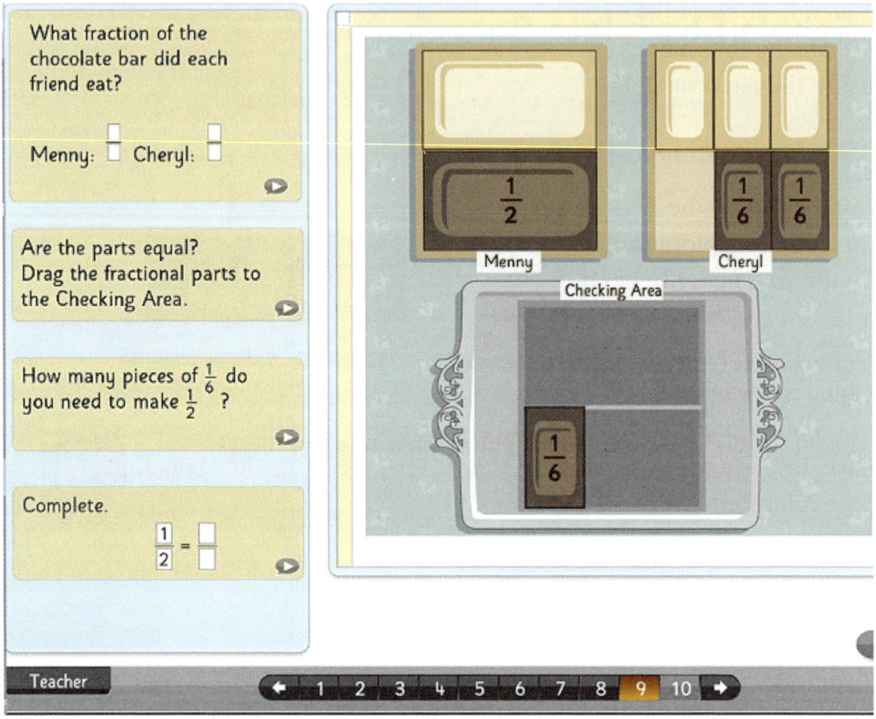

Fig. 2. Interactive application — Area model.

the connections between 1/2 and 3/6. Students see in a tangible way that 3/6 equals 1/2. During this time, Rachel is free to walk around to "visit" her students, assess their working process, and learn about what they have discovered. Rachel explores her students reasoning. She also identifies a few students who have generalized the idea, and came to the understanding that there are other equivalent fractions, such as 4/8, 5/10, etc.

Rachel asks a pair of students to draw a picture to illustrate how 4/8 equals 1/2, and another pair to draw a picture showing how 5/10 equals 1/2. The two student pairs were asked to send their drawings to the gallery, which is a common space that showcases student work.

After about 10 minutes, the teacher stopped the activity and returned to class mode.

Based on her assessment (as part of teaching), Rachel continues to the next step of the lesson. In our case, she decides to lead a discourse toward generalization of the idea. Her students are ready to grasp that 2/4, 3/6, and 4/8 are another names for 1/2 and are able to explain why 1/2 equals 2/4 by using the Area model.

2.3. *Lesson summary*: *Which fractions are equivalent to 1/2?*

In this part of the lesson, Rachel is on stage. She leads a discourse based on student exploration and opens the gallery to demonstrate the work sent by the students. She invites the two pairs to the whiteboard to explain why $4/8 = 1/2$. The students take the role of the teacher and demonstrate their skills to express their thoughts (verbal communication), while using the visual illustration.

Part of the class is able to internalize the principle that fractions can have different names and know how to generate equivalent fractions. For this particular case, most students can now grasp that 1/2 has different names. They can suggest other fractions equivalent to 1/2 and even note what is common to all of them. The students are ready to make the decision about who is right, Menny or Cheryl, and to consolidate and practice their new understanding by working independently.

2.4. *Lesson consolidation*: *Independent learning and practice*

Just like any regular class, this 4th grade group is heterogeneous. Rachel needs differentiated content to provide each student with suitable activities.

The T2K environment includes different activity types and levels. For example, there are tasks designed to demonstrate the key concepts; tasks for practicing a specific skill; tasks oriented for closing gaps of prior knowledge, performing tasks, enrichment tasks, etc. All these and others are available for three differentiated performance levels. This provides a partial response to class heterogeneity.

Rachel assigned each student to work independently. The students open their computers. They work individually or in pairs. Every student or student pair receive activities that is suitable to his or her current performance level and preferred style of learning.

During practice time, while all students are engaged in their own tasks, Rachel works with a group of six students who need personal attention. She gives them a short intervention using manipulatives. At the same time,

Rachel's monitoring the "dashboard" of her portable computer. She sees the progress and performance in real time of each student and the entire class. After a few minutes, an alert is triggered indicating that a student is not progressing as planned. Rachel goes over to him as a timely intervention and then returns to the small group.

3. Discussion

The teaching-learning scenario described above sheds light on how the T2K program addresses three major principles: the Socio-constructivist approach, Formative assessment, and Teacher-centric solution.

Socio-constructivist approach: The lesson starts with an engaging and relevant scenario. The chocolate-bar video provides the realistic base-line context for knowledge construction. The class discussion that follows enables the teacher to informally assess the level of the students and to create a trigger for the exploration phase. In the exploration phase the students work hands-on with the help of an interactive and dynamic fractions applet. In this stage every student starts to develop conceptual insights. In the gallery phase, the teacher conducts a discussion that leads the students to initial knowledge construction. Next the new knowledge is consolidated by personalized practicing.

Formative assessment: The teacher had conducted a formative assessment during the entire time of knowledge construction and consolidation. The collaborative gallery enabled a quick assessment of student outcomes. During practice time, the teacher utilized a real-time dashboard, which helped her to identify students who were having difficulties and to approach them. The students received feedback during practice time and were able to self-assess themselves.

Teacher-centric solution: The teacher's role is essential in facilitating the knowledge construction process while addressing student diversity. The teacher is empowered in various ways. For example, she can use her resources to provide conceptual learning (and not only for practicing) as to promote 21st century skills, such as Communication, Collaboration, Critical Thinking, and Computer Literacy.

4. Research

The T2K teaching and learning environment was reviewed by the Henrietta Szold Institute in Israel in 2009. Participants were 423 students from five schools: 168 4th grade students (90 students in experimental classes and 78 benchmark students); and 255 5th grade students (112 in experimental classes and 143 benchmark students). Twelve teachers and two school principals took part in the experiment and also participated in the study.

Comparative examination of student achievements in standardized tests showed that in general, Time to Know students in the second year of the experiment made better progress throughout the year in subjects that were available in the T2K program, as compared with the control class. Further observations showed that T2K classes were characterized by a high level of student concentration, a reduction in disciplinary problems, and a relatively low percentage of time spent on organizing the lesson. This pattern is reinforced in the second year of the experiment (Henrietta Szold Institute, 2010).

In 2011, the T2K environment was also tested in four primary schools from the Grand Prairie Independent School District (GPISD) near Dallas (Rosen and Beck-Hill, 2012). A total of 476 students participated in data collection (4th grade: 129 experimental, 77 control students; 5th grade: 154 experimental and 116 control students) and 20 teachers (12 experimental and 8 control).

The results of the increased achievement scores, based on the TAKS tests administered before the program and after a year-long participation, demonstrated that 4th grade experimental students significantly outperformed the control students in mathematics scores ($M = 597.6$ compared to 673.9). During the same period, the findings showed only a small increase in the control group (math: $M = 611.6$ compared to 660.1). Fifth grade experimental students significantly outperformed the control group in mathematics scores ($M = 654.7$ compared to 700.6). The findings showed only a small increase in the control group during the same period ($M = 646.4$ compared to 674.1). Findings indicated consistent and highly positive findings of the efficacy of a constructivist one-to-one computing program regarding student mathematical achievement, differentiation in teaching and learning, higher student attendance, and decreased disciplinary actions.

References

Black, P. and Wiliam, D. (2006). *Inside the Black Box: Raising Standards Through Classroom Assessment*, Granada Learning.

Griffin, P., McGaw, B. and Care, E. (2012). *Assessment and Teaching of 21st Century Skills*, Dordrecht: Springer, p. 36.

Henrietta Szold Institute (2010). The Time to Know Research Report.

Palincsar, A. S. (2005). 12 social constructivist perspectives on teaching and learning. An introduction to Vygotsky, **285**.

Rosen, Y. and Beck-Hill, D. (2012). Intertwining digital content and a one-to-one laptop environment in teaching and learning: Lessons from the time to know program, *Journal of Research on Technology in Education*, **44**(3), 225–241.

About the Authors

Dovi Weiss holds a PhD in Mathematics Education from Tel Aviv University. He is head of Digital Pedagogy and director of the Master's Program in Educational Technology at Kibbutzim Academic College of Education. He is also the pedagogical founder of Time to Know and its Chief Scientist.

Tali Wallach holds a PhD in Mathematics Education from the Weizmann Institute of Science and an MSc in Operation Research from the Technion — Israel Institute of Technology. She served as director of Mathematics Curriculum Development at Time to Know while taking an essential part in building its pedagogical innovative vision. Tali is a faculty member at the Achva Academic College of Education.

Part III

Issues and Innovations Related to Mathematics Education at Middle and High School (Grades 7–12) in Israel

Chapter 13

Exhausting Students' Potential in Mathematics: A Comprehensive Approach to Promoting Both Struggling and Promising Students

Orit Zaslavsky[*,#,¶], **Liora Linchevski**[‡], **Noga Hermon**[§], **Drora Livneh**[‡] **and Iris Zodik**[*]

[*]Faculty of Education in Science and Technology, Technion — Israel Institute of Technology, Haifa, Israel
[#]Steinhardt School of Education, New York University, USA
[‡]School of Education, The Hebrew University, Jerusalem, Israel
[§]Kibbutzim Academic College of Education, Tel Aviv, Israel
[¶]oritrath@gmail.com

During the years 2000–2010 we led a national project called *Mitzui u-Metzuyanut in Mathematics* (the meaning of which is "Exhausting Students' Mathematical Potential") that began as an experimental project in 56 middle schools and developed into a nation-wide mandatory program in every middle school in Israel. The project aimed at implementing a special program in mathematics for low-achieving students, hand in hand with a special program for high-achieving students. The project's main goal was to enhance the mathematics learning opportunities for middle school students (7th–9th grades) at both ends — both promising students who are mathematically capable and struggling students who are mathematically disadvantaged. Participating schools were required to implement the program in full — for both ends. The project was launched and supported by the Ministry of Education, and was conducted as a unique collaboration between the Ministry of Education and two academic institutes (The Technion — Israel Institute of Technology in Haifa, and The Hebrew University in Jerusalem). In this chapter we describe the design and implementation of the project as well as its impact on students, teachers, and the school system.

1. Project Background

The project was established in response to a call of the Ministry of Education. During the 15 years preceding the project, the recommendation of the Ministry of Education, and the most common way of organizing mathematics instruction at the middle school, was in heterogeneous classes with no tracking in 7th grades, and in more homogeneous classes, to which students were tracked according to their ability, in 8th and 9th grades. However, two–three years before the project was launched, teachers and educational policy leaders began to doubt whether this teaching organization provides equal learning opportunities for all students. Moreover, further examination of the system indicated that this structure did not meet the special needs of students on both ends — the promising and excelling as well as the struggling and underachieving.

As a result, the department of secondary schools in the Ministry of Education decided to initiate a project that would address this problem at the school level, through the existing mathematics teachers who were already teaching in school, within the hours allocated to teaching mathematics. The idea was to address the special needs of the students on both ends without disadvantaging the rest of the students who were not the target population of the project. From the very beginning, the leading team members from both academic institutions were heavily involved at the level of planning and decision making in close collaboration with the leading persons of the Ministry of Education in charge of secondary schools.

1.1. *Goals*

The goals of the project stemmed from the standpoint that equity in mathematics education requires addressing students' differential needs by special teaching approaches and appropriate learning material and resources (NCTM, 2000). The unifying goal was for each student in the project to do his or her best to exhaust his/her potential in mathematics.

Israel holds national secondary school graduation exams in mathematics at three levels (basic, intermediate, and advanced). Each exam comprises 2–3 questionnaires. At the time, most of the projects that aimed at helping under-achieving students to do better in their mathematics studies focused on the 50th to 60th percentiles. On the other hand, most projects that aimed at fostering excellence in mathematics focused on the 95th to 100th percentiles. In our project, the target populations were defined differently, in a special way. This difference stemmed from the project's goals. The high-achieving population (HA) was defined as students in the 85th to 100th percentile, while the under-achieving population (LA) was defined as students in the 17th to 32th percentile. As a rule, the latter do not reach the grade level that takes the secondary school graduation exam.

For the LA population, the goal was to be able to pass the basic-level graduation exam in mathematics at the end of 9th grade. Note that these students normally would not even be given the chance to take the test, not to mention as early as 9th grade (mostly students take it at the 10th grade or later). For HA students the goals were to motivate them to study mathematics at an advanced level, and to develop their mathematical thinking and creativity, by providing rich and challenging learning opportunities.

An added value of the project, which was set forth as an additional goal, was to empower the participating teachers.

1.2. *Design*

The program's design included the development of a special curriculum for each population (LA and HA), consistent with the official school mathematics curriculum, learning material for students, resource material for teachers, corresponding pedagogical practices, and ongoing in-service professional development activities with teachers that required immediate implementation in their classrooms. In many ways, the program dictated several changes in teacher practice.

The principles on which the HA program was based concur with the approaches found in the professional literature dealing with the need to nurture promising students in mathematics and the desirable ways of addressing this need. Specifically, the program adapted three dimensions described by Sheffield (1999): Depth and complexity within the existing curriculum, enrichment through extra-curricular activities and topics, and increased pace. The HA program constantly included non-routine and unfamiliar problems, for which students were encouraged to search for multiple approaches to solving them. The emphasis on considering alternative approaches was done in order to help students develop fluency (i.e., the number of different correct answers, methods of solution or new questions posed), flexibility (i.e., the number of different categories of answers, methods, or questions, e.g., algebraic, geometrical, graphical), and originality (i.e., solutions, methods, or questions that are unique and show insight).

The LA program set forth specific measurable mathematical learning goals, at a level that this population had never before been faced with. The design principles of the program addressed typical student cognitive obstacles, difficulties in acquiring the mathematical language and basic mathematical concepts, and affective obstacles. It required that teachers learn innovative didactic and pedagogical strategies. The main guiding principles of the LA program included identifying of mathematical concepts and skills within the middle school mathematics curriculum that were critical for the learning of mathematics in higher grades, identifying mathematical concepts and skills that could facilitate the progress in continuing study of mathematics but were not included in the curriculum, and determining the absolutely necessary concepts and skills that could promise success in the basic secondary school graduation exam.

One of the main decisions of the project's team was to conduct workshops and other professional development activities for both teacher populations together — teachers who were assigned to teach the LA students as well as teachers who were assigned to teach the HA students. In addition, there was an attempt to have participating teachers teach both ends. The intention was that all the participating teachers would be exposed to a wide range of theoretical considerations and teaching strategies, as well as to diverse accounts of colleagues who would share their classroom experiences as part of the professional development activities. The idea was for the teachers to significantly enhance their professional knowledge and their ability to judiciously implement the best practice for their classrooms. All the participating teachers taught also classes that were not included in the project (i.e., "mainstream" classes). This way, a considerable amount of additional students benefitted indirectly from the project.

In addition to exposing teachers to the special requirements of the program, regional meetings were held to provide professional support for implementing the program and sharing experiences. Three times per year, we conducted tests that reflected the program's goals, and provided the teachers with an "objective" measure of their students' progress. The leading mathematics supervisors of the state's Department of Education were fully involved in the entire process of designing and implementing the project's tests to ensure consistency with the official requirements of the national school mathematics curriculum at each grade level.

1.3. *Participating schools*

The participating schools were from the Hebrew-speaking sector as well as the Arabic-speaking sector. In all middle schools, students studied mathematics 5 hours per week. In each school we identified out of the entire 7th grade cohort about fifteen low-achieving students that were not likely to reach high school with the background and success necessary for continuing their studies even towards basic-level secondary school graduation exam in mathematics. Emphasis was put on identifying low-achievers that were capable of doing better given the proper opportunities. We also identified in each school about twenty-five 7th graders who were promising students that were either high-achievers or with high potential and motivation to study an enhanced mathematics program. Thus, in each school two special groups were formed. A special program was designed for each group. For the entire 3 years (grades 7–9), each group studied mathematics separately: the LA group 5 hours per week, while the HA group got 2 additional hours, i.e., a total of 7 hours per week.

The selection criteria for the project's students were different for each population. For the HA groups, the main criterion was student motivation, as the program required 2 additional weekly lessons in mathematics, and much more work than in a regular mathematics class. In addition, we encouraged teachers to identify students who exhibited original mathematical thinking in class, even if their formal achievement in mathematics was not at the top level, as often promising students are bored and do not exhaust their potential. The final decisions were done together with the school counselor and mathematics teaching team.

For the LA groups, a special diagnostic tool was developed, in order to ensure that the right students were identified, that is, struggling students who had reached a very minimal achievement level in mathematics, but who could — with an appropriate didactic approach, be able to meet the requirements of middle and high school mathematics, at least at the basic level, and who otherwise would most likely drop out of school before reaching 10th grade. The diagnostic tool and selection process were rather complex, and included several written tests that measured not only student basic mathematical knowledge (at the 4th grade level) but also their ability to deal with higher order thinking. Teachers were also heavily involved in this process, as they were trained to conduct classroom observations to identify students who were likely to benefit from this program. This selection process made sure that schools did not include in the LA groups students on account of discipline issues with which teachers found it hard to deal in their regular classes.

The school administrators selected teachers that seemed suitable for teaching the LA group, the HA group, or both. The implementation of the two programs for the project's students required special preparation of the practicing teachers to teach according to these challenging programs.

2. Indicators of Success

The indicators of the project's success varied with respect to the target populations. Within the context of the project, we identified several manifestations of change in teachers' practices, knowledge and beliefs.

For HA students, the most overwhelming indication of the success was what we consider "voting with one's feet". By this we mean, that in spite of the fact that students studied 2 additional hours per week, worked harder, and did not receive any formal reward for their work — over 90% of the students that enrolled in the program when they were in 7th grade continued persistently till the end, when they graduated from 9th grade. In the Israeli culture this is compelling, as students do not tend to invest more in studying time than officially required. In addition, teachers shared with the program's team members on a regular basis their enthusiasm from their students' motivation and performance.

The success of the LA program was measured mostly by the high percent of students who passed the basic-level secondary school graduation exam in mathematics at the end of 9th grade, a goal that at the launching of the project seemed so ambitious and was hard to believe that it would be possible to fulfill. For example, in 2006 the average score of the (1251) LA students on the basic-level exam was 90 (out of 100). Of these students, 57% scored 95–100. This rate of success is typical of the LA students in other years as well. These results were compelling, as they showed the feasibility of making a significant change in the future of the target population, who otherwise would drop out of the system and would not have the opportunities to graduate from high school. It turned out that the program increased LA students' self-efficacy and their motivation to invest time and energy to succeed in their mathematics studies.

This project turned out to be revolutionary also from the standpoint of teachers. Teachers claimed that they had been "transformed" in ways that made a difference and impacted not only the LA and the HA students. Many of them claimed that they adopted approaches to teaching and types of problems used in the LA and HA programs to other classes in which they teach. Ongoing reports of official mathematics supervisors, who visited classrooms and met with teachers in their schools, indicated that they were impressed by the value and impact of this project. It became evident that by empowering teachers, the whole system benefitted.

Based on the success of this project, the Ministry of Education decided first to expand the program to more schools, in spite of the budgetary implications, and then further — to declare that it was mandatory in *every* school, nation-wide, to have a program that addresses the needs of both the LA and HA student.

References

National Council of Teachers of Mathematics (NCTM) (2000). *Principles and Standards for School Mathematics*, Reston, VA: Author.

Sheffield, L. J. (ed.). (1999). *Developing Mathematically Promising Students*, Reston, VA: National Council of Mathematics, pp. 121–132.

About the Authors

Orit Zaslavsky, a Mathematics Education Professor, was an active member of the Faculty of Education in Science and Technology at the Technion for over 20 years, during which she was director of large-scale projects, which included professional development programs for secondary mathematics teachers, a program for fostering students' excellence in mathematics, and curriculum design. She has retired from the Technion and is now at New York University.

Liora Linchevski, a Mathematics Education Professor, was the head of the S. Amtizur Unit for Research in Mathematics Education, at the Hebrew University of Jerusalem. She directed several large-scale projects, which focused on mixed ability classes and on under achieving students. She is now developing materials for students and teachers at the primary school level.

Noga Hermon served as a Mathematics Education Senior Mentor at the Ministry of Education of Israel, and the coordinator of the project "Mitzui u-Metzuyanoot in Mathematics". She was an instructor at Tel Aviv University, in the field of research methods and evaluation, and is currently an instructor at Kibbutzim Academic College of Education.

Drora Livneh was a lecturer at the School of Education of the Hebrew University of Jerusalem, and at the Mathematics Department in David Yellin Academic College of Education. She served as a leading team member of several projects that focused on mixed ability classes and under-achieving students, including the development of mathematics textbooks.

Iris Zodik is an associate lecturer at the Faculty of Education in Science and Technology at the Technion. She is currently leading the "Academia Classroom" project at the Technion and co-leading the Kadar Center. She has led the development of mathematics textbooks as well as a program aimed at fostering excellence of high-achieving students.

Chapter 14

Middle School Mathematics Curriculum Based on the Power of Open Technological Tools: The Case of CompuMath Project*

Rina Hershkowitz# and Michal Tabach‡

Weizmann Institute of Science, Tel Aviv University, Rehovot, Israel
#Rina.hershkowitz@weizmann.ac.il
‡tabachm@post.tau.ac.il

This chapter uses the CompuMath project as an example of a comprehensive curriculum development project to integrate technological tools into everyday school mathematics. After sketching its border lines, we provide a glimpse from the classroom. We end with some longer term implications of the project.

1. CompuMath Project: Overview

In 1993, the Israeli Ministry of Education decided was made to intensively support the use of technological tools in teaching and learning several topics. The *CompuMath* project was founded then under the responsibility of the Mathematics Group in the Science Teaching Department, The Weizmann Institute. The project was led by Rina Hershkowitz, and included, among others, the following team members: Dany Ben-Zvi, Tommy Dreyfus, Alex Friedlander, Nurit Hadas, Tzippi Reznick, Baruch Schwarz, Hanna Stein, and Michal Tabach. The project is an example of a curriculum development and research project in which not only theoretical frameworks, mathematical contents and computerized tools were taken into account, but also lessons learned from development and from research in previous projects.

*This study was partially supported by Israel Science Foundation grants 438/15.

At the beginning of the project the team members had some experience with curriculum development, which its main characteristics were encouraging investigations and argumentations by middle school students as individuals, groups, and whole class communities — a working inquiry classroom (Artigue and Blomhøj, 2013) — in line with a socio-cultural views about mathematics and its learning.

However, we did not know how this experience would play out in a technology-rich environment. We could only hypothesize the potential of computerized tools. That is, we lacked experience with the systematic integration of technological tools and with consistent use of their potential. As designers, we had not yet acquired the needed instrumentation knowledge (Trouche, 2004). Based on our own use and familiarity with computerized tools, we chose to integrate the use of open, technological tools (e.g., graphical tools, spreadsheets, dynamic geometry software, and Computer Algebra Systems), and to benefit from their potential in learning mathematics. A discussion about criteria for choosing computerized tools can be found in Hershkowitz *et al.* (2002, pp. 661–663).

The development team had to deal with many facets of theory, research, and practices of development and implementation: practices were fed by theory and research, and vice versa. The team functioned as a cell eager to live and develop, whose life was in large part determined by its interaction with an unknown outside world. Through this interaction, the curriculum development activity constantly redefined its own components. The development was followed by implementation in several experimental classes and by research. Research became a crucial component in the curriculum development activity, in design–research–redesign cycles. The project included every topic in the middle school curriculum: algebra, geometry, functions, statistics, and probability (see Appendix for more details). It was a comprehensive process with three stages:

(i) The *predesign* stage involving considerations before starting the actual development and research work (e.g., in algebra, Arcavi, Friedlander and Hershkowitz, 1990).
(ii) The *initial design–research–redesign* stage involving a design of isolated activities and their implementation in a few classrooms, accompanied by class research on learning and teaching practices (e.g., in geometry, Hadas, 2001; Hadas, Hershkowitz and Schwarz, 2000).

(iii) The *expansion* stage comprising the creation of coherent sequences of redesigned activities, published as sequences of textbooks for students and teachers — a complete curriculum and its implementation, including dissemination of the project's curricular aims and "spirit" on a large scale (e.g., in Statistics, Ben-Zvi, 2001).

Next, we bring one example from stage (ii) — research that accompanied implementation of one activity in an experimental 9th grade classroom.

2. The "Growing Rectangles" Example

The next example demonstrates what constitutes evidence when solving problem-situations, or the status of computer displays. We show that certain socio-mathematical norms in our rich environment are not always the result of interactions between students and teacher, as suggested by Yackel and Cobb (1996). Our example leads students to interact with a tool and activity specially designed to convey mathematical ideas.

This example (Fig. 1) is taken from the sixth week of an introductory year-long course on functions in 9th grade. At that stage, students had practiced the passage between representations, actions within representations such as walking on a computerized graph with the pointer and reading the coordinates of the points on the graph. They had also practiced the substitution of the x-value of points on the symbolic representation to obtain the corresponding y-value. Thus, students had intensively practiced actions and displays within specific representations and passing to other representations to view corresponding actions and displays. This practice was obtained as students were engaged in activities such as the Growing Rectangles, in which operations and actions were given context for meaning while investigating problem situations.

The students are first invited to make conjectures without using the computerized tool, then to use it to check them. Students tried to figure out which family will eventually take the lead over the years by using intuition and/or by computing "by hand" the areas of the three families of rectangles for a few values. After allowing a little time for investigation, the teacher initiated a whole class discussion on possible conjectures concerning the changes in areas. Most students guessed that family **B** will take the lead, and family **C** will always remain last. One student, Zivit, argued that **C** will

Each of the following three families of rectangles has its growth pattern:

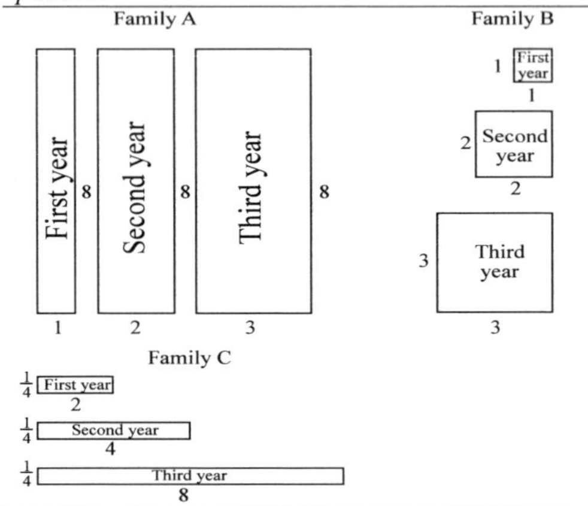

*In family **A**, the width grows each year by one unit; the length remains constant at 8 units. In **B** the width and the length of the rectangle grow each year by one unit. In **C** the length doubles each year, and the width remains equal to 1/4.*

Try to make conjectures comparing change in area, perimeter etc. For example, answer the following questions: In which family will the area be largest and when? When will the area of each family exceed 1000? Check your conjectures with the tools available and explain!

Fig. 1. The "Growing Rectangles".

take over sooner or later and said:

> *The rectangle in family **C** doubles each year and in family **A** it grows by 8 consistently, so we will reach a situation where **C** will overtake **A**.*

The teacher confronted students on this issue and asked all students to vote on Zivit's **C** conjecture against the formerly held **B** conjecture. Many students voted for **B**, two voted in favor of Zivit's conjecture and about ten students chose to abstain. Then the class turned to work in small groups using graphical calculators. The teacher asked the students to write a report on their work. In many groups, students translated the situation into symbolic representations (with some difficulties in translating the area function in **C**, $y = 1/4 * 2^x$), and then obtained the graphs with their graphical calculators (see Fig. 2).

After the groups had written their reports, the teacher discussed the mathematical findings and strategies with the class. Students reported that on the eighth year ($x = 8$), the three rectangles have the same area, and

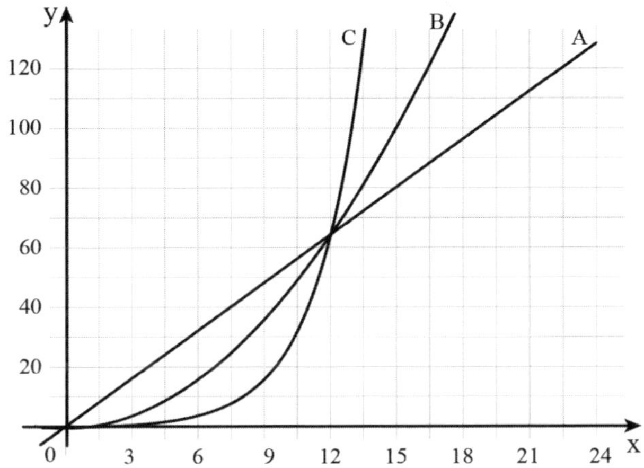

Fig. 2. The three graphs.

that from that day on, family **C** takes the lead from family **A**. Family **B** remains in between. The evidence provided by the different representations was accepted and no student declared the computer wrong. Rather, they tried to reinterpret the situation. Most of the students were seemed to be satisfied by the conclusion that "**C** grows faster than the other families and it catches them up."

In this example, students started to realize that an interaction with the computerized tool, while solving problems, may lead to unexpected results, which confronts former intuitive conjectures. The result (the artifacts embedded in the representations) demands reinterpretation. Evaluating mathematical conjectures with tools in order to confirm or refute them becomes normative in the class. It was surprising to find that when displays or actions lead to a conflict with previous beliefs, it was agreed that the computer displays and actions expressed the "same thing" (Resnick, Schwartz and Hershkowitz, 1994). We interpret this "trust" in the computer as evidence for students' belief in software potential of translation between representations.

For many students, this norm is not negotiated in the sense that the teacher takes an active role in its establishment. Rather, analogical thinking between representations' practices in activities around problem-situations with multi-representational tools, establish tacit agreement that displays from different representations constitute evidence about problem-situations. If a conflict arises between beliefs and the computer displays and actions show, the student will try to revise these beliefs. The socio-mathematical norm of what constitutes evidence in problem-situations was formed here as a consequence of the interaction with the tool. This activity triggered learning about linear, quadratic and exponential growth.

3. To Conclude

The design and research related to the *CompuMath* project started by focusing on cognition, but integrated more and more aspects of classroom practices and of socio-mathematical norms. We see this project as an example of how operative needs lead to research that has a scope and value transcending the operative needs that triggered it. Such research may initiate new educational activity cycles (Hershkowitz *et al.*, 2016). For example, the research in the project showed that several different approaches were possible to many topics, and this necessitated research with a somewhat larger curricular range. The transition from design and research on isolated activities to a curricular continuum necessitated decisions on taking specific approaches to specific topics. The use of the potential of open technological tools added "New Dilemmas in Teaching 'Old' Mathematical Topics" (Yerushalmy, 2009). In particular, the research findings by Stein (2002) informed the team that taking two different approaches to equations — interweaving functional and algebraic considerations — is beneficial for learning: Students become flexible in their understanding of what an equation is and how it may be solved in different technological environments. This was the case even with respect to nonlinear equations, which students had not yet learnt to solve (Friedlander and Stein, 2001).

The CompuMath project was funded by the Ministry of Education for about ten years. In a sense, this project was established in a time where the infrastructure was not ready for an implementation on a large scale. As in other parts of the world, the use of the technological potential of digital tools did not become a formal mainstream approach in mathematics class in the country. For example, in many countries, secondary school graduation examinations do not allow students to use technological tools. Yet, we can see its "footprints" in several channels. For example, a curriculum designer which is very active during the last several years, sees the following influences: (1) relating to the ways in which activities are expressed in todays' learning materials. In many cases we can see a central situation around which a whole lesson or a sequence of lessons is designed; occasionally it also includes the use of the potential of digital tools; (2) students' groupwork as an existence proof for students autonomy and agency in the learning process (Ben Zvi, 2001); (3) the role of the teacher shifted from "the authority of knowledge" in the class towards "provider of opportunities" to learn; and (4) the interweaving of needed skills into the inquiry-based activities, structured around central situations.

In addition, we see the enrichment in teaching approaches to various topics along the curriculum, e.g., the approach for teaching functions in the middle school mathematics curriculum. A noticeable point relates to research. Some trends which originated in the project are active today, e.g., Abstraction in Context (Hershkowitz, Schwarz and Dreyfus, 2001; Dreyfus, Hershkowitz and Schwarz, 2015).

References

Arcavi, A., Friedlander, A. and Hershkowitz, R. (1990). L'algebre avant la lettre, *Petit x*, **24**, 61–71.

Artigue, M. and Blomhøj, M. (2013). Conceptualizing inquiry-based education in mathematics, *ZDM, Mathematics Education*, **45**(6), 797–810.

Ben-Zvi, D. (2001). Children as explorers: Exploratory data analysis by junior high school students in a computer assisted environment, Unpublished dissertation, The Weizmann Institute of Science, Rehovot.

Dreyfus, T., Hershkowitz, R. and Schwarz, B. (2015). The nested epistemic actions model for abstraction in context — Theory as methodological tool and methodological tool as theory, in A. Bikner-Ahsbahs, C. Knipping and N. Presmeg (eds.), *Approaches to Qualitative Research in Mathematics Education: Examples of Methodology and Methods*, Dordrecht: Springer, Advances in Mathematics Education series, pp. 185–217.

Friedlander, A. and Stein, H. (2001). Students' choice of tools in solving equations in a technological learning environment, in M. van den Heuvel-Panhuizen (ed.), *Proceedings of the 25th Conference of the International Group for the Psychology of Mathematics Education*, Utrecht, The Netherlands, Vol. 2, pp. 241–248.

Hadas, N. (2001). Proving as a way to explain and convince in dynamic geometry learning environment, Unpublished dissertation, The Hebrew University, Jerusalem.

Hadas, N., Hershkowitz, R. and Schwarz, B. B. (2000). The role of uncertainty in constructing and proving in computerized environment, *Educational Studies in Mathematics*, **44**(1–2), 127–150.

Hershkowitz, R., Dreyfus, T., Ben-Zvi, D., Friedlander, A., Hadas, N., Resnick, N., Tabach M. and Schwarz, B. B. (2002). Mathematics curriculum development for computerized environments: A designer–researcher–learner-activity, in L. D. English (ed.), *Handbook of International Research in Mathematics Education* Mahwah, New Jersey: Lawrence Erlbaum Associates, pp. 657–694.

Hershkowitz, R., Schwarz, B. B. and Dreyfus, T. (2001). Abstraction in context: Epistemic actions, *Journal for Research in Mathematics Education*, **32**, 195–222.

Resnick, Z., Schwartz, B. B. and Hershkowitz, R. (1994). Global thinking "between and within" function representations in a dynamic interactive medium, in J. P. da Ponte and J. F. Matos (eds.), *Proceedings of the 18th PME Conference*, Vol. IV, University of Lisbon, Lisbon-Portugal, pp. 225–232.

Stein, H. (2002). Changes in the equation concept and its solution methods in a technological learning environment, Unpublished MA thesis, The Weizmann Institute of Science, Rehovot [in Hebrew].

Trouche, L. (2004). Managing complexity of human/machine interactions in computerized learning environments: Guiding students' command process through instrumental orchestrations, *International Journal of Computers for Mathematical Learning*, **9**(3), 281–307.

Yackel, E. and Cobb, P. (1996). Sociomathematical norms, argumentation, and autonomy in mathematics, *Journal for Research in Mathematics Education*, **27**(4), 458–477.

Yerushalmy, M. (2009). Technology-based algebra learning. Epistemological discontinuities and curricular implications, in B. B. Schwarz, T. Dreyfus and R. Hershkowitz (eds.), *Transformation of Knowledge Through Classroom Interactions*, Routledge: London and New York, pp. 56–64.

About the Authors

Rina Hershkowitz is a consultant to the Science Teaching Department at the Weizmann Institute of Science. She was the head of the CompuMath project.

Michal Tabach is currently the head of the Department of Education in Mathematics, Science and Technology, at the School of Education, Tel Aviv University.

Chapter 15

Mathematics at the Virtual School: Why? Why not? Who? What? And So What?

Yaniv Biton[*,#,¶], **Osnat Fellus**[‡,‖], **Dafna Raviv**[#,**], **David Feilchenfeld**[#,##] **and Boris Koichu**[§,‡‡]

[*]*Shaanan Religious Academic College of Education, Haifa, Israel*

[#]*Center for Educational Technology (CET), Tel Aviv, Israel*

[‡]*University of Ottawa, Ottawa, Canada*

[§]*Weizmann Institute of Science, Rehovott, Israel*

[¶]*YanivB@cet.ac.il*

[‖]*osnat.fellus@gmail.com*

[**]*Dafnar@cet.ac.il*

[##]*DavidF@cet.ac.il*

[‡‡]*boris.koichu@weizmann.ac.il*

In September 2012, the Center for Educational Technology together with the Israeli Ministry of Education and the Trump Foundation, launched the first Israeli Virtual High School (VHS). The goal was to increase the number of students who study advanced-level mathematics and physics by addressing a two-tiered problem in Israel's periphery: the shortage of qualified teachers who can teach advanced-level mathematics or physics, and the low numbers of students who opt for these studies — too low to justify funding dedicated classes for them at their home schools. This chapter describes the design, instruction, and support mechanisms of the VHS and presents its growth since its inception in 2012. It then presents preliminary findings gathered from a questionnaire disseminated in 2015 among the first graduating cohort of the VHS and their teachers about their experience.

1. Introduction

The term *virtual learning environment* has been used interchangeably with *distance education*, *e-learning*, and *web-based instruction* (Rice, 2006). Following Rice (2006) and Crean, Gaines, Paul, and Rukobo (2014), we refer to a *virtual learning environment* as an educational system that: (i) provides technology-mediated information spaces, (ii) facilitates synchronous and asynchronous teacher–student and student–student interactions that are independent of geographical proximity, (iii) harnesses pedagogically rich teaching approaches, and (iv) promotes nation-wide or district-wide curricula.

Research highlights several affordances and limitations that are engendered in virtual learning environments. For example, Barbour (2011) shows that virtual schooling has positive effect on rural school students with respect to enhancing their motivation. Clark and Berge (2003) argue that virtual learning environments generate wider accessibility for learning, provide high-quality learning opportunities, and are instrumental for improving learning skills. Some of the limitations identified in the literature include the overwhelming amount of work required from students and learners' distraction (Blau and Caspi, 2008; Kock, 2007), as well as student distress (Hara and Kling, 2000).

Comparative studies aimed at eliciting students' reflections on their experiences in traditional and virtual learning environments are rather inconclusive. While some studies indicate that students are more satisfied with their learning experiences in virtual learning environments than in traditional ones (e.g., York, 2008), other studies (e.g., Driscoll *et al.*, 2012) show no significant difference between the two settings. Against this backdrop of contradicting findings, Barbour (2017) points out that still little is known about K-12 virtual learning environments. We, thus, use Barbour's (2007) call to explore the design, delivery, and support mechanisms of virtual programs and introduce the VHS as a unique part of the Israeli high school mathematics education landscape. The chapter also presents empirical findings that shed light onto the first VHS cohort's perceptions of design, pedagogy, and support of the VHS.

2. The Israeli Virtual High School

The VHS classes are integrated in the schedule of the participating high schools and substitute the regular classes of the selected subjects for the VHS students. The instruction model employed in the VHS consists of the following weekly activities: four to six synchronous lessons with a teacher and 25 to 30 other students who log in the VHS at different high schools dispersed throughout Israel; two to three synchronous classes of virtual tutoring, during after school hours, that take place in groups of up to four students working with a university student who majors in a STEM-related field; homework assignments that are submitted electronically; and online quizzes and tests. The number of students studying in the Israeli VHS grew almost tenfold in five years starting from about 100 in 2012 to almost a thousand students in 2017. The demographics of the VHS student population are notable, as they comprise diverse sectors including the Bedouin sector, the Orthodox-Jewish sector, the Arab sector, and the religious and secular Jewish sectors. Fifty-six percent of the VHS students are females.

The distribution of the VHS students' socio-economic status (SES) suggests a close representation of the SES levels in the general Israeli public (see Fig. 1). A closer examination of these demographics shows that

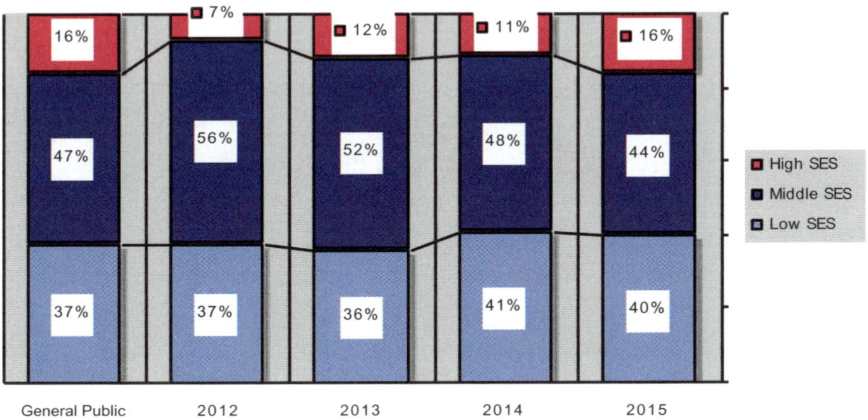

Fig. 1. SES distribution by clusters.

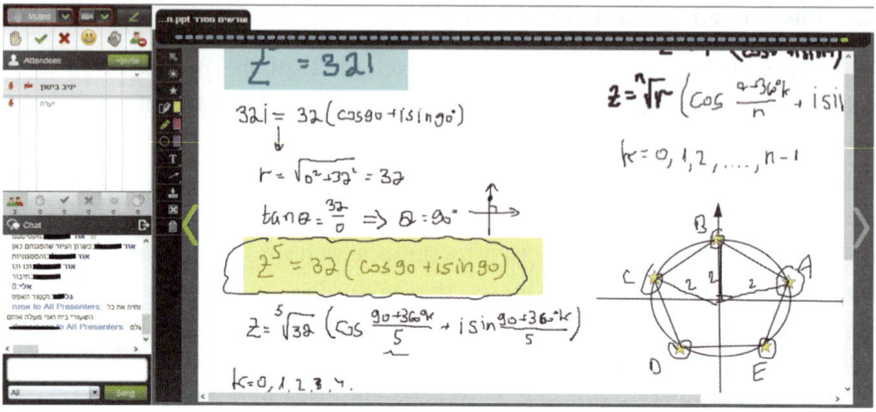

Fig. 2. White screen & tools of communication available on the platform.

students from the lowest SES level have a slightly larger representation in comparison to their size in the general public. This is hardly surprising given the goal of the VHS to reach out to all students who are interested in learning advanced-level mathematics.

The online VHS platform provides content and support through highly diverse media, real-time feedback, and ongoing support that is adapted to the needs of each student through, inter alia, the work of tutors. To participate in synchronous classes, the students log in the VHS during the school day in their respective schools. The VHS teacher shares content through the platform, which includes videos, ready-made presentations, and a great variety of interactive tools. For example, the "white board" feature (see Fig. 2) — a dynamic writing platform allows both teachers and students to share ideas and carry out discussions. Integrated tools of interaction (such as *chat*) and available software (such as *GeoGebra*) are also ample and diverse.

The Israeli VHS is unique in the sense that it is fully online, which means its design (i.e., the content available through the VHS platform); pedagogy (i.e., the teaching methods); and the support mechanisms (i.e., home school support, work with tutors, recorded classes, teacher-student, and student-student interaction) are all carried out through the VHS platform.

3. Feedback from the VHS' First Cohort and their Teachers

Given the unique design of the growing Israeli Virtual High School, CET's Research and Evaluation Unit (REU) gauges and guides the school's work and progress. As part of the evaluation process, the REU disseminates questionnaires and conducts interviews with the students, the teaching staff, and the home school principals. This section is based on the data collected from questionnaires (22 teachers and 86 students) and discussions carried out on professional-development days organized for the VHS teachers. A content analysis was employed through the use of the SWOT model (Strengths, Weaknesses, Opportunities, and Threats) (see Helms and Nixon, 2010) to pin down collectively identified strengths and challenges. In the next sections we present an aggregated summary of the findings, in accordance with the chosen three themes: design, pedagogy, and support.

3.1. *Design-related input*

Some of the items in the questionnaire specifically dealt with different design-related aspects of the VHS. While 77% of the students pointed out that the synchronous classes with their teachers contributed to their achievements, 86% of them noted that the synchronous sessions with their tutors did so. Next, 77% of the students indicated that the digital presentations contributed to their achievements, but when asked about the use of recorded lessons and GeoGebra, 58% and 35%, respectively, identified those as contributing to their learning. This finding implies the need to underscore the utilization of the available tools by teachers, tutors, and students of the VHS. Data analyzed through the lens of the SWOT model show that teachers perceive the design of the VHS as one that allows them to flexibly use the platform's tools and diversify their teaching. Another strength, which was highlighted by the teachers as an important element in the design of the VHS through the SWOT model, was working closely with the tutors and mentoring them.

3.2. *Pedagogy-related input*

Overall, students' feedback on their pedagogy-related experience, i.e., teaching practices as they are experienced by the students, is very positive.

It underscores notions such as the high quality of the teaching staff and the opportunity to learn in a way that addresses students' needs. As one student wrote, "The teacher taught us in a way that made it difficult to forget... He was committed to making sure we did well." The students' feedback is complemented by the data collected from the teachers, who reported that they allocated time for students' questions (91%) and carried discussions over problems and solutions (77%). Teachers of mathematics (70%) reported that they used simulations or dynamic applications. Note the gap between the students' report on the extent to which dynamic simulations such as GeoGebra contributed to their learning (35%) on the one hand, and teachers' report on using digital tools, including GeoGebra, in their teaching (70%), on the other hand. This gap might indicate that students are not yet adequately cognizant of the potential of, and added value in, using such tools. In regard to other pedagogically relevant insights, reflecting on his teaching in the VHS using the SWOT model, one teacher noted, "I let them 'invent' almost anything during the class. We 'rewrite' the presentation together from scratch." Such pedagogical practices reflect the teachers' efforts in creating opportunities for learners in the VHS to be actively involved in making sense of mathematical ideas.

3.3. *Support-related input*

Both teachers and students highlight the remarkable support system of the VHS. The following is a representative statement: "[What I like the most is] knowing that there is always someone you can turn to, that you can receive constant support and that you can experience innovative learning." Similarly, teachers not only expressed their enthusiasm about working in an environment that provided enhanced support, but also explained how they themselves supported students' learning. One teacher wrote, "We address students' needs beyond the time allotted for teaching in the VHS." In terms of points-for-improvement, one student reflected about the support of his home school, noting that "[there is a need to] put more emphasis on the responsibility the home schools have in supporting their VHS students by providing an adequate learning environment, such as computers and quiet classrooms."

4. Summary and Implications

The Israeli VHS was launched to address the steadily decreasing number of students who choose to study advanced-level mathematics. The purpose of this chapter was to introduce the first Israeli VHS, which is aimed at improvement of access to advanced-level mathematics instruction for students who live in Israel's periphery. An examination of the data collected yielded several important insights in regard to the high-quality teaching, cutting-edge technology, and unique online interactions among the teaching staff and the students in the VHS. This chapter captures only a speck of the rich learning environment in the VHS. Further development of the VHS requires close attention to context-specific challenges. Addressing these challenges both effectively and efficiently is paramount to the continued success of the Israeli VHS.

The title of our chapter reflects a combination of questions. Throughout the chapter, we provided answers, however concise, to these questions. To answer the "so what?" question posed in the title, we believe that the Israeli VHS is taking us another step in promoting equity of access and opportunity. The promises associated with harnessing a VHS in mathematics — *if fulfilled* — have the potential to substantially increase the number of students who choose to take advanced-level mathematics — because now they can.

References

Barbour, M. K. (2007). Teacher and developer perceptions of effective web-based design for secondary school students, *Journal of Distance Education*, **21**(3), 93–114.

Barbour, M. K. (2011). The promise and the reality: exploring virtual schooling in rural jurisdictions, *Education in Rural Australia*, **21**(1), 1–15.

Barbour, M. K. (2017). K-12 online learning and school choice: Growth and expansion in the absence of evidence, in R. A. Fox and N. K. Buchanan (eds.), *School Choice: A Handbook for Researchers, Practitioners, Policy-Makers, and Journalists*, New York: John Wiley & Sons, pp. 421–440.

Blau, I. and Caspi, A. (2008). Do media richness and visual anonymity influence learning? A comparative study using Skype™, in Y. Eshet, A. Caspi and N. Geri (eds.), *Learning in the Technological Era*, Ra'anana, Israel: The Open University of Israel, pp. 18–24 (in Hebrew).

Clark, T. and Berge, Z. (2003). Virtual schools and e-learning: Planning for success, in *19th Annual Conference on Distance Teaching and Learning*, Retrieved on June 4, 2017 from: https://pdfs.semanticscholar.org/c949/3f2687230bbcb66b48175a17339c3d37ddf7.pdf

Crean D. A., Gaines, J., Paul, A. and Rukobo, E. Z. (2014). *Virtual Learning Program Rubric*. Philadelphia, PA: Center on Innovations in Learning.

Driscoll, A., Jicha, K., Hunt, A. N., Tichavsky, L. and Thompson, G. (2012). Can online courses deliver in-class results? A comparison of student performance and satisfaction in an online versus a face-to-face introductory sociology course, *Teaching Sociology*, **40**(4), 312–331.

Hara, N. and Kling, R. (2000). Student distress in a web-based distance education course, *Information, Communication & Society*, **3**(4), 557–579.

Helms, M. M. and Nixon, J. (2010). Exploring SWOT analysis — where are we now? A review of academic research from the last decade, *Journal of Strategy and Management*, **3**(3), 215–251.

Kock, N. (2007). Media naturalness and compensatory encoding: The burden of electronic media obstacles is on senders, *Decision Support Systems*, **44**(1), 175–187.

Rice, K. L. (2006). A comprehensive look at distance education in the K-12 context, *Journal of Research on Technology in Education*, **38**(4), 425–448.

York, R. O. (2008). Comparing three modes instruction in a graduate social work program, *Journal of Social Work Education*, **44**(2), 157–171.

About the Authors

Yaniv Biton is a faculty member at Shaanan Religious Academic College of Education and a staff member in the team of mathematics education at the Center for Educational Technology, Tel Aviv. His research focuses on assessment and technology in mathematics education.

Osnat Fellus is a PhD candidate at the Faculty of Education, University of Ottawa. Her PhD work focuses on learning and teaching with a specific concentration in theories of identity in mathematics education.

Dafna Raviv is the head of the Virtual High School at the Center for Educational Technology, Tel Aviv.

David Feilchenfeld is a mentor of mathematics teaching in Jerusalem, a faculty member at the David Yellin Academic College of Education in Jerusalem, and the head of the development team in the field of mathematics for the Virtual High School at the Center for Educational Technology, Tel Aviv.

Boris Koichu is a faculty member at the Weizmann Institute of Science. His research focuses on mathematical problem solving and problem posing.

Chapter 16

Nurturing Students with High Mathematical Potential

Abraham (Avi) Berman*,‡ and Roza Leikin#,§

*Department of Mathematics,
Technion — Israel Institute if Technology, Haifa, Israel

#Department of Mathematics Education, RANGE Center,
Faculty of Education, University of Haifa, Haifa, Israel

‡berman@technion.ac.il

§rozal@edu.haifa.ac.il

In this chapter, we describe different types of programs for students with high mathematical potential in Israel. Nurturing mathematically advanced students requires solving complex problems associated with making decisions about the scope of the programs and the choice of appropriate educational approaches. We focus our attention on classes for students with high mathematical potential, university programs for mathematically advanced school students, mathematics Olympiads and competitions. We provide some representative examples of such programs.

1. Introduction

It is generally agreed that the human capital in the State of Israel is the primary source of the country's advancements in science and high technologies. For this reason, societal awareness of the great importance of nurturing students with high mathematical potential has grown in the country. Due to the political and cultural character of Israeli society, there is no school for mathematically gifted students in Israel (Nevo, 2004). Rather, there are various other programs including classes for mathematically advanced students, mathematics clubs, and mathematics competitions.

In what follows, we describe different types of programs and provide concrete examples. Due to the space limitations of the chapter, we do not describe all the programs that exist in Israel, but rather give some representative examples of such programs — not necessarily according to their importance. Some of the programs are presented elsewhere in this volume and thus we do not describe them in this chapter (e.g., Avissar-Zeldis — this volume, Arad and Fridman — this volume).

2. Classes for Students with High Mathematical Potential

Overall there are about 20 secondary schools (grades 7–12) in Israel in which students can study in self-contained advanced mathematics classes for students who passed a national examination of giftedness conducted nation-wide among 3rd grade students. In middle school all the students in these classes study an enhanced mathematical curriculum together, whereas starting from 10th grade students are allowed to choose the level of mathematics that they prefer to study. In addition to the three ordinary levels (i.e., basic, intermediate, and advanced), classes for gifted students offer an enhanced high level mathematical program.

For example, the Leo Baeck Secondary School offers a special program, which integrates elements of the "unified modern mathematics" curriculum (Columbia Mathematical Curriculum, 1968). The curriculum relies on two principles: spiral learning and connectivity between Number Theory, Algebra and Geometry. It was translated to Hebrew by Akiva Skidel and experimentally implemented by Benjamin Mayer and Nitsa Hadar in 1970.

A famous graduate of this program is Maria Chudnovsky. Maria studied in a special mathematics school in St. Petersburg, and then after immigrating to Israel, took part in the Columbia Program in Leo Baeck. She participated in enrichment classes at the Technion, and was accepted to the Technion program for exceptionally excellent students while still a high school student. She completed her Master's studies at the Technion and completed her doctorate in Princeton, where she proved, with her advisor and two other mathematicians, the Perfect Graph Conjecture that was open for 40 years. One of her many achievements is being awarded a MacArthur "Genius" Fellowship. Nowadays she is a professor of mathematics at Princeton.

For many years, mathematicians from the Technion, the Israel Institute of Technology, entertained the idea of creating a special mathematics school. This initiative lead to the establishment of two mathematics classes in Haifa, the Technion Class in Ironi Hey (Municipal School Number 5) and the Mathematics Class in the Reali School, which is a private school.

The Reali, Leo Baeck and Ironi Hey are examples of schools with special programs in mathematics. Other examples include ***The Israel Arts and Sciences Academy*** (IASA) and ***Leyada*** (an abbreviation in Hebrew for "near the university"), a school founded in 1935 near the Hebrew University in Jerusalem. Leyada has a long tradition of fostering excellence and its graduates include two Nobel Prize winners. Elon Lindenstrauss (graduated 1988) was awarded the Fields medal in Mathematics in 2010.

The Technion classes and Mathematics classes in Reali include students from the top 1–2% of the general population. In 1999, in order to increase learning opportunities for mathematically promising students, the Technion established TELEM — Technion Lessons in Mathematics — classes. The main goals of the program are to enable students to study advanced-level mathematics, to develop higher order mathematical thinking and to promote mathematical excellence. The curriculum is enhanced by mathematical investigations related to topics from the regular mathematical curriculum. The program also includes a mathematical enrichment course one hour per week (Mednikov, Gurevich and Leikin, 2001).

3. University Programs for Mathematically Advanced Students

The Future Scientists Program was established by President Shimon Peres to promote mathematics, science and technology. It is funded by the Maimonides Foundation and managed by four Israeli universities (Tel Aviv University, the Hebrew University, the Technion and Ben Gurion University), and by the Ministry of Education. It consists of three programs: Odyssey, Alpha, and Idea. *Odyssey* identifies and recruits 9th graders with extremely high capabilities in mathematics. This program includes meetings with leading people in science and technology in Israel; a source of motivation and inspiration for the participants, who are expected to lead science and technology in Israel in the future. *Alpha* spans three years and includes exposure to the world of scientific research and work on a personal research project (in science or in mathematics) with a mentor. The research project can be in place of the advanced-level secondary school exam (which is an option given to all students in Israeli high schools, in all topics studied in school). *Idea* aims at promoting giftedness in the Humanities.

All Israeli universities allow high school students (grades 10–12) who excel in high school level mathematics to take regular courses towards a B.Sc. degree in mathematics. With sufficiently high grades in these courses, they can later be accepted to the universities as regular students without a secondary school graduation certificate.

The Open University (OU)

The OU enables talented high school students to enroll in academic courses and to begin their studies towards a BA degree in parallel to their high school studies. Studies at the OU are only suitable for students who are prepared to invest about two hours daily in academic studies and to contend with the difficulties of the self-study method. It is important to note two things: that the courses were originally developed for adult students, and that they may be accepted in place of secondary school graduation exams.

Technion

In addition to TELEM classes, and the *Future Scientists and Inventors Program* mentioned above, the Technion offers special activities for gifted students, and for students with high mathematical potential in particular.

The Noam Center for Talented Youth operates under the auspices of the Department of Mathematics. Noam center (Hebrew acronym for Excelling Youth) hosts up to 20 students in a 2-week Number Theory Camp for high school students (grades 9–12). The program combines study of number theory and the RSA encryption system with various fun social activities.

The Mathematical Riddles Course at the Technion Mathematics Department deals with mathematical problems and puzzles. Most of the problems are at the level of problems given at the Mathematics Olympics for high school students. Some originate in actual research. The course requires the active involvement of the participants in group and individual problem solving. The Noam center also takes part in organizing the Grossman Mathematics Olympiad (see the Olympiads section of this paper) and in the publication of a virtual mathematics magazine — Net-Gar.

Net-Gar is a web mathematics journal for high school students (and for everyone who loves mathematics). It succeeds the paper magazines, *Etgar-Gilyonot Mathematica* (lit. Challenge — Mathematics Volumes) published between 1985–2001, and its namesakes *Gilyonot Mathematica* and *Etgar*, published before the birth of the State of Israel.

The Technion Youth Department runs an international summer research program — SciTech — for gifted students. SciTech is a 3–4-week program which combines scientific research in several disciplines, including mathematics, with cultural and social activities. It brings together outstanding students from North and South America, Europe, Asia, Israel and many other countries. SciTech participants work on a research project supervised by Technion researchers.

Tel Aviv University

The *Beno Arbel Outstanding Young Students Program* is operated by the School of Mathematical Sciences of Tel Aviv University. The program, founded in 1970, was named after the late Prof. Beno Arbel, who was its chair for many years. It allows mathematically gifted secondary school students to finish their undergraduate studies (and sometimes even a master's degree) in parallel to their high school studies (Arbel, 1991). The program is modular and enables outstanding students to work at the pace that is most appropriate for them.

Weizmann Institute of Science

Mathematics by Mail is operated by the Davidson Institute of Science Education at the Weizmann Institute of Science. It offers talented and curious students (grades 3–9) extra-curricular activities in recreational mathematics that develop their creative thinking and logic. Its activities include studying advanced topics over the course of the school year. In addition, participants receive a weekly newsletter with additional riddles and challenges. Many mathematics topics can be discussed in the weekly chat with the Mathematics by Mail team, researchers at the Weizmann Institute of Science.

The Davidson Institute also takes part in organizing the *Prof. Joseph Gillis National Mathematics Olympiad* (see the section Olympiads in this chapter).

Ben Gurion University

Kidumatica — The Youth Mathematics Forum was established at the Center for Science and Technology Education at Ben-Gurion University of the Negev, in order to advance the overall level of mathematical education in the Negev. Kidumatica is a play on words in Hebrew meaning "advancing mathematics." Students ages 10–17 from all over the Negev take part in Kidumatica. After-school activities include workshops, interactive lectures, competitions and mathematical sports that are enjoyable for the pupils, who are introduced to many mathematical topics, based on cognitive parameters related to mathematical thinking and creativity (Amit and Neria, 2008).

University of Haifa

The ETGAR Program at the University of Haifa offers a B.Sc. degree in Computer Science to high school students who excel in mathematics. The program is unique in that in the first years the students take the courses in small study groups designed just for them. In addition, the program provides mentoring for each student by older students and support from educational consultants.

RANGE — The Interdisciplinary Center for Research and Advancement of Giftedness and Excellence was established at the University of Haifa with the aim of advancing innovative interdisciplinary research associated with creativity, giftedness and excellence in education in various fields such as mathematics, science, art and neuroscience.

4. Mathematics Olympiads and Competitions

The main national mathematics competitions in Israel are the Grossman Olympics and the Gillis Olympics. The winners receive scholarships and are invited to register for studies in universities without additional requirements.

The Grossman Mathematics Olympics is organized by the Department of Mathematics at the Technion. It is named after the late Prof. Yirmiyahu Grossman and is the oldest mathematics competition in Israel. The competition is open to high school students and soldiers doing compulsory military service. It offers participants challenging and non-standard questions in elementary mathematics.

The Prof. Joseph Gillis National Mathematics Olympiad is a nationwide competition in mathematics designed for high school students, held at the Weizmann Institute of Science. Step one of the competition consists of a test sent to all interested schools. Step two is a competition held at the Davidson Institute of Science Education, which is open to all interested students. Students who pass the second stage with distinction become the Israeli national team in mathematics, and are invited to training camps at the Weizmann Institute. Six outstanding members of the team represent Israel at the International Mathematical Olympiad.

Zuta Olympiad (*Mini Olympics*) is a national competition in mathematics for students in grades 7–9. Its purpose is to locate excellent students nationwide. Students who pass the competition with honors become the young Israeli national team in mathematics, and are invited to training camps at the Weizmann Institute.

Additional mathematical competitions that should be mentioned are the *Israeli Open Mathematics Competition* and the *Computer Science Olympiad.*

5. Summary

Nurturing students with high mathematical potential requires solving complex problems — and not only in mathematics. First, there is currently no test for "mathematical giftedness" and no clear definition of mathematical giftedness. Second, opening the programs to a broader population also addresses the dilemma between selective vs. inclusive policy. Finally, the dilemma of choosing appropriate teaching approaches is raised. Just as challenging mathematical problems can be solved in multiple ways, so too the problem of how to nurture these students has many solutions.

Recently, the Advisory Board in Mathematics Education of the Israeli Ministry of Education appointed a Program Committee for high school mathematics education. The committee includes research mathematicians, mathematics education researchers, teacher educators and teachers. The Program Committee proposes teaching mathematics at four levels: the three current levels and an *advanced-plus* level. Addition of the advanced-plus level of mathematics should provide mathematically gifted students with a curriculum that allows them to realize their very high mathematical potential. According to the proposal, the advanced-plus level will integrate acceleration, deepening and enrichment, and will allow students and teachers to enjoy teaching and learning mathematics. We hope that this plan will be realized.

References

Arbel, B. (1991). *Problem-Solving Strategies*, Tel Aviv, Open University Press (in Hebrew).

Amit, M. and Neria, D. (2008). "Rising to the challenge": Using generalization in pattern problems to unearth the algebraic skills of talented pre-algebra students. *ZDM*, **40**, 111–129.

Columbia mathematical curriculum (1968). http://math.buffalostate.edu/~med600/SignedNumbrs/UnifiedMatematicsProject.pdf.

Mednikov, L., Gurevich, I. and Leikin, R. (2001). *Enrichment Materials for the Development of Mathematical Thinking: For Gifted Students*, Units 1–12. Technion, Haifa (in Hebrew).

Nevo, B. (2004). *Recommendations Steering Committee on the Educations of Outstanding and Gifted Students in Israel*, The Division for Outstanding and Gifted Students, The Ministry of Education, Jerusalem, Israel.

About the Authors

Abraham (Avi) Berman is professor emeritus of Mathematics at the Technion — Israel Institute of Technology, Haifa. He earned his PhD at Northwestern University in 1970. Prof. Berman's research interests include Nonnegative Matrices, Spectral Graph Theory, Teaching Linear Algebra and Mathematical Giftedness. He is the author of 6 books and 4 edited books and had supervised more than 50 post-doctoral, doctoral, and master students. His past service to the Technion includes holding The Israel Pollak Chair; serving as the head of the Department of Education in Science and Technology, the Center for Pre-University Education, and the Interdepartmental Committee on Applied Mathematics and serving on the Senate Steering Committee and the Board of Governors Council. His past public service includes serving as the head of the Israeli Society for the Promotion of and Research on Giftedness and Creativity;

guest editor of LAA, ELA; editorial board of SIMAX an IJRUME and editor of *Etgar — Gilyonot Mathematica* (high school mathematics journal in Hebrew). Currently he serves on the Academic Board of the Israel National Museum and on the Board of Directors of the Hebrew Reali School in Haifa. He has been involved in the organization of the following past and present conferences: *The Haifa Matrix Theory Conferences, The Hamilton Workshops on Nonnegative Matrices and the Jerusalem Conferences on Research in Mathematics Education*. His memberships include: ICMI, ILAS, IMU, MCG, and WFNMC.

Roza Leikin is a professor of Mathematics Education and Education of the Gifted at the Faculty of Education, University of Haifa. Her research and practice embrace mathematical creativity and ability in teachers and students. She is the head of the Department of Mathematics Education (since October 2016) and director of the RANGE Center in the University (since 2008). She served as the head of the National Advisory Committee for Mathematics Education of the Israeli Ministry of Education (2013–2017) and the President of the International Group for Mathematical Creativity and Giftedness (2012–2017). She has published more than 150 research papers.

Chapter 17

The Bar-Ilan University — ICAMS[a] Program for the Advancement of Mathematically Talented Youth[b]

Zvi Arad[*,#,§] and Elisheva (Gerstein) Fridman[‡,¶]

[*]Bar-Ilan University, Ramat-Gan, Israel
[#]Netanya Academic College of Education, Netanya, Israel
[‡]ICAMS-Israel Center for the Advancement of the Mathematical Sciences, Jerusalem, Israel
[§]aradtzvi@netanya.ac.il
[¶]eli@yuni.co.il

This chapter describes the Bar-Ilan University — ICAMS Program for the Advancement of Mathematically Talented Youth. It gives the rationale and a general description of the program, a quantitative summary of the program's outcomes, as well as some pedagogical details of the program. It also gives an overview of the development of the program over the years.

1. Rationale and General Description of the Program

1.1. *Rationale of the program*

The Bar-Ilan University — ICAMS Program has chosen **Enrichment** and **Accelerated Learning** from among several possibilities available for talented/gifted students. This is based on the recognition that cognitive and learning ability and the rates of cognitive, emotional and social development

[a]The Israel Center for the Advancement of the Mathematical Sciences (ICAMS) is a non-profit organization which works towards the advancement of mathematically talented youth.
[b]A joint project of ICAMS and Bar-Ilan University with financial support of Mr. Victor Bentata and his family.

are not necessarily dependent on the rate of development of the chronological age of the student, and indeed, vary among any given age group.

Talented/gifted students do not need as much introductory material and repetition to learn. They think and learn quicker than their peers. They also seek broader and deeper study.

The most effective **Accelerated Learning** program is one which provides the student with the material they are to learn at the pace, depth and breadth appropriate **for them.**

The Bar-Ilan University — ICAMS Program also includes a substantial element of self-study and learning. This is a tool which must be mastered by young outstanding students in order to progress in the world of science and technology and other areas which require a strong foundation in mathematics.

The program also recognizes that talented/gifted mathematics students can actually learn and accomplish a great deal at an extremely young age. Therefore, the earlier the student achieves knowledge of mathematics the greater the window of opportunity.

The program consists of three stages:

Stage 1: The Program for Mathematical Enrichment;
Stage 2: The Program for Mathematically Talented Youth: Accelerated Learning;
Stage 3: Academic Studies.

Stage 1: The Program for Mathematical Enrichment (two-year program)

This is a two-year program for sixth and seventh graders and consists of thirty 2-hour lessons, after school hours, over the school year.

Admission: Any fifth or sixth grade student may take the entrance exam which is designed to determine their suitability to the program based on testing knowledge taught at school as well as analytical thinking. Applicants are accepted only after passing the entrance exam.

The basic goals include: development of "mathematical thinking", enrichment of mathematical knowledge beyond what is taught in the primary school mathematics curriculum, identifying mathematically talented youth and encouraging them to study mathematics.

These studies also help to prepare the interested students for participation in the more rigorous Stage 2 program of Accelerated Learning. During the course of the Program for Mathematical Enrichment, the students develop mathematical thought processes and are provided enrichment in many areas of mathematics.

Stage 2: The Program for Mathematically Talented Youth: Accelerated Learning (three-year program)

This is a three-year program for students in grades 8–10, and consists of 120 academic hours given on a weekly basis in the afternoons, after regular school classes, over the course of the school year.

Admission and Integration*:** Every student applying to Stage 2, including those who have successfully completed the Program for Mathematical Enrichment, must pass the entrance examination for Stage 2. The entrance exam is given at the end of seventh grade. It evaluates mathematical aptitude as well as basic skills of the candidates. Students who achieve the required grade on the exam are accepted to the program and begin their studies in 8th grade. This is a very high pressure course of study, and requires a great investment of time and energy, even for the talented student. Most of the students have not had to contend with such a situation in their studies until now. Some students are not emotionally suited to cope with this situation and choose to withdraw. Some students, despite doing well on the entrance exam, are found to be not suited scholastically and are advised to withdraw by the pedagogical advisors. Finally, there are some well suited students who decide after beginning the program that their priorities lay elsewhere. Students who succeed in 8th grade are ***integrated into the program.

This program has two main goals: The first goal is to prepare the students to take the advanced-level secondary school graduation exam in mathematics at the end of 10th grade. This exam is normally taken at the end of 12th grade. The syllabus for the Accelerated Learning Program is based on the high school mathematics curriculum, contains topics that are normally studied during the course of regular high school studies, and includes mathematical topics that are required for academic studies in the natural as well as social sciences.

The second goal is to enable and prepare the students who successfully complete the secondary school graduation exam in 10th grade to enroll in regular academic courses in mathematics, physics or engineering at Israeli universities and colleges during 11th and 12th grades, while completing their high school studies (i.e., Stage 3). An important part of the preparation is to educate and encourage the students towards self-learning. This is accomplished by requiring specially chosen assignments and projects for homework.

There are uniform exams for students in Stage 2 (grades 8–10) four times a year — midterms and final exams each semester. Each of these exams is given at the same time and graded by a special staff, not the individual teachers. These exams give an indication of the success of the teachers as well as the students.

At the end of every year, each student is evaluated in order to determine whether or not they are suited to continue in the program. This process is especially crucial during 8th grade because it determines whether the student will be integrated into the program.

Stage 3: Academic Studies

Students who have completed the Stage 2 program, which includes the advanced-level secondary school graduation exam at the end of 10th grade, and are interested in pursuing academic studies in mathematics, can enroll in university courses during grades 11–12 and successfully complete over half of the requirements for a Bachelor's degree in mathematics, physics or engineering during this time. These students are granted a one-year deferment of their military service (two years for engineering students) so that they can complete their degree during the year following 12th grade. A special course schedule has been designed for these students so that they can combine academic studies with their regular high school studies.

Students who are not interested in such intensive studies, so as to complete their degree one year after high school, can enroll in various mathematics courses offered in the mathematics department. They earn academic credit for those courses and when they register to a university these credits are recognized.

The Department of Mathematics at Bar-Ilan University accommodates students in both of the above tracks by offering personal guidance and assistance in arranging their schedules.

Once students begin academic studies, they are under the academic supervision of the relevant department. These students who enter the military with a Bachelor's degree are generally assigned to advanced technology or intelligence units where they make a great impact.

2. Administration and Location of the Program

The Stage 2 program is under the academic auspices of Bar-Ilan University and under the administrative auspices of The Israel Center for the Advancement of the Mathematical Sciences (ICAMS).

The activities of the Program take place in over 30 locations throughout Israel.

The main administrative office of the Program is located on the campus of Bar-Ilan University in Ramat-Gan. However, because of the widespread locations of the activities, some administrators work from home or at other locations. All the administrators have access to a network where all the necessary information is accessible.

3. Pedagogical Aspects of the Program

The level and the depth of the curriculum are at a much higher level than in the school system. This is due to the homogeneity of these classes. The rate of study is intensive. It is these aspects that draw the talented students and are of interest to them. The program demands independent thought and effort.

The curriculum is covered in less time than in the high school system. The focus of the accelerated learning for these students is mathematics, with the structure following the standard curriculum. However, the material is covered in a deeper fashion than in standard high school classes and at times in a different arrangement.

When students begin the Stage 2 program in 8th grade, they find the program to be very challenging. Many begin the program without having developed proper study habits beforehand. In the earlier grades they were the "best in the class" and did not develop the habit of self-study. The demands are suddenly much greater than in lower grades. Therefore, in the first year of the program the teacher concentrates on helping the students to develop proper study habits. The students find learning material on their own and discover the benefit of reading ahead before the lessons. With the increased knowledge before the class, the sessions frequently are more dynamic, as much give and take and discussions take place.

The curriculum stresses connections between various subject matters studied, for example, analytic geometry, vectors, and trigonometry.

The quality and experience of the teachers is crucial to the success. The staff of the program is constantly searching for appropriate teachers. Many training sessions are conducted for the teachers on an ongoing basis. Some very successful teachers come from among scientists or experts in hi-tech industries who enjoy teaching, and consider it a contribution to society to be part of this project. Mathematics graduate students are also eager to take part in this program. In addition, there is a large number of high school teachers who teach in the program after regular school hours.

Many Israeli students take part in local and International Mathematical Olympiads, with a fair degree of success. In recent years, a large number of medal winners among the Israeli medal winners were students or graduates of the program. In 2016, Israeli students participated in 6 international Olympics in the exact sciences. Altogether, 31 Israeli students participated in the Olympics, 19 of them studied in the enrichment and/or the Stage 2 program.

Stage 2 has drawn some international attention, and in 2014 a delegation of Chinese specialists in mathematics education spent a week in Israel to study the program and to determine whether such a program can be adapted for Chinese students.

4. The Development and Outcomes of the Program

This program was started over 35 years ago. The founders of the program were Professors Zvi Arad and Bernard Pinchuk, members of the Department of Mathematics at Bar-Ilan University.

The program is available to students from Ashkelon in the South to Karmiel in the North.

The following chart displays some quantitative information about Stages 2 and 3 over the past three years.

Class of	No. of students beginning 9th grade	No. of students taking advanced-level exam	Average grade (out of 100)	No. of 11th grade students continuing academic studies
2012–2014	433	302	96	214
2013–2015	409	323	97	223
2014–2016	446	308	98	201

Acknowledgments

We would like to thank Prof. Bernard Pinchuk for his assistance. We also thank Mrs. Miriam Beller for her technical and editorial assistance.

About the Authors

Zvi Arad is a well-known mathematician who served as President and Rector of Bar-Ilan University. He was also a member of the Council for Higher Education of the Stage of Israel. Profs. Arad and Bernard Pinchuk established the "Bar-Ilan University-ICAMS Program for Mathematically Talented Youth". Arad is currently President of the Netanya Academic College of Education.

Elisheva (Gerstein) Fridman has a BSc and a Teaching Certificate in mathematics from Bar-Ilan University. She taught at the Bar-Ilan ICAMS Program from 1993 to 1999 and since 1996 serves as the Director of the Program.

Chapter 18

Mathematical Excellence: The Mofet Way

Tamara Avissar-Zeldis
Achva Academic College of Education, Arugott, Israel
Mathematics Department, Mofet Network, Israel
tamaraMofet@gmail.com

This chapter introduces the Mofet Network. Starting with its history, it continues with the description of the findings obtained in a research on Mofet's achievements. The various pedagogical departments of Mofet are presented, focusing on the Mathematics Department. Our vision and goals are defined, along with some examples of tasks we give our students in order to achieve our stated goals.

1. The Mofet Network: History and Research

Mofet (the acronym for Mathematics, Physics and Culture, in Hebrew) is a non-profit public institution, which promotes personal excellence in the studies of mathematics, physics, computers, English, and social science. Mofet is considered a unique anthropological phenomenon in Israeli society. It aims at bringing about a fundamental change in education, focusing on developing thinking skills while promoting essential social values, Students are obliged to take an active place in social activities at their schools and communities, approximately 60 hours each year.

Mofet was first suggested by a group of teachers from the former Soviet Union in the 1970s. In 1992, led by Zeev Geizel and Yakov Mazganov, they opened an educational center in the Shevach School in Tel Aviv, which offered extracurricular activities after school hours. This ultimately brought about the establishment of the Mofet Association. There was substantial demand for the afternoon enrichment programs offered by Mazganov and his colleagues, which encouraged Mazganov and Geizel to establish an association with the objective of setting up a network of schools for mathematics and physics in Israel.

The opening of special morning classes, in 1994, for highly motivated students, immigrants from the former Soviet Union, marked the beginning of the switch from afternoon enrichment studies to regular classes in the formal education framework. This was a crucial step in the association's development. In 1997, Orna Schneiderman became the Israeli director of the Mofet Association. Schneiderman, who understood Mofet's potential for implementation in Israel's mainstream, realized that the system could work for any student and therefore should be disseminated amongst the non-immigrant Israeli population as well. Since the 2015/6 school year, the program is intended for grades 7–12.

Horowitz compared the achievements of Mofet students with students in other classes (Horowitz, Shamai and Ilatov, 2014, pp. 68–94). She lists four main points:

(1) Achievements in the secondary school graduation exams were particularly high for Mofet students. This supports schools' policy on empowering low-achievers and reinforcing high-achievers. This finding is further validated by the low percentage of dropouts from these schools.
(2) Mofet students are highly motivated, with broad horizons, knowledge, and the tenacity to channel their actions toward the realization of their academic goals. They are very busy with their studies, yet many add more courses to their curriculum. Findings indicate these students to be inquisitive and curious, not afraid of challenges, motivated and talented, and with good leadership capability. This means they are capable of, and motivated toward, higher level learning. The findings indicate the Mofet Program's contribution to the development of educational and personal excellence.

Before we go on with more findings, it should be stated that analysis of the attitudes and conceptions of Mofet students, compared with the views and conceptions of students in regular classes, shows that Mofet students have higher motivation, sometimes higher self-esteem, and a more positive attitude towards school. They see a purpose in schooling, have a high feeling of belonging and contribute to their schools. They tend to feel less estranged and less fearful of the academic world.

Furthermore, Mofet students see their future roles positively, whether within school or in society. Across time, it seems that some students in other classes emulate Mofet students' attitudes, conceptions, and their philosophy of life regarding school. This is called the *Halo effect* of the Mofet class on school climate.

It is interesting to see the mathematics teachers' attitude in schools. The attitude was especially positive and there is almost no difference between Mophet students and other students. It can be assumed that this is a result of the "mathematics learning culture" created in these schools.

(3) Mofet teachers did not consider themselves as an exclusive group, because many taught both ordinary and Mofet classes, contributing to the "Halo" effect.
(4) Integrating Mofet into schools was a catalyst for positive change in the target schools. It was not just another new curriculum, but another pedagogical concept that seemed to affect school cultures in a positive conceptual change. Most teacher teams saw the potential for a significant meaningful change.

2. Scope of Mofet Activity Through the Years

Figure 1 shows the scope of activity of Mofet in mainstream education (middle and high schools). Mofet is implemented within a broad integrative

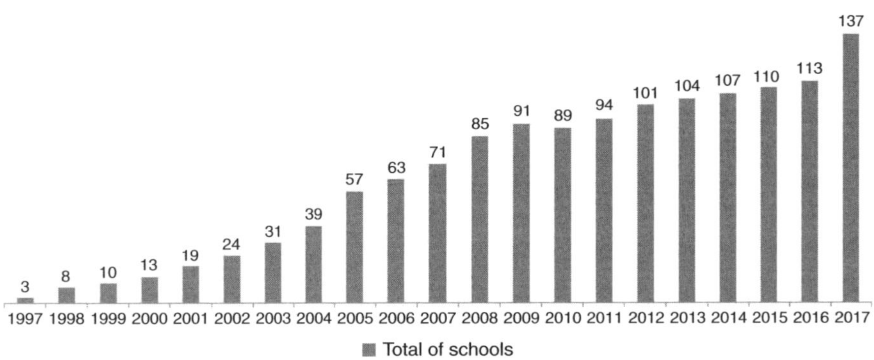

Fig. 1. The scope of activity of the **Mofet** program.

nationwide scope: 20,000 students study in 120 educational institutions within 60 municipalities. It is specifically implemented in the peripheral areas of society, including Jewish secular and religious sectors and the Druze and Arab sectors.

3. Mofet: Structure, Goals, and the Mathematics Department

Mofet currently has five pedagogical departments — Mathematics, Physics, Computer Science, English, and Empowerment. The teaching staff is highly qualified, with advanced degrees (MA and PhD).

3.1. *Mofet's main goals*

Mofet's vision and main goal is that its graduates will be those who lead research and development of science and economy, thus basing a strong foundation for an innovative, stable, and reasonable society in Israel.

3.2. *The mathematics department*

The mathematics department's proclaimed goals are to expose students to the richness and beauty of the mathematics discipline, and encourage their curiosity and motivation to contend with mathematical challenges. Developing logical, creative, critical thinkers with the tenacity and knowledge for solving problems will prepare students for research, development, innovation, and a knowledge-intensive industry.

There are many theories regarding the question of how to define learning improvement. In Mofet we define it by two major criteria: (1) the ability to bring students beyond the drill-and-practice level into the problem-solving level. By this we mean implementing techniques and concepts for solving more complex problems, using an original approach; (2) the level of enthusiasm, inquisitiveness, and ambitiousness we can maintain among students in mathematics classes.

The department's staff compiled a unique teaching program, which is based on the middle school mathematics curriculum. This program is used for the development of supplementary learning materials for enrichment and deepening.

The team provides individual instruction and group teachers' training courses for school staff. The mathematics department team constructed comparative tests and analysis of their results as a tool for evaluation, culminating in a biennial Mathematics Olympic Game in middle school.

Our curriculum emphasizes the implementation of each technique, concept or rule that we teach. We use a wide range of problems aimed at applying the subject studied in many different situations and in a creative way (Kieran, 2006).

First example: In 8th grade, after teaching factoring, we challenge our students by the following task:

Find the maximal value of the expression:

$$4b(5a - b) - (5a - 2)(5a + 2).$$

This question is unfamiliar to the students and they have to figure out how to approach it. The way is to connect to previous studies concerning the square function where they encountered the maximal value of an expression and use factoring.

So the answer is

$$\begin{aligned} 4b(5a - b) - (5a - 2)(5a + 2) &= 20ab - 4b^2 - 25a^2 + 4 \\ &= 4 - (5a - 2b)^2. \end{aligned}$$

And thus the maximal value is 4.

Second example: At the same stage, we ask students to find how many solutions, if any, the following equation has, without solving it:

$$a^4 - 2a^3 + a^2 - 1 = 0.$$

Students are accustomed to solving equations, and are confused by the request to refrain from finding the solutions. They find it hard to even begin answering the question. They know they should reduce the powers by factoring, but how? The correct answer is

$$\begin{aligned} \underbrace{a^4 - 2a^3 + a^2} - 1 &= (a^2 - a)^2 - (1)^2 \\ &= (\underset{\Delta<0}{a^2 - a + 1})\,(\underset{\Delta>0}{a^2 - a - 1}) = 0. \end{aligned}$$

So we have two different solutions.

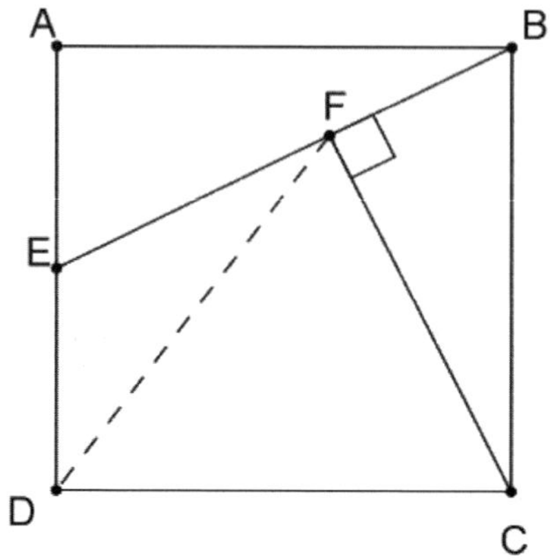

Fig. 2. A question in geometry.

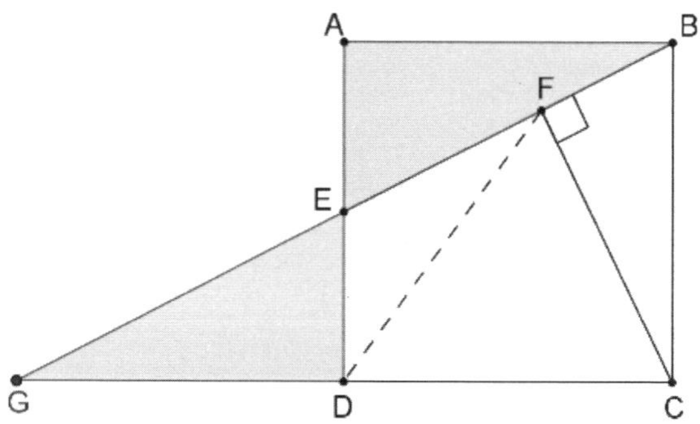

Fig. 3. A geometric solution.

Third example: For the 9th grade. Given a square ABCD with a side of 1 cm. Let $AE = ED$ and $CF \perp BE$. Prove that $DF = 1$ cm; solve in as many ways as you can find (see Fig. 2).

We show a sample of three different solution for this question.

Geometric solution: We extend BE through E and DC through D. They meet at point G (see Fig. 3). Then we get

$$\Delta ABE \cong \Delta DGE \Rightarrow GD = DC = 1\,\text{cm}\ \Delta GFC = 90^\circ$$

$$\Rightarrow DF = DC = 1\,\text{cm}.$$

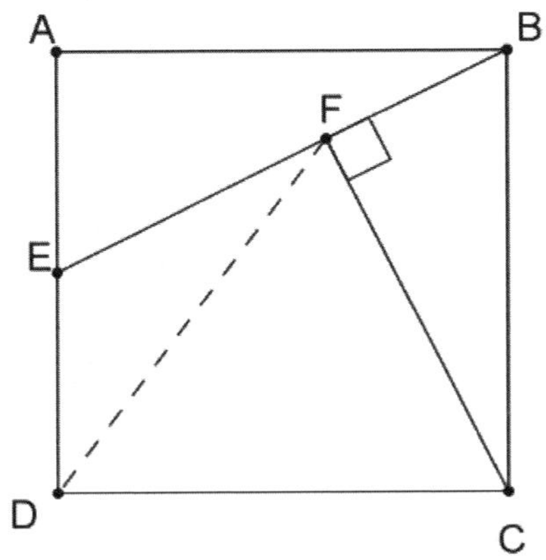

Fig. 4. A trigonometric solution.

Trigonometric solution: Using the law of cosines (see Fig. 4),

in ΔABE,

Tan $\Delta ABE = 0.5 \Rightarrow \Delta ABE = 26.565^\circ$

in ΔBFC,

$\Delta\ FCB = 26.565^\circ \Rightarrow FC = 0.89442\,\text{cm}$

in ΔDFC

$\Rightarrow DF = 1\,\text{cm}$.

Analytic geometry solution: Let (see Fig. 5)

$$B(1, 1),\ E\left(0, \frac{1}{2}\right) \Rightarrow l_{BE} : y = \frac{1}{2}x + \frac{1}{2}.$$

Let

$$C(1, 0),\ BE \perp CF \Rightarrow l_{FC} : y = -2x + 2 \Rightarrow F\left(\frac{3}{5}, \frac{4}{5}\right) \Rightarrow |OF| = 1.$$

4. Summary

In this chapter, we briefly introduced the Mofet Network, using the Mathematics Department as an anchor. For more information, the reader is invited to visit the Mofet website: www.reshetmofet.org/en.

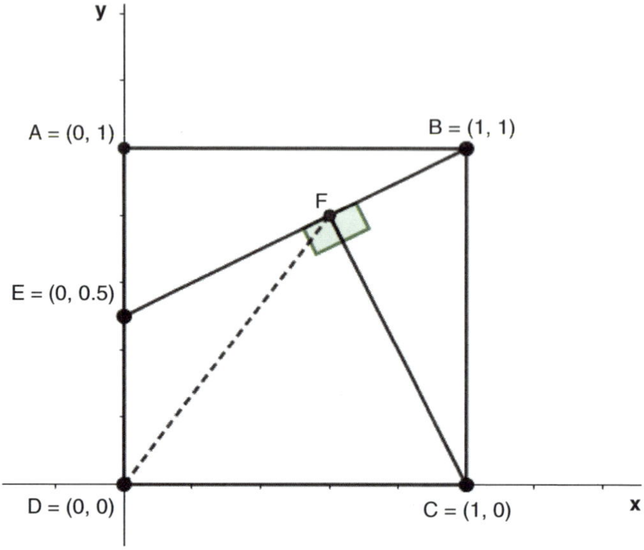

Fig. 5. An analytic solution.

References

Horowitz, T., Shamai, S. and Ilatov, Z. (2014). *From an Immigrants Association to a National Education Network*. Lanham, MD: University Press of America.

Kieran, C. (2006). Research on the learning and teaching of algebra, in A. Gutiérrez and P. Boero (eds.), *Handbook of Research on the Psychology of Mathematics Education: Past, Present and Future*, Rotterdam, The Netherlands: Sense Publishers, pp. 11–49.

About the Author

Tamara Avissar-Zeldis received her PhD from Tel Aviv University. She is a Teacher and Pedagogic Instructor at Achva College. She has 30 years of experience in mathematics teaching for high school students, and 15 years of experience in educating mathematics teachers.

Chapter 19

The Advancement of Mathematics Studies in the ORT Israel Educational Network — Policy and Implementation

Lea Dolev* and Eli Eisenberg#

ORT Israel, Israel

**leadolev@gmail.com*

#elieizenberg@gmail.com

ORT Israel is the country's leading network of comprehensive schools geared to teaching advanced technology and sciences. Its Research, Development, and Training Administration works with and for the ORT schools, to create innovative pedagogical projects that develop teachers, enhance students' skills to raise their matriculation eligibility and achievements, and contribute to bridging gaps between different population groups. ORT Israel's educational policies are implemented in all its schools, including for the study of mathematics. The policy for mathematics focuses on developing a nurturing environment, enabling students to persevere with their mathematics and science studies at the advanced level throughout grades 10–12, and attain higher scholastic achievements. The change in ORT's mathematics education policy involves pedagogical aspects which strengthen motivation, and organizational aspects highlighting the importance of the department chair and leading teachers in the guidance and professional development of the teaching staff. The goal of this policy is to support the aspirations of principals, teachers and students for a high-quality secondary school graduation certificate that will pave the way to tertiary education in advanced fields suited to the 21st-century employment environment.

1. Introduction

The ORT Israel (acronym for *Association for Vocational Crafts*, in Russian) network has 208 educational institutions in over 70 localities, many in

Israel's periphery, and is geared to teaching advanced technology and sciences. Its comprehensive schools impart knowledge, skills, and values through innovative and cutting-edge methodologies in high-tech and science education, while also emphasizing the development of social and humanistic values.

ORT Israel's highly innovative systemic pedagogical projects apply a variety of teaching-learning methodologies in many disciplines, the purpose of which is to develop our teachers, enhance students' skills to raise their secondary school graduation certification eligibility and achievements, and contribute to bridging gaps between different population groups. Based on qualitative and quantitative criteria and research, ORT Israel has proven itself as an organization that delivers results matching its objectives.

With one out of every ten graduates of the Israeli education system being an ORT student, the network has determined educational policies that are implemented in all its schools, including a policy for high-quality teaching of mathematics, which is the focus of this chapter.

2. Educational Policy for Mathematics Teaching

How we define high-quality mathematics teaching is highly important when it comes to formulating a policy aimed at enhancing students' mathematical knowledge. This definition is not absolute, but rather is grounded in a mindset of what is important to know in mathematics and what the objectives of mathematics teaching are.

ORT Israel's policy for mathematics teaching focuses on developing a learning environment that nurtures every student rather than merely using ability grouping. A nurturing environment is created through an adaptive teaching approach aiming to achieve a common instructional goal with learners whose prior achievements, aptitude, or learning styles, differ (Ikwumelu, Oyibe and Oketa, 2015). An adaptive teaching approach allows the teacher and the school as an organization to develop considerations, methods, and tools that enable students to persevere with their mathematics and science studies at the advanced level throughout grades 10–12 and attain higher scholastic achievements. Teachers must have pedagogical expertise that integrates knowledge of the discipline and teaching methods and means, as well as familiarity with the students and their thinking processes (Gavish, 2004).

ORT Israel leads in what the professional literature calls "third-order change". Research literature on strategies for educational change and innovation (Chen, 2004) describes several possible levels of change: first-order change — changing one sub-system in a school; second-order change for a basic unit structure (i.e., a school); and third-order change, which aims to change an entire system (e.g., the ORT Israel network).

Policy focuses on the initial ability grouping of students as a basis for change: paving the way for students to improve, while making every effort to keep students in the higher levels through extra tutoring and support to prevent a "drop out" to lower levels. To achieve this, a change must occur on several levels: defining what disciplinary knowledge is important to learn, what steps should be taken to teach it, the extent and methods of support provided both to students and teachers, and the organization and integration of all necessary changes.

The starting point is formulating a policy focused on improving mathematics teaching, with a full commitment of the management of ORT Israel and of each school. The Research, Development, and Training (RD&T) Administration deals with curriculum development and the professional development of the teaching. RD&T works with and for the ORT schools to enhance teaching and develop study content for all disciplines studied in the ORT network (Fig. 1).

In the case presented here — the change in education policy for mathematics — the RD&T Administration worked on both organizational and pedagogical aspects:

- The organizational track empowers department chairs and leading teachers in the professional development of the teaching staff, in terms of their disciplinary knowledge and the nurturing of their students' motivation and skills.
- The pedagogical track includes: (1) strengthening students' motivation as an integral part of teaching, by cultivating their sense of belonging, efficacy, and autonomy; and (2) developing an approach to cognitive empowerment in presenting and practicing topics (Deci, Ryan and Williams, 1996).

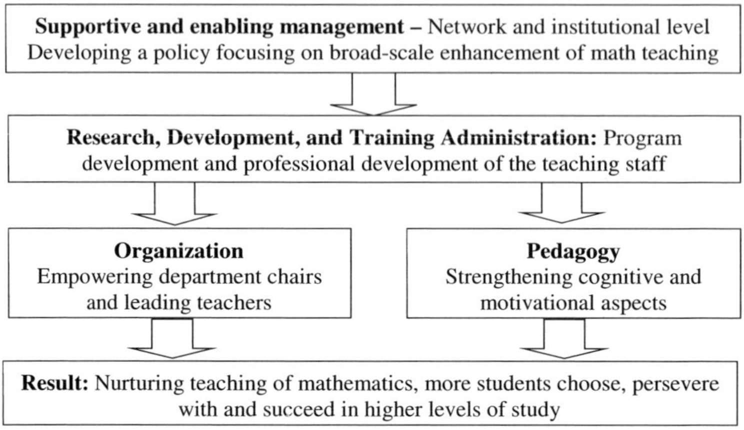

Fig. 1. ORT Israel's policy guidelines for mathematics teaching.

These aspects are not imposed "top-down", but are created through iterative processes among RD&T teams, department chairs, and teachers in the field. This is the tremendous advantage of the network: on the one hand there are the defined goals and targets of the organization to which all schools are committed, but on the other hand, schools, department chairs, and teachers are not alone as they strive to attain the goals, because they are active players in the development and use of supportive pedagogies, and are backed by the RD&T Administration throughout the process.

2.1. *Organizational aspects — department chairs and leading teachers*

ORT Israel's RD&T Administration launched the "Coordinating for 5" program in 2015, aimed both at retaining students studying mathematics at the intermediate and advanced levels and increasing their number.

The program, which began in 10 schools and then expanded to 20, works in two channels. The first channel revolves around the department chairs and supports them through empowerment of their staff-management skills, on target-oriented data-driven work (Michaeli and Somer, 2004). Over the years, the department chairs have received personal accompaniment by the RD&T Administration. There is also an annual professional development day attended by most (85%) of the ORT mathematics department chairs, at which they are introduced to innovations in mathematics teaching and management.

The second channel revolves around the department chairs together with the leading teachers, forming a community for shared discussion and consultation, to create leaders of excellence in mathematics teaching. This community will be able to co-construct pedagogical knowledge that links experience in the field with theory and contemporary research (Birenbaum, 2009; Blanga, Lendler-Pardo and Shachar, 2011). This is a community of 30 leading teachers and 35 department chairs in the "Coordinating for 5" program. Teachers have a 4-hour meeting once every three weeks, while department chairs meet twice a year (one of these meetings is a 2-day seminar) for peer learning, work on data-driven management, leadership skills, and innovations in teaching pedagogies.

2.2. *Pedagogical aspects*

The pedagogical aspects comprise strengthening student motivation and cognitive empowerment of students.

2.2.1. *Strengthening student motivation in mathematics lessons*

The RD&T Administration perceives teachers as supporters of students' motivation. Three resources that can enhance student motivation are the sense of autonomy, the sense of self-efficacy, and the sense of belonging. Hence mathematics teachers must support student motivation through the following means:

(1) Student-teacher dialogue — highlights students' ability to choose and to reach personal goals and outcomes that are not easy to attain.
(2) Class discourse, norms, and climate — reinforce students' sense of belonging to a group and feeling safe and protected. Class discourse or climate, in which the students have a sense that they can change and influence matters, supports individual motivation as well.
(3) The mathematical tasks and their delivery — teaching the material, presenting the tasks, and the individual and class conduct while solving them, must all be linked to the need for autonomy, efficacy, and belonging. Each task has aspects which the teacher should utilize to enhance student motivation.

2.2.2. *Cognitive empowerment of the student in mathematics lessons*

We believe that teaching literacy is more than simply providing generic literacy skills and that effective teaching of literacy skills can only take place within a disciplinary context, so as to familiarize students with ways to create knowledge that are specific to each discipline. In this manner, learning will be more authentic, substantive, and meaningful to real life (Olshtain, 2012).

The principles underlying the concept of mathematical literacy in the ORT Israel network today are as follows:

- organize and form mathematical thinking through communication;
- communicating mathematical thinking clearly and consistently to peers and teachers;
- use mathematical language to express ideas accurately and precisely.

Following these principles requires cultivating skills which are relevant for expression, communication, and coping with wording in mathematics. In our training programs, teachers work on ways to enable students to create their own mathematical wording: they work on the variety of opportunities in which students express themselves with mathematical language in reference to mathematical content. The approach and tools linked with literacy were developed mainly for AKADEM — a program for literacy, learning community, and ongoing dialogue, implemented in 20 ORT schools.

Integrating technology into mathematics teaching: Interactive technological tools may change how students think about mathematical concepts and how teachers teach them. The use of technological tools in class brings the mathematical concepts to life and enriches mathematical thinking (NCTM, 2000). According to a common perception of mathematical literacy, the use of technological tools encourages mathematical communication since it enables the design of tasks in which mathematical concepts have visual and dynamic representations. These, in turn, constitute the basis for active learning through discussion and independent work.

Our training program for mathematics teachers was developed and implemented with a group of six mathematics teachers from different

ORT Israel schools, who undertook upon themselves to study the issue, implement it in their classes, and lead the in-service course for their fellow teachers. The course has been running since 2012. So far about 25% of all ORT mathematics teachers have taken the course and are successfully integrating technology into their teaching.

3. Conclusion

Between 2001 and 2015 there have been significant changes in the features of the secondary school graduation exams in mathematics among ORT Israel graduates. The most significant change is the drop in the number of students who did not take the exam — from almost one third down to only 15% — and a parallel rise in the number of graduates passing the basic-level exam. The percentage of graduates examined at the advanced level rose from 26% to 29%. Even though changes have been minor, in raw numbers this means 100 additional graduates with a high quality secondary school graduation certificate. This trend of change shows that the change policy is, indeed, bearing fruit.

The organizational and pedagogical aspects of the policy for improving mathematics teaching in ORT Israel are in different stages of implementation in schools. However, a combination of actions will eventually be implemented in all the network's schools. This will lead to an enhanced teaching of the discipline itself, enhanced school discourse, and a student-teacher dialogue regarding the goal of having more students choosing, persevering, and succeeding in mathematics studies at the intermediate and advanced levels. The goal is to support the principals, teachers, and students to aspire to a high-quality secondary school graduation certificate that will pave the way to high-level tertiary studies in meaningful advanced fields, suited to the 21st-century employment environment.

References

Birenbaum, M. (2009). Assessment for learning and characteristics of a school professional community and classroom culture that empowers it, in Y. Kashti (ed.), *Assessment, Jewish Education, and the History of Education: Anthology in Commemoration of Prof. Arieh Levi*, Tel Aviv University School of Education and Ramot Publications (in Hebrew), pp. 77–100.

Blanga, A., Lendler-Pardo, G. and Shachar, M. (2011). Teachers' professional learning communities: A review commissioned as background material for the work of the Committee for Language and Literacy, http://education.academy.ac.il (in Hebrew).

Chen, D. (2004). *Between Vision and Reason: About the Strategies of Change and Innovation in Education*, Tel Aviv: Ramot (in Hebrew).

Deci, E. L., Ryan, R. M. and Williams, G. C. (1996). Need satisfaction and the self-regulation of learning, *Learning and Individual Differences*, **8**(3), 165–183.

Gavish, V. (2004). The relationship between "external" agents of change and "internal" agents of change in schools undergoing a process of change, in S. Guri-Rosenblatt (ed.), *Professional Education of Teachers from a Multi-disciplinary Perspective*, Open University (in Hebrew).

Ikwumelu, S. N., Oyibe, O. A. and Oketa, E. C. (2015). Adaptive teaching: An invaluable pedagogic practice in social studies education, *Journal of Education and Practice*, **6**(33), 140–144.

Michaeli, N. and Somer, O. (eds.) (2004). *Leading Teachers as Agents of Improvement in the Education System*, Jerusalem: The Initiative for Applied Education Research, Israel Academy of Sciences and Humanities (in Hebrew).

National Council of Teachers of Mathematics (NCTM) (2000). *Principles and Standards for School Mathematics*, Reston, VA: Author.

Olshtain, E. (2012). An outline of the discussion, in I. Pollack (ed.) *Middle School Literacy: Between Language and Disciplinary Literacy — Workshop Report*, Jerusalem: The Initiative for Applied Research in Education, Israel Academy of Sciences and Humanities, pp. 6–7 (in Hebrew).

About the Authors

Lea Dolev is head of Mathematics Studies, Research & Development and Training at ORT Israel. She received her PhD from Technion, Institute of Technology, Haifa, MA degree from Tel Aviv University and BSc from Haifa University, Israel.

Eli Eisenberg is a Senior Deputy Director General for Research & Development and Training at ORT Israel. He received his PhD, MSc and BSc degrees from the Technion — Israel Institute of Technology, Haifa.

Chapter 20

Promoting Advanced-Level Mathematics in Diverse Populations in the Amal Educational Network

Ronit Ashkenazy and Anna Vaknin*

AMAL Educational Network, Tel Aviv, Israel

**annav@amalnet.k12.il*

This chapter focuses on how the Amal Educational Network, a leader in promoting scientific and technological excellence, responds to the needs of students and teachers, both as groups and as individuals. This is accomplished through the implementation of three programs: a program to encourage girls to pursue advanced studies in STEM subjects, while raising teachers' awareness gender-specific aspects of learning; a program to tap the potential of high-achieving students in intermediate-level mathematics, while educating teachers on how to respond to students' individual needs on the emotional, functional, and learning levels; and a program to improve teaching in the Bedouin sector in southern Israel by recruiting local teachers, providing individualized support for each teacher based on the their needs, and partnership in the unique community of mathematics teachers employed in the Bedouin sector. Each of these programs has served as a model for implementation in additional Amal schools.

1. Historic and Pedagogical Background of Amal

This chapter describes Amal and three unique programs it implements for the teaching of mathematics, with the goal of providing educational solutions suited to the diversity of students and teachers.

Amal is a large and well-established education network in Israel, with an enrollment of approximately 40,000 students in all sectors of the population — Jewish, Arab, Druze, Bedouin, secular and religious. Amal has been in operation since 1928, leading in scientific and technological excellence, with more than 100 educational facilities spread across the country, from middle schools to adult education.

In recent years, Amal has focused efforts on increasing the number of students who pass the advanced-level secondary school graduation exam in mathematics. The quality of the secondary school graduation certificate is on the national agenda, with emphasis placed on advancing the study of mathematics, physics, and computers, and a corresponding allocation of resources. Teams of mathematics teachers are investing time in thinking of ways to increase the number of advanced-level students. Extensive study hours are invested in programs sponsored by the Ministry of Education, aimed at creating a technological and scientific reserve, with mathematics as a priority. In addition, Amal operates an enrichment program with the main objective of providing professional enrichment for teachers, with training in disciplinary-specific and other pedagogical skills, diagnosis and monitoring.

2. Rationale

Over the years, Amal has strived to provide mathematics education to a wide range of students. Diversity is expressed in differences between populations (e.g., gender and language) and between individuals within a population (e.g., learning styles and types of difficulties). According to Gardner (1996),

> A school focused on the individual allows understanding and optimal development of the cognitive profile of each student... The ideal school focuses on two complementary principles: providing a broad solution and emphasizing difference, and adapting the learning processes accordingly.

In this spirit, we have identified students with an unrealized potential for learning advanced-level mathematics. We have focused our efforts on populations that in the past were less likely to pursue and advanced-level secondary school graduation exam in mathematics, while taking into consideration and addressing their unique needs.

3. Special Programs

In addition to general programs intended for all Amal teachers and students (seminars, enrichment hours, and ongoing support), Amal focuses on helping three students subgroups realize their potential: female students;

high-achieving students in intermediate-level mathematics; and Bedouin students in southern Israel. In this section, we will elaborate on the rationale for each of these programs.

3.1. *Gender-aware pedagogy*

We attach great importance to the study of science, technology, engineering, and mathematics (STEM) at the highest level by girls, and believe that choosing these subjects will open up a wide range of opportunities for their successful integration into the employment market. As part of the program to encourage girls, we operate two complementing intervention tracks: one aims to change attitudes among high school teachers (in collaboration with the Trump Foundation), and another targets 9th grade girls, just before they elect which subjects they want to study at the advanced level. This track is based on research and experience in the education system in Israel and the world that focuses on addressing conscious and emotional obstacles that deter girls from pursuing advanced STEM studies.

In the 2013–2014 and 2014–2015 school years, the program included 500 9th grade female students from 10 Amal schools, who had demonstrated the potential to succeed. The students were exposed to the exciting world of science and technology, meetings with women leaders in the world of science and industry, and workshops on career development and future employment paths. Particular attention was paid to integrating content that had been tested and found to be relevant for girls, such as health and the environment, as well as ethical issues that require physical and technological solutions, which according to research findings are the areas of science most preferred by women.

In order to create lasting and viable systemic change, it is not enough to create motivation among girls, a process of changing perceptions among teachers is also necessary. Teachers need to be made aware of the different learning styles of girls and boys, as well as the fact that teaching in the traditional approach is more suited to the learning style of boys than girls. For example, individual learning is more compatible with boys' learning style, while group work is more compatible with girls' learning style. Teaching that emphasizes competition is often not suitable for girls and even blocks them. Theoretical teaching based on algorithms only is more suitable for boys while teaching that links learning to everyday life is more suitable for girls.

> It is recommended to encourage mathematics and science teachers to change their teaching methods in order to create an open classroom climate with less competitiveness, and to employ pedagogies that encourage understanding as a way to reduce the gaps between boys and girls (Zohar and Sela, 2002).

The training and guidance of the teaching staff contributes to the program's uniqueness and success. Focused work with teachers, based on observing lessons in the classroom, brings about a substantial change in teachers' faith in the ability of their female students.

The process of working with teachers includes several stages. First, teachers are exposed to facts and research in the field. They observe lessons, and monitor behavior patterns and teaching methods with regard to girls in their classes. Once the data has been collected, they focus on case studies, analyze them, focus on the issues requiring change, evaluate ideas, and propose changes. Next, each teacher sets up a task and a unit of study or activity in keeping with the principles learned, and implements it in the classroom. In the final stage, a workshop is held where teachers reflect on the process and what they have learned from it. The process is supervised by a teacher who has proven success in advancing girls' achievements in mathematics. A significant part of the work with the teachers involves personal mentoring.

One of the teachers who participated in the program reported,

> "I learned the most by watching my own lessons. Until then I hadn't noticed how much boys love competition, this is why they are always the first to jump in. Although the girls are more hesitant to volunteer, this does not mean they don't know the answer. In addition, I give more teamwork assignments now, girls like to learn in groups. At the same time, I try to link teaching to everyday life topics and make them relevant."

As a result of these two tracks, we have seen an increase in the number of girls who choose to study mathematics and science at an advanced level. The percentage of girls in advanced-level mathematics classes has increased from one-third to one-half in schools where the program has been implemented. Both programs, which operate in parallel and with close coordination, advance the girls, while helping teachers change the climate in the classroom, and, when necessary, changing unconscious attitudes.

3.2. *Pedagogy adapted to students' needs*

At the root of this program is the assumption that a high percentage of high-achieving students in intermediate-level mathematics have untapped potential. From the previous program for gender-focused instruction, we learned that tailored guidance, adapted for teachers by master teachers or teacher-educators, is the key to advancement. Providing individual solutions based on the needs of teachers, and support in dealing with difficulties and constant reflection, lead them to success.

We chose teacher-educators based on their experience and their ability to bring different populations to succeed in their studies at the highest level. In its initial stage, we focused our intervention program on five Amal multidisciplinary schools with a high percentage of high-achieving students in intermediate-level mathematics.

In this program, we raise teachers' awareness of the diversity in students' learning styles, cognitive and emotional characteristics, and the need for teaching methods adapted to the needs and abilities of each student. We nurture teachers' skills for recognizing and handling the student's mathematical, emotional and functional obstacles. Diagnosis and treatment are conducted through questionnaires, formative assessment tasks, and dialogue between teachers and students. The teacher-educator serves as a model for teachers learning about diagnosis and treatment, by managing conversations with students, evaluating the characteristics of their learning, and building a tailored program suited to each student's needs. The student is a full partner in the process of building a personal plan, goals and schedule.

At the same time, in schools where the program is run, activities aim to instill a new pedagogical culture around the study of mathematics, in cooperation with all stakeholders: management, counselors, educators, program directors, mathematics teachers, and students. This change is reflected in school processes such as: defining new criteria for assigning students to advanced-level mathematics at the beginning of 10th grade; writing and implementing a procedure for transitioning between the different levels of study in mathematics; and the recruitment of educators and counselors to provide students with emotional support. In each school where the program has been implemented, the number of students opting to study advanced-level mathematics has significantly increased. At the onset of the program, only 12% of students in the participating schools studied mathematics at the advanced level. Two years later, 16% of 12th grade students, 24% of 11th grade students, and 30% of 10th grade students opted to study advanced-level mathematics.

In its second year of operation, this program has served as a model for disseminating information to other schools in the network and has generated great interest among the coordinators of mathematics and mathematics teachers in Amal high schools. We plan to introduce this model into additional Amal schools in the 2017–2018 school year.

3.3. *Unique activities in the Bedouin sector*

Amal operates 11 schools in the Bedouin sector in southern Israel. This population has unique characteristics. Amal faces multiple challenges in this sector that require appropriate solutions.

The first challenge is to create a suitable and stable, local teaching force. Until recently, teachers in the Bedouin sector in the south came from the north, and left after acquiring teaching experience. This led to the need to train new teachers each year and to hire inexperienced teachers on a regular basis. In recent years, emphasis has been placed on preference for the employment of local teachers. We have taken initiatives to recruit local teachers, for example, we contact students of teaching at Kaye College and Ben-Gurion University and employ them as teaching assistants. In addition, we encourage graduates of our schools who are studying toward a mathematics teaching certificate to return to their villages and even to the school where they studied in their youth. In our experience, local teachers have a greater commitment to the student population, school and community where they grew up.

Today, after several years, the number of local teachers of mathematics has increased in all schools in the Bedouin sector in the south. As part of the integration process, these teachers receive guidance from a local teacher-educator, who speaks the language of instruction and who teaches either in the same school or in a nearby school. The form of guidance varies according to the needs in the field. Guidance typically includes mutual observation of the teacher-educator and teacher in the classroom, as well as assistance in creating lesson plans and tests, and in analyzing the test results.

The second challenge is to raise student motivation for the study of mathematics, while bolstering confidence in their abilities. In addition to traditional solutions, such as additing hours and study groups to reinforce skills and bridge gaps in preparation for high school, we conduct unique activities with students, parents and teachers. According to Dr. Ruth Lusting *et al.*, parental involvement by "Academic Socialization", and the associated parents' involvement, contribute to raising student motivation and confidence in their ability to cope with the challenges posed by the study of advanced-level mathematics and physics.

We hold meetings for parents and students, where we present the advantages of studying advanced-level mathematics, and explain what is required of the parents to enable their children to succeed. We raise the awareness of both students and their parents regarding the investment required and the importance of parental support and encouragement.

The lack of suitable textbooks in Arabic presents an additional challenge for students, who need to study mathematics in Hebrew, which is not their mother tongue. Amal has initiated a series of professional development days and seminars for teachers, the products of which are educational materials in Arabic. A culture of mutual cooperation has been created, whereby the teachers routinely translate materials and share them with their colleagues. In addition, to support the unique needs of these teachers, we have established a community of dedicated, mathematics teachers in the Bedouin sector in the south, a sub-community of Israel's mathematics teachers who prepare students for the advanced-level examinations.

4. Summary

Implementation of excellence in teaching mathematics requires programs tailored to both teachers and students, in keeping with and addressing the differences in each of the populations. For this purpose, there is a need to map and analyze needs and create customized programs. The core idea is to tailor learning according to the needs of both teachers and students, while focusing on populations identified as having untapped potential.

References

Gardner, H. (1996). *Multiple Intelligences*, Hebrew edition, Branco Weiss Institute, Jerusalem, Israel (in Hebrew).

Zohar, A. and Sela, D. (2002). Her physics, his physics: gender issues in Israeli advanced placement physics classes, *International Journal of Science Education*, **25**, 245–268.

About the Authors

Ronit Ashkenazy is the Pedagogical Director at Amal, where she leads innovative pedagogy, entrepreneurship and excellence. She is formerly STEM Projects Manager, Ministry of Education. She pursued degrees in DSc, Biotechnology; MBA; MSc, Medical Science; BSc and Teacher Certification in Biology.

Anna Vaknin is the head of Mathematics Studies at Amal. She pursued her BSc degree in Mathematics and Computer Science.

Chapter 21

Problem-Solving Forums on Social Networks that Accompany the Learning of Mathematics in Israeli High Schools

Boris Koichu[*,||], Yaniv Biton[#,], Olga Gendler[‡,§,##] and Nelly Keller[§,¶,‡‡]**

[*]Weizmann Institute of Science, Rehovot, Israel
[#]Shaanan Religious Academic College of Education, Haifa, Israel and CET, Tel Aviv, Israel
[‡]ORT Kramim School, Karmiel, Israel
[§]The Technion, Haifa, Israel
[¶]Ulpana Sgula School, Kiryat Motskin, Israel
[||]boris.koichu@weizmann.ac.il
[**]YanivB@cet.ac.il
[##]olgag@kramim.ort.org.il
[‡‡]nellyk@tx.technion.ac.il

We present three examples of online problem-solving forums that accompany the learning of mathematics in Israeli high schools. The first example introduces a Facebook forum intended to prepare students from different schools for the secondary school graduation exam in mathematics. The second example concerns a WhatsApp asynchronous forum opened by a novice mathematics teacher for her 10th grade. The forum has gradually developed into a multifunctional space for students seeking assistance and sharing solutions to homework problems. The third example is a synchronous Google+ forum in an 11th grade initiated by an experienced mathematics teacher. This forum functioned as a platform for collaborative problem solving after school hours. Trends in development of self-organized online activity are discussed by the end of the chapter.

1. Introduction

Online activity is an inalienable part of social functioning of young people. The youth chat with friends, seek information, and upload on social networks artifacts related to events of importance to their lives (Boyd, 2008). Studying mathematics is part of schoolchildren's lives. Accordingly, it should not be surprising that online forums at social networks accompany the learning of mathematics in many classes in Israel. Students' online activity related to their studies is still insufficiently documented and explored, but its scope and diversity seem to rapidly grow (Borba *et al.*, 2016). The goal of this chapter is to document some of this activity and articulate some of its gains for enhancing the learning of high school mathematics. This goal is pursued based on presentation of three examples, with special attention to mathematical problem solving.

2. First Example: Mathematical Marathon at Facebook[a]

Four days before a secondary school graduation exam in mathematics in 2013, the Center for Educational Technology (CET) opened eight groups on Facebook; four groups were in Hebrew and four in Arabic. The exam consisted of four forms. The forms (numbered 804–807) differed in level of difficulty and involved various mathematics topics. Each Facebook group was devoted to a particular form. Overall, 614 students and 16 teachers participated. The students chose a group appropriate to their needs and raised any question they wished. The teachers, who were trained in online tutoring and in Facebook tools, responded to the students' questions. The groups worked 12 hours a day for four days. Each participating teacher was on duty for four hours a day. Table 1 presents the scope of the activity in each group.

About half of the enlisted students appeared at the forums, and the rest chose to act as silent observers. As a rule, students posted requests for help in solving a particular problem from a textbook or a problem book. In many cases, the students uploaded their attempted solutions and asked the teacher to find a mistake. There were also highly specific questions related to particular solution steps. The teacher on duty immediately acknowledged the receipt of the question and after about 10 minutes uploaded a response to the forum. A typical response consisted of a photo of a hand-written solution or of a hint.

In the words of one of the participating students, "It [the forum] gave me a chance to handle exercises that I couldn't solve. Not to give up like

[a]This section summarizes a case study reported in Biton, Hershkovitz and Hoch (2015) and Biton *et al.* (2015).

Table 1. Numbers of participants and problems inquired about in each group.

	The number of the enlisted participants	The number of problems inquired about
Group 804	191	171
Group 805	59	12
Group 806	245	184
Group 807	119	68

I usually do, but to get the solutions from a teacher. This really helped me." In the words of one of the participating teachers, "I really liked that the students asked relevant questions related to the solutions, and didn't give up until they understood."

There were also occasions of student–student interactions. An interesting online exchange between two previously unfamiliar students was documented. It occurred during the short period of time when one student uploaded his question but not yet received a reply from the teacher. To that student's delight, he got the explanation from a peer participant before the teacher responded.

In summary, Biton, Hershkovitz and Hoch (2015) identified a number of meaningful learning opportunities that stem from both teacher–student and student–student interactions. The named opportunities were related to: formulating questions in writing, critical reading of teacher-provided solutions, scaffolding student problem solving, exposure to the peer questions and evaluating peers' solutions.

Most of the students were very enthusiastic when reflecting on their experience. For example, one student wrote: "I would be very happy to get this kind of help throughout the year. It is above and beyond what a student can expect for success". It is of note that in 2013 studying mathematics on Facebook was a novelty for most of the participating students and teachers, and that now participation in online forums that support studying mathematics is quite a routine for many Israeli schoolchildren, as the following two examples show.

3. Second Example: A Multifunctional WhatsApp Forum

During her first year of teaching in high school (2014–2015), OG (the third author) joined a research project entitled "Heuristic and engagement aspects of learning through long-term collaborative mathematical problem solving."[b] One of the goals of the project was to enhance mathematical problem solving in a learning environment combining a classroom and an online components. Seven high school classes took part in the project. The pedagogical idea was to encourage the students to approach challenging mathematics problems in a classroom and continue solving them after school hours on an online forum. The realization of this idea appeared to be not simple at all (Keller and Koichu, 2017; Koichu and Keller, 2017).

Notwithstanding various quandaries associated with being a first-year high school teacher, OG managed to gradually incorporate into her lessons a desired problem-solving component. However, OG's online problem-solving forum did not become active despite considerable effort. A similar situation was observed in OG's additional class in 2015–2016. It is important to note that OG was convinced in the potential usefulness of online forums, based on intensive interactions with some other participants of the project.

After deeply reflecting on the accumulated experience, OG opened a multifunctional rather than a problem-solving forum for her new class in 2016–2017. The main lesson learned from the first two attempts was that the forum can be active on a long run only if it addresses some existing needs of the students and develops gradually, along with the development of problem-solving activity in a classroom. The data below characterizes the first four months of the forum in OG's 10th grade class. The class consisted of 17 students who studied intermediate-level mathematics.

When the forum on WhatsApp was open in September 2016, 11 students enlisted and eight students actually appeared on the forum. OG introduced the forum as a virtual space for handling anything related to the learning of mathematics. For example, the students sought administrative information that they missed at the lessons (e.g., in which room the next mathematics lesson would be, what homework was assigned, and when it was due). At the beginning, students' questions were mostly answered by OG, but in a short while she began readdressing some of the student questions to other students. In about two months, students began answering each other questions voluntarily, that is, without OG's interventions,

[b]The project was conducted under the auspices of the Technion, where Boris Koichu, the PI of the project, worked until February 2017.

Table 2. Student activity in OG's WhatsApp forum in September–December 2016.

	Sept.	Oct.	Nov.	Dec.	**Overall**
Requests for assistance in solving homework problems	5	8	17	12	**42**
Requests for information (mainly administrative issues)	8	4	10	17	**39**
Verbal responses to the others' questions	5	7	28	13	**53**
Solutions to problems posted as pictures or scanned documents	2	8	24	69	**103**
Questions to the teacher	6	27	32	22	**87**
Overall	**26**	**54**	**111**	**133**	**324**

and the forum became rich with student–student interactions. The number of the forum participants grew gradually, and in December all 17 students enlisted and became active on the forum. The key events that enhanced the forum were four problem-solving lessons in which the students used their smartphones, instead of the traditional whiteboard, as a tool for sharing their problem-solving ideas. Table 2 shows the scope and types of the student activity on the forum.

The forum gradually extended and evolved. As one can see from Table 2, the portion of administrative questions in the overall activity decreased from 31% in September to 13% in December. Uploading attempted and full solutions to homework problems became the most widespread type of the student activity (8% in September, and 52% in December). It is of note that forum activity was unequally distributed among the students: 265 out of 324 entries to the forum (82%) belonged to nine particularly active students. However, unlike the Facebook forum described in Section 2, not every student question or uploaded solution deserved a response. A sizable part of the forum was a repository of attempted solutions without evidence of collaborative problem solving; only a few (about 10) attempts to solve problems together were documented.

In sum, it would be safe to say that the forum has not yet developed as a collaborative problem-solving space during the first four months of its existence, but it was tremendously instrumental for gradually developing, in OG's class, the norm of sharing work in progress and the norm of perceiving peers (and not only the teacher) as a source of useful information and assistance. These norms can be seen as precursors to collaborative problem solving.

4. Third Example: A Problem-Solving Google+ Forum

NK (the forth author) participated in the aforementioned research project since 2013, with two of her classes that studied advanced-level mathematics. Different aspects of the development of the forum in the first class (10th grade, 17 students) are reported in Keller and Koichu (2017) and Koichu and Keller (2017). In brief, after three months of relatively unstable activity, the forum, which was opened on Google+ and then moved to WhatsApp, flourished. The forum consisted of nine problem-solving episodes of 100 to 300 posts each. As in the previous examples, about half of the students were active on the forum and about half were silent observers. More than 95% of the interactions were among the students; the role of the teacher was restricted to offering the students challenging problems and making encouraging remarks. The forum functioned for three years until the class graduated in 2016.

The data below were collected in the second NK's class, in 2016. We choose to focus on that class (11th grade, 19 students) in order to represent the online activity at its mature stage, that is, at the stage when the "rules of the game" were well-established. It is of note that the class had opened a WhatsApp forum at least one year before they met with NK. The class continued using this forum also in the 2016–2017 school year, and granted NK access. Generally speaking, the WhatsApp forum was slightly more active than the forum in OG's class (to recall, the use of a forum for studying mathematics was a novelty for OG's students), and functioned in similar ways. On average, 10 posts appeared on the forum during ordinary days. Relatively small portion of the posts were related to administrative issues (e.g., "Do we have a lesson tomorrow?" or "What was the homework?"). Most forum discussions concerned the homework assignments (e.g., "somebody please upload a sketch of a derivative from problem 96" or "can somebody send the answers to exercises 73 and 74?") The students also occasionally helped each other with theoretical matters (e.g., one student wrote that she did not understand under which manipulations a sign of an algebraic inequality changes, and another student explained it).

Similar to OG's forum, NK's forum was particularly active (80–150 posts a day) when students studied for tests. For example, the online activity towards preparation to one of the tests involved 15 students who made 133 contributions. Most of the contributions (74%) consisted of requests for assistance and queries related to the status of particular solution steps and ideas; most of the questions were answered by the peer students.

Encouraged by students' willingness to collaborate online and by her successful experience in the first participating class, NK suggested to open an additional forum on Google+, a forum for collaboratively solving challenging problems. The forum was opened as an online synchronous platform for weekly meetings after school hours. Each meeting lasted two hours and as a rule was devoted to solving one problem. Consider an example.

One of the problems[c] that NK offered on the Google+ forum evoked 236 contributions by 16 students. The contributions were more or less equally distributed among six discussion threads. Five threads developed around five different problem-solving ideas; most students contributed to more than one thread. Fifty eight percent of the student contributions were in the form of questions and queries of different types, but 38% can be classified, in terms of Clark, James and Montelle (2014) as "group synergy". Clark, James and Montelle (2014) introduced the group synergy category in order "to capture the combination and confluence of two or more group members' problem-solving moves that could only occur when solving problems as a member of a group" (p. 10). In other words, the students meaningfully collaborated, continued and developed each other's ideas and obtained the solutions that they could not obtain if working on the problem alone. In sum, a Google+ forum in NK's class can indeed be called a platform for collaborative problem solving.

5. Concluding Remarks

This chapter presents evidence that online problem-solving activity can augment the study of mathematics in high school. The presented examples

[c]The problem (adapted from Sharygin and Gordin, 2001) was as follows: "Two non-intersecting circles are given. From the center of each circle, two tangent segments to the other circle are constructed. The points of intersection of the tangent lines with each circle define two chords. Prove that the obtained chords are equal." This problem was also used on the asynchronous forum of the first NK's class. On that forum, discussion of the problem lasted for 4 days (see Koichu, 2015).

suggest that such an activity can gain considerable response from students and, most importantly, can be self-organized. The use of social networks as a medium for studying mathematics appears to become more and more widespread in Israel.

Some learning opportunities associated with student–student and teacher–student online interactions around solving challenging mathematical problems are presented. In addition, the role of the online forums for gradual development of norms of collaborative problem solving is illustrated. It is of note that long-term online interactions focusing on the students' learning needs and having informal nature, as it is the custom in the social networks, bears remarkable potential for enhancing problem solving also in the classroom setting.

Still, it looks like most of the self-organized problem-solving forums develop as question–answer forums at which the students share attempted solutions and seek assistance. Special pedagogical effort is needed in order to transform informal online discussions into genuine platforms for collaborative problem solving. Our hope is that in time the self-organized online activity will develop also in this direction. Certainly, social networks provide new opportunities for enhancing mathematics study, but these opportunities should be properly seized.

Acknowledgments

The research described in Sections 2 and 3 of this chapter was supported by the Israeli Science Foundation (grant 1596/13, PI Koichu). Opinions expressed here are those of the authors and are not necessarily those of the Foundation.

References

Borba, M. C., Askar, P., Engelbrecht, J., Gadanidis, G., Llinares, S. and Aguilar, M. S. (2016). Blended learning, e-learning and mobile learning in mathematics education, *ZDM Mathematics Education*, **48**(5), 589–610.

Boyd, D. (2008). Why youth ♡ social network sites: The role of networked publics in teenage social life, in D. Buckingham (ed.), *Youth, Identity, and Digital Media*, Cambridge, MA: The MIT Press, pp. 119–142.

Biton, Y., Fellus, O., Hershkovitz, S. and Hoch, M. (2015). Using the lens of complexity theory to examine a second chance at school mathematizing through Facebook, *Paper Presented at the 2016 Annual Meeting of the American Educational Research Association*, Chicago, IL. Retrieved from the AERA Online Paper Repository.

Biton, Y., Hershkovitz, S. and Hoch, M. (2015). Peer learning in mathematics forum on Facebook: A case study, in K. Krainer and N. Vondrova (eds.), *Proceedings of the Ninth Congress of the European Society for Research in Mathematics Education*, CERME 9, Prague, Czech Republic, pp. 2473–2479.

Clark, K., James, A. and Montelle, C. (2014). "We definitely wouldn't be able to solve it all by ourselves, but together...": Group synergy in tertiary students' problem-solving practices, *Research in Mathematics Education*, **16**(2), 306–323.

Keller, N. and Koichu, B. (2017). A dialogue about integrating a class environment and an online environment in mathematics education, in Schwarz, B. B., Rosenberg, H. and Asterhan, C. S. C. (eds.), *Breaking Down Barriers in Education? Teachers, Students and Social Network Sites*, MOFET, Israel, pp. 156–180 (in Hebrew).

Koichu, B. (2015). Towards a confluence framework of problem solving in educational contexts, in Krainer, K. and Vondrová, N. (eds.), *Proceedings of the 9th Conference of the European Society for Research in Mathematics Education*, Charles University in Prague, Czech Republic, pp. 2668–2674.

Koichu, B. and Keller, N. (2017). Implementation enterprise through the lens of a theory of diffusion of innovations: A case of online problem-solving forums, *The 10th Conference of the European Society for Research in Mathematics Education*, Institute of Education, Dublin City University, Ireland. Available at: https://keynote.conference-services.net/resources/444/5118/pdf/CERME10_0562.pdf.

Sharygin, I. F. and Gordin, R. K. (2001). Collection of problem in geometry. 5000 problems with solutions. (Sbornik zadach po geometrii. 5000 zadach s otvetami). Moscow: Astrel (in Russian).

About the Authors

Boris Koichu is a faculty member at the Weizmann Institute of Science. His research focuses on mathematical problem solving and problem posing.

Yaniv Biton is a faculty member at Shaanan Religious Academic College of Education and a staff member in the team of mathematics education at the Center for Educational Technology, Tel Aviv. His research focuses on assessment and technology in mathematics education.

Olga Gendler is a mathematics teacher at the high school ORT Kramim. She is a member, as a teacher and a researcher, of a research project focusing on collaborative mathematical problem solving.

Nelly Keller is an adjunct lecturer at the Technion and a mathematics teacher in Ulpana Sgula high school. Her research interests are related to unusual learning forms. She currently focuses on developing learning spaces which unite classical classroom learning and problem solving in social networks' forums.

Chapter 22

Designing Mathematics Textbooks: The Case of the *Integrated Mathematics Curriculum Program*

Alex Friedlander*, **Ruhama Even and Naomi Robinson**

Weizmann Institute of Science, Rehovot, Israel
**alex.fried30@gmail.com*

The *Integrated Mathematics Project* (*Mathematica Meshulevet*) is a comprehensive middle school mathematics curriculum program developed in response to the introduction of a new national curriculum. The chapter presents and discusses two salient aspects of the curriculum program: (1) integrating procedural knowledge and higher order thinking skills, and (2) addressing the diverse needs of the Israeli student population within the track for which the program is intended.

1. Introduction

In response to the introduction of a new national middle school mathematics curriculum in 2009 by the Ministry of Education, we have developed a comprehensive middle school mathematics curriculum program — *Integrated Mathematics* (*Mathematica Meshulevet*). The curriculum program emphasizes problem solving, thinking, and reasoning for all students, as well as connections among mathematical concepts, topics, and domains. The development team was headed by the authors, and comprised experienced curriculum developers and teachers. The textbooks were developed in standard/expanded and in limited scope versions, along with a supplementary collection of activities *Excelling Rehovot* (*Metzuyanut Rehovot*) for high-achieving students.

The development of each textbook comprised an experimental and a final version. The latter was edited by scientific, pedagogical, language, gender, and graphic editors, and had to be approved by the Ministry of Education in a three-cycle review process that took about 10 months.

The textbooks were written in Hebrew, and then translated to Arabic, as well as converted to a version suitable for the ultra-Orthodox Jewish community. A teacher's guide and other accompanying resources (e.g., assessment items, additional activities, and applets) come to about 10,000 pages.

Developing the complete series of textbooks required seven years of intensive work. The *Integrated Mathematics* curriculum program is currently used in more than 250 schools throughout Israel, and we provide counseling and support for teachers all over the country via in-service courses and an interactive website hosting nearly 200,000 visits per year (https://stwww1.weizmann.ac.il/math-rehovot/).

Each student textbook includes about 30 units; each unit is organized as a series of four or five lessons. The written lessons suggested in the textbooks generally have a similar structure that includes four main parts. First, an introductory problem is presented, whose aim is to motivate students (i.e., create a need for learning a new concept), "set the scene" (i.e., present a context) or provide a "larger picture" of a new concept or tool to be acquired during the lesson. Then three sets of tasks are presented: tasks that lead to the formalization of the new concept or tool, tasks that require students to use and apply various aspects of the formalized concept, and tasks that consolidate the newly acquired knowledge. In addition, tasks that require higher-order thinking, such as reflecting on the learned concept, making connections with other concepts, comparing different solutions, or detecting errors in hypothetical students' solutions are presented throughout the textbook.

The *Integrated Mathematics* curriculum program has several unique characteristics. This chapter focuses on two salient aspects: (1) integrating procedural knowledge and higher-order thinking skills, and (2) addressing the diversity of the Israeli student population.

2. Integrating Procedural Knowledge and Higher-order Thinking Skills

The learning of algebraic procedures in middle school algebra is usually perceived as an algorithmic activity, achieved by performing sequences of short drill-and-practice tasks, having little to do with conceptual learning or with higher order mathematical thinking. In contrast, one of the characteristics of the *Integrated Mathematics* curriculum program is that it integrates the domains of algebraic procedures, higher order thinking, and conceptual learning within some of its procedural tasks. Adopting the integrative approach described and advocated by Friedlander and Arcavi (2012), the *Integrated Mathematics* textbooks also require students to apply higher order thinking in the context of procedural tasks.

Figure 1 presents three examples of integrative procedural tasks from the standard/expanded scope version. In each example, besides possessing procedural knowledge, students must possess some of the following skills:

- *Reversed thinking*: Examples 1–3 require students to utilize backward thinking, and reconstruct a procedure already performed but missing —

Example 1

Find an equation with the solution $x = 3$.

Find an equation that has algebraic expressions on both of its sides, and its solution is $x = 3$.

Fill in the blank in the equation $3x + \square = 2x - 5$ so that its solution will be $x = 3$.

(Grade 7 — Part C, p. 58)

Example 2

In each case, fill in the blank, so that the resulting expression can be simplified to 1, and define the corresponding domain.

$$\frac{(x-2)(\_\_\_\_)}{x^2-4x+4} \qquad \frac{(x-1)(\_\_\_\_)}{x^3-x^2} \qquad \frac{5x(\_\_\_\_)}{4x^2-4} \qquad \frac{(x+1)(\_\_\_\_)}{6x^2+18x}$$

(Grade 9 — Part B, p. 28)

Example 3

Solve:

(a) $1 + 2x = 3^2$ (b) $1 + 3x = 4^2$ (c) $1 + 4x = 5^2$

(d) $1 + 5x = 6^2$ (e) $1 + 6x = 7^2$ (f) $1 + 7x = 8^2$

Find a similar equation with the solution $x = 11$, and an equation with the solution $x = 21$.

(Grade 7 — Part B, p. 100)

Fig. 1. Examples of integrating procedural knowledge and higher-order thinking skills.

to find or reconstruct equations with a given solution in Examples 1 and 3, and to reconstruct expressions equivalent to a given number in Example 2.
- *Generating examples*: In Example 1, students are encouraged to explore and review their understanding of solving equations by generating several equations with the same solution. Note that providing a variety of examples or counterexamples and comparing them with those provided by others also stimulates divergent thinking.
- *Global comprehension*: Examples 2 and 3 require students to consider the algebraic structures of an expression or an equation as entities that can be identified and operated upon as units — rather than as a collection of "atomic" components such as numbers, letters, and operations.
- *Generalizing and justifying*: Example 3 requires students to consider commonalities within a given set of equations, and to identify a relation between the numbers in such an equation and its root. At a more advanced stage of learning algebra, they should be able to use parameters to express the structure of these equations, and to generalize and justify algebraically the emerging pattern.

3. Addressing the Diversity of the Israeli Student Population

Parallel versions of the *Integrated Mathematics* learning materials were developed to address the diversity of the Israeli student population with regard to (1) levels of mathematical achievement (limited, standard/expanded, and high), and (2) strands within the Israeli educational system (Hebrew speakers, Arabic speakers, and ultra-Orthodox Jews). Next, we elaborate and exemplify the guiding principles used in producing the various versions, highlighting the similarities and differences between them.

3.1. *Addressing the needs of students with different achievement levels*

As mentioned earlier, the *Integrated Mathematics* curriculum program offers two parallel versions of textbooks for each grade level:

- A standard/expanded scope version intended for students with medium or high achievements.
- A limited scope version intended for students with low achievements.

In addition, the *Excelling Rehovot* supplementary collections of activities are offered for students with high achievements.

In line with the national curriculum and with our past experience in research and curriculum design (Karsenty, Arcavi and Hadas, 2007), both the standard/expanded scope and the limited scope versions cover the same mathematical topics. The two versions of the textbooks are also similar with regard to many of their design principles. Thus, *all* students — those with low, medium, or high achievements — are provided with tasks that require (1) working at various cognitive levels (lower-level replication of learned knowledge and procedures, medium-level applications of learned knowledge, and higher-level investigations), (2) employing both concrete and abstract thinking, and (3) working at various degrees of innovative thinking. Also, the tasks included in the textbooks of both the standard/expanded and the limited scope versions promote meaningful learning, mathematical literacy, and mathematical reasoning.

The main differences between the standard/expanded and the limited scope versions of the textbooks are rooted in (1) the extent of guidance (scaffolding) provided, (2) the verbal load (reading comprehension) required, (3) the cognitive load (the volume of data and constraints) required while working on a mathematical task, (4) the procedural complexity of the tasks, and (5) the complexity of meta-cognitive thinking (e.g., monitoring own work, detecting errors, comparing given solutions, and interpreting results).

3.2. *Addressing the demographic diversity of the Israeli student population*

As mentioned, the standard/expanded and the limited scope versions of the textbooks were originally developed in Hebrew, as were the *Excelling Rehovot* collections of activities intended for students with high achievements. After approval by the Ministry of Education for use at school, the various versions of the textbooks and the *Excelling Rehovot* collections were translated to Arabic and converted to a version suitable for the ultra-Orthodox Jewish community.

In addition to the obvious language conversion, the translation to Arabic included cultural adaptations of aspects related to the phrasing and the context of some of the mathematical tasks. For example, when Hebrew children's names were used in word problems in the original textbooks, they were replaced with names commonly used by Arabic speakers in Israel. The contexts of some mathematics problems were also modified. For example, when a Jewish holiday served as the context for a mathematics problem in the original textbook, it was replaced by an Islamic or Christian holiday.

Conversion to a version suitable for the ultra-Orthodox Jewish community did not require language conversion as in the translation to Arabic. However, for approval for use in these schools, some cultural, contextual, and even content-bound adaptations were necessary, e.g., changing commonly used contemporary names of Israeli children in word problems to traditional (mainly Biblical) Jewish names. Similarly, some decorative or illustrative pictures from the original textbooks — such as illustrations of people or some animals — were omitted whenever the pictures were considered unacceptable by the ultra-Orthodox Jewish community. With regard to content-bound adaptations, all computer-based tasks and web-related references had to be omitted due to religious restrictions.

Sometimes the context of a problem had to be modified or even completely changed. For example, in the original textbook, introducing the tree diagram as a tool for calculating the probabilities of a complex event (Fig. 2(a)) takes place in a love story involving several characteristics

(a) Once upon a time there lived a king and his beautiful daughter. The king wished to marry off his daughter to a prince, but she fell in love with a low-class young man. The king decided to leave the decision regarding the marriage to a combination of chance and his daughter's wisdom: He asked his daughter to choose between Rooms A and B in a labyrinth (see drawing). If the young man manages to get to the princess, he will be allowed to marry her. But if not, he will have to deal with a ferocious tiger that will wait in the other room. At each crossroad, the young man has to make a random choice of his path. Which room should the princess choose? Explain.

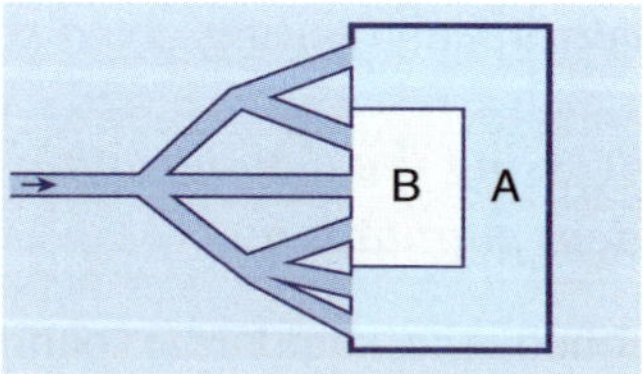

(b) At a Purim party, girls and their mothers played a labyrinth game (see drawing). Each mother chooses between Rooms A and B of the labyrinth. If her daughter reaches her, she will get a prize. At each crossroad, the girls make a random choice of their path. Which room should the mothers choose? Explain.

Fig. 2. Change of context from the original text (a) to the ultra-Orthodox version (b).

considered morally wrong by the ultra-Orthodox Jewish community (e.g., falling in love, a daughter choosing her husband-to-be instead of her parents utilizing matchmaking). In the ultra-Orthodox Jewish version, the scenario of the love story was replaced by a game of mothers and daughters (see Fig. 2(b)).

4. From Design Principles to Design Concerns

Underlying the two design principles described in this chapter (integrating procedural and conceptual knowledge, and addressing the diversity of the Israeli student population) are a wide range of concerns that had to be addressed by the design team on a daily basis, throughout the development process. Next, we present a brief list of selected challenges that arose regarding the two principles previously described.

- What would be a suitable distribution of tasks requiring various cognitive levels (lower and medium-level replication of learned knowledge and procedures, and higher-level investigations) and what should this distribution be for high-achievers?
- What would be a desirable balance between tasks that require innovative thinking versus more routine tasks?
- What would be a desirable balance between the level of support provided (scaffolding) and the verbal load required in complex tasks; particularly for low-achieving students?
- What part of the procedural tasks should be integrative, and what part should be allotted to "plain" algorithmic exercises?
- What should be the technical difficulty level of an integrative procedural task?

Additional challenges were faced by the design team regarding other design principles, not discussed here.

References

Friedlander, A. and Arcavi, A. (2012). Practicing algebraic skills: A conceptual approach, *Mathematics Teacher*, **105**, 608–614.

Karsenty, R., Arcavi, A. and Hadas, N. (2007). Exploring informal mathematical products of low achievers at the secondary school level, *Journal of Mathematical Behavior*, **26**(2), 156–177.

About the Authors

Alex Friedlander is a researcher, teacher, and teacher educator. His interests lie in curriculum design and research on classroom practice.

Ruhama Even holds the Rudy Bruner Professorial Chair of Science Teaching. Her research focuses on: (a) teacher professional development, (b) curriculum development and analysis, and (c) the interplay of factors shaping students' opportunities to learn mathematics.

Naomi Robinson is a curriculum designer. Her interests concern adapting curriculum materials to the diverse needs of Israeli students.

Chapter 23

Introducing All High School Students to the True Nature of Contemporary Mathematics: Meeting the Challenge

Nitsa Movshovitz-Hadar

Kesher-Cham, R&D Center for Reviving Mathematics Education,
Technion — Israel Institute of Technology, Haifa, Israel
nitsa@technion.ac.il

This chapter describes an on-going attempt at decreasing the gap between the ordinary school mathematics curriculum and contemporary mathematics, in order to give all high school students a taste of what contemporary mathematics is about.

1. Introduction

Over 100,000 items are added each year to the Mathematical Reviews database of the American Mathematical Society (2017). "The overwhelming majority of works in this ocean contain new mathematical theorems and their proofs." (Sevryuk, 2006, p. 101.) This chapter describes an on-going project that was set up to enable high school students to have a glimpse into this "overwhelming ocean" and get at least a bird-eye view of the rapidly growing area of mathematics, in order to achieve a better perspective of what mathematics is all about.

2. The Challenge

In many countries, including Israel, school mathematics curricula do not reflect the ever-growing, open-ended, and application-rich true nature of mathematics. Nor do they expose students to mathematicians' struggle with solving open problems, and establishing new results. Consequently, mathematics is perceived as a body of facts and procedures that are at best loosely connected, a deterministic, black and white, cut and dried domain with little if any place for personal investment, creativity, or intellectual joy. In many studies since the 1980s (e.g., Schoenfeld, 1992; Szydlik, 2013), these beliefs about the nature of mathematics were found to be widespread among school students, implying that students refrain from choosing to study mathematics beyond the compulsory years. This issue has developed into an economic, social and national concern in many countries (e.g., in the UK, see Kounine, Marks and Truss, 2008).

One of the hypotheses underlying the Mathematical News Snapshots (MNS) project was that these typical widespread yet wrong perspectives are largely due to the wide gap between school mathematics curriculum and contemporary mathematics, a gap that is reasonably attributed to the hierarchical accumulative nature of mathematics.

How might it be possible to communicate the true nature of contemporary mathematics, to ALL high school students, without harming their progress in learning the heavily loaded mandatory curriculum, and despite the gap between their knowledge and the background needed for in-depth understanding of contemporary mathematics? Furthermore, can the ordinary high school mathematics teacher be expected to be able to give the ordinary high school student a taste of contemporary mathematics without harming the progress in teaching the heavily loaded mandatory curriculum, and despite the gap between the effort it takes to be a full-time mathematics teacher and the effort it takes to keep up-to-date in contemporary mathematics?

3. Meeting the Multifaceted Challenge

To respond to the above questions, Movshovitz-Hadar (2008) suggested interweaving mathematical news snapshots (MNSs) in the ordinary teaching of high school mathematics. As defined by Amit and Movshovitz-Hadar (2011), each MNS is an intermezzo with three main characteristics:

(i) a mathematical result published in the professional literature in the past 3–4 decades;
(ii) exposition of about 30 minutes focusing on the news, elaborating on its history, the main underlying ideas and the people involved in its creation;

(iii) taking into account students' limited background, preferably linking it to some topics in the mandatory curriculum.

The MNSs are not meant to be complete and in-depth, but merely a "glass-bottom boat tour" over Sevryuk's overwhelming ocean (*ibid*) without getting wet...

4. Realizing the Suggested Solution

4.1. *Developing mathematical news snapshots*

Our goal is to have 30 MNSs ready for teachers to choose from, 10 for each of the three years of high school. The development of each MNS takes about 100 hours, starting from choosing an appropriate piece of news, through developing the outlines of "the story", preparing the Power Point presentation, testing it and revising it before handing it out to the teachers. Presently, 24 MNSs are ready; a few have been translated into English and presented in international congresses (ESU 5 2007 in Prague, ESU 6 2010 in Vienna, ICME 12 2012 in Seoul, HICE 13 see Movshovitz-Hadar, 2015). Their topics are:

Benford's law;
Fermat's last theorem;
Goldbach's (weak) conjecture;
Julia set and the invention of Fractals;
Kepler's conjecture;
Map coloring;
Mathematics match making;
Möbius strip and its applications;
Musical mathematics;
Non-round wheels;
Non-transitive dice;
Origami and its applications;
Pentagonal tiling;
Prime numbers;
Random walk;
Solving Rubik's cube;
Sudoku — not just for fun;

The art gallery problem;
The cake-cutting problem;
The digits of Pi;
The double bubble problem.
The Fibonacci myth;
The Katalan Conjecture

4.2. *Teacher preparation*

The MNS project was launched in 2008 as an action research carried out by a single teacher interweaving an early version of 9 MNSs in four classes of various levels (Amit, Movshovitz-Hadar and Berman, 2011). It evolved through clinical work in three high schools for 2 years followed by a call for participation in an online preparation for the sake of scalability. In Summer 2016, a longitudinal study of the effect of interweaving MNSs for 3 years attracted 47 beginning teachers and 14 continuing teachers. This study is currently ongoing. Students' as well as teachers' responses are documented and we hope to establish the impact of this intervention in a report due in 2019, which will call the Ministry of Education to adopt MNSs as a part of the mandatory high school mathematics curriculum.

Acknowledgment

The MNS project enjoyed a grant from the Israeli academy of Science in 2012–2015 (Grant no. 144/2011), and is presently supported by the Ministry of Science (Grant no. 3-12942). Thanks are due to Varda Zigerson and Atara Shriki who are the leaders of the school implementation and the accompanied research respectively to my co-PI Prof. Avi Berman, to the PostDoctoral fellows: Ruthi Segal and Boaz Silverman, and to the administrative coordinator Orna Imber.

References

American Mathematical Society (2017). *MathSciNet by the Numbers*, http://www.ams.org/mathscinet/help/about.html?version=2, and http://www.ams.org/mathscinet/help/byTheNumbers.html (Retrieved May 27, 2017).

Amit, B. and Movshovitz-Hadar, N. (2011). Design and high-school implementation of Mathematics-News-Snapshots — An action research into 'Today's News is Tomorrow's History', in Krongellner and Tzanakis (eds.), *History and Epistemology in*

Mathematics Education: Proceedings of the 6th European Summer University, Austria: Verlag Holzhausen GmbH/Holzhausen Publishing Inc., pp. 171–184.

Amit, B., Movshovitz-Hadar, N. and Berman, A. (2011). Exposure to mathematics in the making — Interweaving Math News Snapshots in the teaching of high school math, in V. Katz, and C. Tzanakis, (eds.), *Recent Developments on Introducing a Historical Dimension in Mathematics Education*, Washington, DC: MAA — Mathematical Association of America, pp. 91–102.

Kounine, L., Marks, J. and Truss, E. (2008). *The Value of Mathematics*, London: Reform.

Movshovitz-Hadar, N. (2008). Today's news is tomorrow's history — Interweaving mathematical news in teaching high-school math, in F. Barbin, N. Vondrora and C. Tzanakis (eds.), *History and Epistemology in Mathematics Education: Proceedings of the 5th European Summer University*, Prague: Vydavatelsky Press, pp. 535–546.

Movshovitz-Hadar, N. (2015). Mathematics news snapshots — Meeting the challenge of introducing ALL senior high school students to contemporary mathematics. Presented at HICE13, the 13th annual Hawaii International Conference of Education, Honolulu, Hawaii. Conference proceedings pp. 2209–2210.

Schoenfeld, A. H. (1992). Learning to think mathematically: Problem solving, metacognition and sense making in mathematics, in D. A. Grouws (ed.), *Handbook of Research on Mathematics Teaching and Learning*, New York: Macmillan, pp. 334–370.

Sevryuk, M. B. (2006). Book reviews, *Bulletin of the American Mathematical Society*, **43**(1), 101–109. Retrieved from: http://www.ams.org/bull/2006-43-01/S0273-0979-05-01069-4/S0273-0979-05-01069-4.pdf.

Szydlik, S. D. (2013). Beliefs of liberal arts mathematics students regarding the nature of mathematics, *Teaching Mathematics Applications*, **32**(3), 95–111.

About the Author

Nitsa Movshovitz-Hadar is an Emerita Professor of mathematics education at Technion — Israel Institute of Technology. She is a Senior Researcher at Samuel Neamann Institute for national policy studies and the Head of Kesher-Cham — R&D center for reviving mathematics education in Israel, which she established in 1986. See also her homepage on Technion website.

Part IV

Issues and Innovations Related to Teacher Education — Training and Professional Development in Israel

Chapter 24

Mathematics Teacher Education in Israeli Colleges of Education: Past Trends, Current State of the Art and Future Plans

Atara Shriki* and Tili Wagner#

**Oranim Academic College of Education, Tivon, Israel*
Kesher-Cham, Technion — Israel Institute of Technology, Haifa, Israel
#Beit Berl Academic College of Education, Kfar-Saba, Israel

In this chapter, we present a brief overview of changes that have taken place in the nature and status of Israeli colleges of education over the past 50 years and the controversy these changes have introduced into the discussion of mathematics teacher preparation. We then present data regarding students in the various programs for mathematics teacher training in colleges of education, and conclude by pointing out some current trends that may be the cause for the rising number of students who apply for mathematics teaching certificates in Israeli education colleges, which may have implications on future plans.

1. Introduction

In Israel, in order to become a mathematics teacher, one must hold a teaching certificate and a teaching license. A **teaching certificate** is an academic diploma awarded by an academic institution — a university or a college of education [CoE] (also known as a teachers' college). Currently, about 90% of Israeli teachers earn their teaching certificate at CoEs, with the remaining 10% at universities. Students at CoEs can choose to qualify either as early childhood educators or as primary or secondary school teachers, while universities provide teaching certificates for secondary school teachers only. CoEs have three teacher training programs: a regular program, a teaching certification program for those holding an academic degree (academic retraining), and an M.Teach program. The course of studies for the students in the regular program is 4 years, for academic retraining — 1–2 years (depending on their background), and for M.Teach — 2 years. In addition, some of the CoEs offer programs designed to support the professional development of in-service teachers, among them M.Ed. programs and various specialized courses. A **teaching license** is granted by the Department of Teacher Education at the Ministry of Education [MoE] subsequent to successful completion of a one-year internship. Mathematics teachers for secondary school may be entitled to one of two types of teaching licenses: A license for teaching 7th–10th grades, traditionally awarded to those who complete a Bachelor's degree in mathematics or mathematics-intensive disciplines; and a license for teaching 7th–12th grades, awarded to those with a Master's or Ph.D. degree in mathematics.

2. Changes in the Status of Education Colleges in Israel and Their Implications for Mathematics Teacher Training Programs

Since the 1970s, the issue of teacher training programs has been on the public agenda. Until the late 1970s, Israeli education colleges were non-academic with a 3-year certification program. An "academization" process then began, and the teachers' colleges became academic colleges providing a 4-year teaching certificate program granting a Bachelor of Education degree (B.Ed.). In 2004, CoEs were accredited to grant a Masters of Education degree (M.Ed.), and in 2009, in order to attract graduates of universities to the teaching profession, the CoEs were also accredited to grant Master of Teaching degrees (M.Teach). Five prominent commissions have left their mark on the process of academization: The Yaffe Commission in 1971, the Etzioni Commission in 1979, the Ben Peretz Commission in 2001, the Dovrat Commission in 2005 and the Ariav Committee in 2006 (Hofman and Niederland, 2012).

The main goal of the **1971 Commission** was to upgrade the colleges to 4-year academic CoEs granting a B.Ed. degree. Nonetheless, the commission maintained that academic training for teachers should be different from academic studies at universities; while the standards of studies in CoEs must be equivalent to those in the universities, the disciplinary studies must be supplemented with education toward social and national values in addition to the practical aspects of teaching.

The **1979 Commission** examined the status of teachers and the teaching profession, and declared that the teacher's role is critical to the future of Israel, and therefore working conditions of teachers must be improved. The commission mentioned academization as a way to improve the quality of new teachers as well as veteran ones, and had several meaningful recommendations, among them: selecting the candidates on the basis of more rigorous standards than those generally accepted; establishing clear criteria for faculty members in teacher training institutions; allowing teachers without academic degrees to attain a full academic education; and encouraging academic staff with a pedagogical orientation to become teacher educators. These recommendations had a significant influence on the academization of teacher training, including the Council for Higher Education's [CHE] new guidelines for academic training of teachers, a measure that laid the cornerstone for the current CoEs. During the second half of the 1990s, in order to meet the growing demand for higher education and to overcome relatively high admission requirements at universities, several new academic colleges ("regional colleges") were established under the academic sponsorship of universities (most of them are autonomous colleges today). The CHE required that some of these new academic colleges open teacher training programs; however, MoE policy stated that teacher training programs should only exist at designated colleges.

The **2001 Commission** primarily addressed institutional and structural issues of the teacher education system and, similar to the 1979 Commission, suggested raising the bar of admission for teacher training programs, in particular the threshold grade on the Psychometric Entrance Test, and advised that a state licensing exam be implemented. While the admission level of the Psychometric Entrance Test was indeed raised, the suggestion of the licensing exam was not accepted. The report of the commission led the CHE to the conclusion that in order to raise the status of teaching, it is the Council that should be the body granting teaching certification and supervising the CoEs. Such a move would mean that the MoE would lose their control over the CoEs; due to the MoE's objection this report had no practical consequences on the teacher training system. The process of academization and the commissions' recommendations sparked a debate over the relative weight of each component included in the mathematics teacher training program. In addition to the general courses and courses dealing with various aspects of education, mathematics teacher training programs for the regular program students in CoEs consist of three main components: disciplinary training (mathematical content knowledge and mathematical knowledge for teaching); didactic-pedagogical training (knowledge of mathematics school curricula, textbooks, teaching methods and pedagogical approaches); and practical training (actual experience in teaching mathematics at school). Without a binding policy regarding the allocation of hours for each component and the nature and level of study of courses in each, such decisions are subject to the approaches of each CoE; as a result, mathematics teacher training programs designed in CoEs differ in scope, level and quality. In fact, this might be considered the Achilles heel of the CoE. In addition, the CoEs began to demand of its academic staff to complete doctoral studies, and new candidates are required to hold Ph.D. degrees in mathematics or mathematics education.

The **2005 Commission** was appointed by the MoE and its task was to assess the entire Israeli education system, to recommend a general reform without exceeding the current budget, and to suggest guidelines for implementing the reform (Inbar, 2006). The commission declared that teaching, like any other job, should be measured according to inputs and outputs. The commission determined that a general reform of teacher training programs should be implemented, and that teachers must receive full academic training in their discipline (B.A./B.Sc.), as well as a teaching certificate which includes broad pedagogical training. This training, according to the report, may take place at colleges that will be upgraded to academic colleges or at universities that will redesign their teacher training courses to better meet the needs of prospective teachers. The opponents to this recommendation maintained that those who intend to teach mathematics in primary school do not have sufficient mathematical abilities to allow them to study towards a bachelor's degree in mathematics, which is why they have not chosen to teach mathematics in secondary schools in the first place. In any case, this controversy has exposed the fact that some consider the CoEs to be "sub-academic" in contrast to the universities, and that the level of CoE students is low. We return to this point in Section 4. Furthermore, the recommendation was to close several CoEs so that the budget saved could be used toward reforming the entire education system. The Minister of Education at that time insisted on implementing the reform in its entirety, and her inflexible attitude resulted in a strong backlash, particularly from the teachers' unions. Nonetheless, the CoEs paid an immediate price in the form of budget cuts.

A year later, the **2006 Commission** that was appointed by the CHE published Layouts for teacher training in institutions of higher education, which were identical for both CoEs and universities (Council of Higher Education, 2006). Unlike the 2005 commission, which viewed teachers as "school workers", this commission used terms such as "professional teacher" or "teacher-educator", and declared that teaching is an activity anchored in both theoretical research and practical-reflective knowledge. The unified guidelines for training teachers suggested reducing total studies for B.Ed. from 108 academic points to 90–96 points, of which 60 points would be for disciplinary content. This commission also defined the common knowledge base that every teacher should hold, regardless of the discipline or educational level, and recommended the study of either two disciplines, or a single extended one (Dror, 2012). The implementation of the Layouts, which is currently under reevaluation, contributed to the enhancement of mathematical knowledge of new teachers.

3. Mathematics Teacher Education in Numbers

Currently, there are 22 CoEs in Israel. Among them, 19 are in the Jewish sector and three in the Arab sector. The Jewish CoEs include nine secular colleges and 10 religious ones. Half of the religious colleges are very small. However, these numbers are expected to decline over the coming years due to the above-mentioned directive to close or merge colleges.

Official documents of the MoE indicate that in the 2016–2017 academic year close to 50,000 students attend CoEs, among them about 4000 specializing in mathematics. Table 1 presents the number of students in CoEs by track, sector and gender (F = Female). The "others" in the "Track" line refers to student whose track was not specified, while the "others" in the "Sector" section relates to Druze, Circassians, Armenians, etc. In total, there are 3061 females out of 3816 students (80.21%). However, in the tracks of academic retraining and M.Teach for grades 7–12, the percentage of females is about 63%. This reflects changes in the composition of the mathematics teaching force in secondary schools.

Table 1. Distribution of mathematics students in CoEs, by track, grade, sector and gender (2016–2017).

Track $N = 3816$		B.Ed./BA $N = 2161$	Academic retraining $N = 929$	M.Ed. $N = 402$	M.Teach $N = 175$	Others $N = 149$
Grades $N = 3816$	1–6 $N = 1660$ (43.5%)	1129 $F = 88.22\%$	362 $F = 94.75\%$	115 $F = 89.21\%$		54
	7–12 $N = 2156$ (56.5%)	1032 $F = 77.13\%$	567 $F = 63.14\%$	287 $F = 77.82\%$	175 $F = 63.43\%$	95
Sector $N = 3816$	Jewish $N = 2553$ (66.9%)	1368 $F = 80.87\%$	725 $F = 73.93\%$	225 $F = 81.78\%$	154 $F = 61.03\%$	81
	Arabs $N = 1096$ (28.72%)	754 $F = 79.17\%$	155 $F = 78.06\%$	169 $F = 82.24\%$	18 $F = 77.78\%$	
	Others $N = 167$ (4.38%)	39 $F = 92.3\%$	49 $F = 89.79\%$	8 $F = 87.5\%$	3 $F = 100\%$	68

An interesting trend over the past 10 years, indicated in Fig. 1, is the massive increase in the number of students specializing in mathematics teaching at CoEs. This figure describes the number of students in their last year of studies, and includes four groups: Regular program (B.Ed./B.A.) students in their 4th year of studies (those who specialized in primary/secondary school teaching indicated by Reg 1–6/Reg 7–12, respectively), and academic retraining students in their last year of studies (similarly indicated by AR 1–6/AR 7–12). During the years indicated, the number of Reg 1–6 students almost tripled (from 125 to 353), and the number of Reg 7–12 increased by about 67% (from 124 to 208). A significant increase occurred in the number of AR students: the number of AR 1–6 increased from 27 to 207 (760%) and the number of AR 7–12 increased from 69 to 397 (580%).

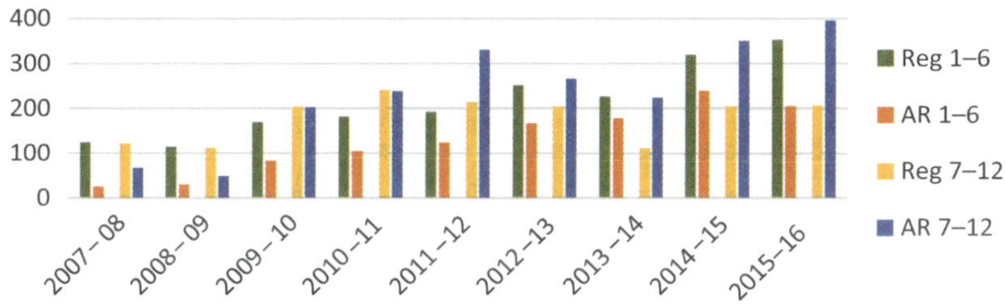

Fig. 1. Number of students in their last year of studies by track and grade, between 2007–2008 and 2015–2016.

Note that these findings indicate that the number of teachers starting teaching mathematics at the secondary school level today that graduated from the AR track is twice the number of those that graduated the regular track. In addition to challenges related to the nature of the qualification, this also has implications on the entire secondary school system; this issue, however, is beyond the scope of this chapter.

4. Recent Trends

In 2015, the Wineburg Committee (Wineburg, 2015) for the evaluation of education and science education study programs appointed by the CHE indicated that many of the lecturers in the Faculty of Education in universities do not feel committed to the education system but are there chiefly for research, and that students do not receive sufficient practical experience in schools. This stands in stark contrast to the picture in most CoEs, where the emphasis in general is on an optimal balance between the practical, the disciplinary and the pedagogical-didactic, and the teaching staff is oriented first and foremost toward teacher training. Nonetheless, on January 2017, the chief superintendent of school mathematics, under the auspices of the Director General of the MoE, issued a directive with regard to eligibility for the teaching license: regular program or academic retraining students holding B.A./B.Sc. degrees in mathematics or some specific engineering fields will be entitled to a teaching license for 7th–12th grades. Academic retraining students below a certain threshold level of previous studies in mathematics will be required to complete additional courses in mathematics, and will then be entitled to a teaching license for 7th–10th grades. Other candidates will be eligible to study toward a mathematics teaching certificate for 7th–10th grades only upon completion of their missing points in mathematics at a mathematics department at one of the universities. Prior to this directive, such candidates were allowed to complete their missing studies at CoEs. This new directive, no doubt, illustrates the goal of the MoE to improve the process of mathematics teacher training; however, in practice this may hold back the COEs rather than advance their academization process. In the meantime, it is difficult to anticipate the impact of this directive on the training of mathematics teachers in colleges of education.

References

Council of Higher Education (2006). *Guiding Frameworks for Teacher Education in Higher Education Institutions in Israel*, Jerusalem: Council of Higher Education (in Hebrew).

Dror, Y. (2012). Teacher education in Israel and their impact on professional development of teacher educators, in M. Ben-Peretz, S. Kleeman, R. Reichenberg and S. Shimoni (eds.), *Embracing the Social and the Creative — New Scenarios for Teacher Education*, Lanham, MA: Rawman & Littlefield Education, pp. 35–56.

Hofman, A. and Niederland, D. (2012). Is teacher education higher education? The politics if teacher education in Israel, 1970–2010, *Higher Education Policy*, **25**, 87–106.

Inbar, D. (2006). Introduction, in D. Inbar (ed.), *Towards an Educational Revolution*? Jerusalem: Van Leer Institute and Hakibbutz, pp. 9–11 (in Hebrew).

Wineburg, S. (2015). Committee for the evaluation of education and science education study programs. Available at: http://che.org.il/wp-content/uploads/2016/04/Education-and-Science-Teaching-QA-Commitee-General-Report.pdf.

About the Authors

Atara Shriki is an associate professor who teaches mathematics and mathematical didactics at Oranim Academic College of Education. She also works as a researcher at the Samuel Neamann Institute for National Policy Research and at Kesher-Cham — R&D center for mathematics education at Technion — Israel Institute of Technology.

Tili Wagner is an assistant professor and head of the Mathematics and Physics Residency teacher education programs at Beit Berl Academic College of Education. Her major fields of research and expertise are science education (development of scientific literacy) and teacher education (unique teacher education models).

Chapter 25

The Rothschild–Weizmann Master's Program for Practicing Mathematics Teachers*

Ruhama Even#, Zvi Artstein and Avital Elbaum-Cohen

Weizmann Institute of Science, Rehovot, Israel

#ruhama.even@weizmann.ac.il

The Rothschild–Weizmann Program is a uniquely crafted master's program at the Weizmann Institute of Science, especially designed for practicing secondary school teachers of mathematics, physics, chemistry, and biology. The chapter describes the rationale and structure of the two components of the mathematics strand of the program: mathematics and mathematics education studies, and presents the view of a program graduate on her studies in the program.

1. Introduction

The Rothschild–Weizmann Program is a unique master's program, without thesis, offered by the Weizmann Institute of Science, especially designed for experienced Israeli secondary school teachers of mathematics, physics, chemistry, and biology. The program, which started in the 2008–2009 academic year, grants an M.Sc. degree in science education. It is a collaboration between the Department of Science Teaching and the respected Faculties of Exact and Life Sciences. Each of the program's four strands — Mathematics, Physics, Chemistry, and Biology — is co-directed by two Weizmann

*The Rothschild–Weizmann Program for Excellence in Science Teaching is supported by the Caesarea Edmond Benjamin de Rothschild Foundation.

Institute faculty members: one from the corresponding scientific faculty and the other from the Department of Science Teaching.

This chapter describes the Rothschild–Weizmann Program in which academic-level mathematics and research in mathematics education comprise the main components. The Program's approach is pertinent to two open questions that received international attention in recent years, namely, the relevance of academic mathematics to teaching mathematics at school, and the relevance of research in mathematics education to practicing teachers.

Admission to the program takes into account the applicants' background. A Bachelor's degree in a relevant scientific field and three years of experience in teaching secondary school mathematics or science are required. Also taken into account is the candidate's potential to influence mathematics or science teaching at their school or region.

The program extends over two years. During each year, the participants spend two days a week on campus. Participants in the program receive study grants and are exempt from paying tuition. They continue to teach at schools in parallel to their studies, although are expected to reduce their teaching load.

Apart from three courses that are taught to all four strand students — cognition, learning, and instruction; assessment and evaluation; and integration of learning technologies — each of the four strands designs and conducts specific courses and activities for its respective students. In the following we describe the rationale and structure of the two components of the mathematics strand of the program. The view of one of the program graduates regarding the contribution and relevance of her studies in the program to her teaching at school concludes the chapter.

2. The Mathematics Strand Program

A major component of the mathematics strand of the program comprises mathematical studies tailored for the students at the program. Another component focuses on important issues and aspects in the academic field of mathematics education, regarded relevant to the teaching of secondary school mathematics. In the following we describe these two components.

2.1. *The mathematics studies*

The emphasis of the courses in mathematics is on providing students with a perspective on the subject matter, perspective that connects what they themselves teach in their respective secondary schools, with modern mathematical research and open issues. Although each course contains graduate materials, the courses would not qualify as graduate courses toward a degree in mathematics. The courses are taught by active researchers, most of them faculty members at the Faculty of Mathematics and Computer Science at the Institute. The goal is to demonstrate that mathematics is a field that is very much alive, and that topics that are taught in high school are related to topics of current mathematics research. Seminars on current research issues were also presented. In particular, the emphasis is not on a prescribed body of definitions, theorems and proofs. Rather, the main purpose of the mathematics component is to broaden secondary school teachers' perspective and understanding of and about the discipline of mathematics and its relevance. As mentioned, research mathematicians, who usually teach only mathematics master's and doctoral students, have designed and taught these courses. In addition to the formal courses, the students have opportunities to get to learn and use computational tools.

The formal mathematics program is comprised of eight courses and a final project. The courses include core courses related to subjects taught in high school (courses 1–5 below), courses devoted to the applications of mathematics (courses 6–7 below), and a course on the history and philosophy of mathematics (course 8 below). Each course is independent of the other courses, and thus they can be offered on alternate years to students of two cohorts together. Below are sample topics covered in each course (at times, small variations of the planned courses were introduced).

(1) *Algebra*: Getting acquainted with algebraic structures and their roles, symmetry, number bases, groups, linear algebra, vectors, matrices, fields, Galois Theory, squaring the circle.
(2) *Analysis*: Understanding the general role of analysis, various definitions of integration, differential equations, operators, Taylor series, transforms, elements of functional analysis, wavelets, calculus of variations.
(3) *Geometry*: From ancient to modern geometry, trigonometry, curves, linear spaces, surfaces, elementary differential geometry, Euclidean

and non-Euclidean geometries, different representations of the same objects, the Erlangen program.

(4) *Mathematics of Computer Science*: The structure of computer programs, deterministic and non-deterministic finite automata, Turing machines, propositional logic, temporal logic, syntax and semantics, semantic tableaux, verification of concurrent programs.

(5) *Statistics and Probability*: Understanding the dissonance between the intuitive and the mathematical approach to uncertainty, basic notions in statistical inference and probability, modeling issues, samples, limit laws, common fallacies of intuitive judgment.

(6) *Applied Mathematics in Science*: Examples arising in life sciences, physics, engineering, and in computational sciences, analysis versus simulations, new understanding of a range of scientific problems that are seemingly unrelated to each other.

(7) *Mathematics in the Social Sciences*: The extent to which mathematics can be applied to social problems, elements of game theory and economics, decision theory.

(8) *Philosophy and History of Mathematics*: Overview of the development of mathematics from ancient times through the Greeks, the logical approach, Platonism versus formalism, the philosophy of current mathematical research.

The final project involves an independent study of an unfamiliar mathematics topic, carried out under the supervision of a faculty member from the Faculty of Mathematics and Computer Science. Work on the project is accompanied by a seminar that involves reviewing mathematics and mathematics education papers, and presenting and discussing intermediate stages of the final projects. The final project is presented both orally and in a written report. It requires students to show some originality, either by solving an adequate mathematical problem, or sorting out and presenting a topic otherwise scattered in the literature. Sometimes included are suggestions for teaching that piece of mathematics to teachers or to advanced high school students. Examples of titles of final projects are: “Elliptic integrals and Beethoven’s metronome”, “The envelopes paradox”, “Complexity and algorithms of multiplication”, “Extensions of fields — compass and ruler construction”, “Dimensional analysis”.

2.2. *The mathematics education studies*

The main purpose of the component of the program that focuses on mathematics education is to broaden secondary school teachers' perspective and understanding of the scholarly discipline of mathematics education. The formal mathematics education program is comprised of six courses, and is substantially based on the MANOR Program.[a,b] The courses include a course that serves as an introduction to the academic field of mathematics education (course 1 below), courses dealing with research and design related to the learning and teaching of central topics in the school curriculum (courses 2–3 below), courses dealing with fundamental problems and topics in mathematics education (courses 4–5 below), and a course linking what is studied in the mathematics courses with classroom instruction (course 6 below). Following are details on each course.

(1) *Introduction to Mathematics Education*: Overview of the field of mathematics education and its development in the last 60 years, focusing on trends and main characteristics of each decade with regard to mathematics curricula, perspectives on learning mathematics, approaches to teaching mathematics and to classroom culture, national and international tests and their implications.
(2) *Research in Teaching and Learning Algebra*: Research, theory, and practice related to historical view on algebra and its development, conceptualizations of school algebra and its central concepts, aspects of algebra learning and understanding, algebra teaching, and algebra curriculum.
(3) *Research in Teaching and Learning Geometry*: Logical, visual, cognitive, and didactic aspects of learning geometry at different age levels, various kinds of definitions and their role in creating different "geometrical worlds", justification and proofs and their role in convincing and explaining, theories of geometry learning, connections between geometry and other mathematical areas.

[a]Even, R. (1999). The development of teacher-leaders and inservice teacher educators, *Journal of Mathematics Teacher Education*, **2**, 3–24.
[b]Even, R. (2005). Integrating knowledge and practice at MANOR in the development of providers of professional development for teachers, *Journal of Mathematics Teacher Education*, **8**, 343–357.

(4) *Mathematics for All*: Teaching and learning of mathematics among sub-groups of learners: gifted and advanced students, low achievers, students with special needs, focusing on cognitive, metacognitive, and affective characteristics of students, teaching approaches that take these characteristics into account.
(5) *Basic Topics in Mathematics Education*: Fundamental topics in mathematics teaching, such as classroom discourse, connecting students' conceptions and curriculum design.
(6) *Mathematics Seminar*: Connecting the mathematics learned in the mathematics courses to the school curriculum and teaching, taking into consideration students with different achievement levels, focusing on principal mathematical notions, such as linearity, function, measure and measurement, invariance, and algebraic structures.

3. View of a Program Graduate

Fifty-five mathematics teachers from eight cohorts graduated from the Rothschild–Weizmann Program so far; two-thirds were women. They came from different parts of the country, from Jewish and Arab sectors, and from Orthodox religious and secular sectors. In the following, Avital Elbaum-Cohen, a first cohort graduate, reflects on the years she spent as a student in the program and its impact on her today, in a first-person point of view. She points out four main insights.

Encounter with university-level mathematics. When I entered the program, being an experienced teacher, I was already very well acquainted with the high school mathematics curriculum, and had a broad perspective regarding its implications and pitfalls. Therefore, the renewed encounter with university-level mathematics allowed me to approach it from a practicing teacher's point of view that enabled me to focus on its relevance for my teaching and to make links that I could not make before, between high school mathematics and university-level mathematics. My personal experience shows that being able to embed high school mathematics within a broader mathematical context undoubtedly contributed to my practice of teaching.

Studying university-level mathematics can be difficult. Before attending the program, being a teacher for many years, I had a blurred memory of my experiences as a university mathematics student. The difficulties that struck me then were seldom mentioned and were mostly remembered in humor. At that time, difficulties in learning mathematics seemed to be those things that other people, namely my students, experience. Coming "back to school" as a student, learning once more difficult mathematics, made me struggle again and let me experience once more how it feels to make an effort to master new and challenging mathematics. Being in a state of not knowing and sometimes failing to understand what I'm being taught at first try, made me realize — for the first time in a long while — what emotional frustration some of my students must endure. This renewed realization that studying mathematics can be strenuous, not only helps me identify with and feel empathy towards my students, but also strongly motivates me to improve my didactical repertoire.

Attitude towards research in mathematics education. The distance between research in mathematics education and the teaching practice from a teacher's perspective is sometimes expressed with negative feelings towards research. Teachers in my close community perceive research as condescending and attempting to impose "scientifically approved" conclusions that sometimes contradict their gut feeling, which is based on their own long professional experience. The Rothschild–Weizmann program provided me, through several eye-opening courses in science and mathematics education, an opportunity to make a fresh start regarding my attitude towards research. I now believe that there were two main reasons for the transformation of attitude I went through. First, most of the educational courses in the program were specially designed for in-service proficient teachers and therefore emphasized subjects in research that are relevant to both the practice of teaching and the subject matter being taught. Second, it seems that in recent years one can detect a tendency of the research community to focus more on the complexities that underlie the practice of teaching, dwelling on questions such as "how to train people to become proficient mathematics teachers" or "what is the knowledge that sets the foundations of the practice of teaching?" Maybe the fact that the research community recognizes the importance of the teacher in the student's process of learning helped me appreciate its relevance to improving the practical aspects of education.

Becoming a part of a new community. Last, studying in the program and becoming acquainted with educational programs initiated both at the Science Teaching Department of the Weizmann Institute and elsewhere, introduced me and my fellow students to the Israeli science and mathematics education community in its widest scope and helped us join it as active and contributing members.

Looking back on my experience as a student in the Rothschild–Weizmann Program I realize it encouraged me to adopt a constant educated reflective stance toward my practice, taking every opportunity I have to improve it.

About the Authors

Ruhama Even holds the Rudy Bruner Professorial Chair of Science Teaching. She is the co-director of the mathematics strand of the Rothschild–Weizmann Program.

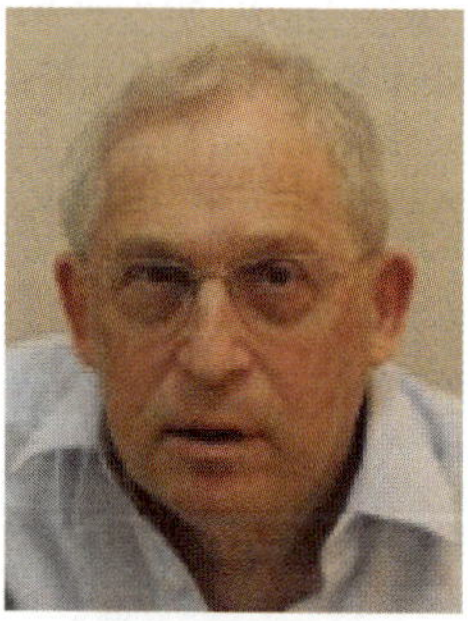

Zvi Artstein is a professor of mathematics at The Weizmann Institute of Science, and was heavily involved in crafting the mathematical studies in the Rothschild–Weizmann Program.

Avital Elbaum-Cohen is a mathematics and physics teacher at Aharon Katzir High School in Rehovot. She is a first cohort graduate of the Rothschild–Weizmann Program, and recently earned a PhD in Mathematics Education from the Weizmann Institute of Science.

Chapter 26

Mathematics Teacher Education for Secondary Schools: Bridging Between University Courses and Field Work

Pessia Tsamir*, Dina Tirosh#, Ruthi Barkai‡ and Esther Levenson§

Department of Mathematics, Science and Technology,
School of Education, Tel Aviv University, Tel Aviv, Israel
**Pessia@post.tau.ac.il*
#Dina@post.tau.ac.il
‡ruthi11@netvision.net.il
§Levensone@gmail.com

This chapter describes the program for prospective secondary school mathematics teachers held at Tel Aviv University. At the university, prospective teachers already have a bachelor's degree in mathematics or a mathematically rich field of study. The main aim of the program is to promote their mathematical knowledge needed for teaching. Prospective teachers attend courses at the university and comply with field work requirements at local schools. The challenge is to bridge coursework with fieldwork, bringing the courses into the field as well as bringing the field into the university. This chapter focuses on the use of cases as tools in fostering this bridging. The chapter begins by describing the rationale and general outline of the teacher education program. It then describes two specific courses, and in particular, how cases were used in those courses. For each course, we offer some findings on what prospective teachers gained from the analysis of cases, as well as their perspectives on the use of cases as learning tool.

1. Introduction

Preparing future mathematics teachers is a complex process involving both academic and practical elements. Academic elements often include university courses aimed at promoting prospective teachers' subject matter knowledge (SMK) and pedagogical content knowledge (PCK) (Shulman, 1986). Practical elements, aimed at applying what was learned in theory, often include field work such as classroom observations and student teaching. Yet, bridging out-of-university practice and within-university academic courses can be a challenge (Zeichner, 2010). While some educators argue that clinical experiences should be central to teacher education (Ball and Forzani, 2009), others point out that prospective teachers' lack of experience may limit what may be learned from field work (Masinglia and Doerr, 2002).

One way to bridge classroom experiences with theoretical courses is to have prospective teachers analyze cases. Cases may convey real events that took place in a classroom or a hypothetical situation based on research. They are used to raise a problem or dilemma inviting readers to suggest ways of responding to the problem (Markovits and Even, 1999). Cases may also be used to enhance prospective teachers' ability to focus on substantive features, rather than general features, of a mathematics lesson (Pang, 2011).

At Tel Aviv University (TAU), we decided to adopt cases as a tool in teacher education and implement it in several academic courses. In the following section, we describe our program in general and the specific use of cases in two different courses. Taking into consideration that one's beliefs and affect may impact on the way that individual engages in learning (Roesken-Winter, 2013), we also report on prospective teachers' views of the ways in which analyzing cases impacted on their learning.

2. The Mathematics Teacher Certificate Program

The mathematics teacher certification program at TAU was founded more than 40 years ago. It is open to those holding an academic degree in mathematics or in a mathematically rich field (e.g., engineering). Where warranted, candidates are mandated to take additional mathematics courses at the mathematics department in order to receive certification. The teacher certification program is characterized by a focus on the specific challenges faced by mathematics teachers and provides a carefully formulated blend of SMK and PCK, based on secondary school related content and relevant theoretical and research-based data. For example, the syllabus for the course "Learning and teaching mathematics in middle school" focuses on mathematical topics taken from the middle school mathematics curriculum, various ways of teaching mathematics, and their didactical aspects. The teaching staff of the program includes tenured faculty members, most of whom have had many years of experience teaching at the secondary school level. In addition, the practice part of the program is under the supervision of outstanding hand-picked teachers.

The current format of the program consists of 48 semester hours made up of four components: methodology courses (6 hours), education courses (16 hours), PCK courses (12 hours), and practice (14 hours). Topics covered in these courses include: the role of affect and cognition in mathematics learning; adapting mathematics instruction to students with different abilities; integrative learning of mathematics in schools; formal, algorithmic, and intuitive aspects of mathematics learning and teaching; the use of technology in mathematics teaching and learning; and mathematics education as a scientific discipline. Subject-matter knowledge is promoted within each course. For example, one semester of the course dedicated to technology in education focuses on geometrical topics such as congruence and similarity. The second semester focuses on functions and calculus.

The practice component of the program has both university and school components. At the university there is a peer learning and teaching course and two workshops — one focused on learning-related issues, and the second focused on teaching-related issues. At the workshops, students meet the accompanying instructor on a weekly basis, report on their experience at school and discuss a variety of issues and problems. Out of the university, students must participate in field practice that includes: observing various mathematics teachers, assisting them, taking part in small group mini-teaching sessions, participating in parents–teacher sessions and teachers' meetings, and exam-designing and grading. Prospective teachers are placed under the tutelage of excellent mathematics teachers.

3. Integrating Cases in Academic Courses

In this section we describe two separate courses, the purpose of using cases in each course, and how cases were used. Participants' perspectives of the use of cases in each course are presented.

3.1. *Research literacy: Mathematics education*

In the Research Literacy course, participants include both prospective and practicing teachers. Participants read classical articles in mathematics education that discuss content, skills, and processes within primary and secondary mathematics classes. Teachers' knowledge necessary for teaching mathematics, different approaches to teaching and learning mathematics, and relevant research aspects are discussed.

In general, the purpose of introducing cases in this course was to strengthen and deepen participants' knowledge of the articles studied during the course. A second aim was to demonstrate how theories and research may be used as a tool when analyzing classroom events. Cases were presented three times during the course. The first two times, participants were given the names and full references of three different articles which had previously been studied and were asked to analyze cases using at least one of those articles. For the third assignment, students were told to identify and explain the main concepts of one article, read a case transcript, and to illustrate those main concepts by referring to specific lines in the case transcript. Each assignment was reviewed by the instructor who gave written feedback to each student.

Two questions were investigated: (1) Do participants' view case analysis as an effective way to deepen their knowledge of mathematics education research articles? (2) When given two slightly different tasks relating cases to articles, do teachers perceive a difference between these two tasks and if so, in what ways? At the end of the course, participants were requested to assess the different methods used in the course. Specifically, they were asked if they felt a significant difference between the contribution of the first two tasks and the third task to their understanding of the articles discussed in the course. Fifty participants responded to the questions. Table 1 presents the frequency of responses.

As can be seen from Table 1, participants viewed both types of case tasks as contributing greatly to their understanding of the research literature discussed during the course. Most participants, however, felt that there was a difference between the tasks preferring to focus on one theory and use only that theory when analyzing a case.

4. Formal, Algorithmic and Intuitive Aspects of Learning Mathematics

In this course, participants are only prospective teachers. The focus of the course is on the algorithmic, formal and intuitive aspects of mathematical reasoning. It addresses research and theoretical frameworks used to analyze

Table 1. Frequency (%) of responses to assessing the difference between tasks ($N = 50$).

Are the tasks different?	Frequency (%)	Reason
There is no difference between the types of tasks	6 (12)	"For both tasks you had to read the articles and understand the main topic and see how the concepts and ideas come up in real life"
There is a difference but both types of tasks contributed equally	6 (12)	"The third task (one article and one case) required a deeper analysis. . . However, it required a lot of time and effort. The first task, when you do not know which article to connect with the case, is more interesting. Maybe both types of tasks could somehow be combined"
There is a difference and I preferred the first two tasks (three articles)	14 (28)	"You can find connections between the different articles and there is the possibility of analyzing the same text (case) from different angles"
There is a difference and I preferred the third task (one article)	24 (48)	"The first part of the task was to identify the main concepts in the article and that . . . assisted us in connecting the article to the case and to reason about the happenings in the case and connect them back to the article"

learners' common errors in advanced mathematics topics, as well as some possible related implications to instruction. In general, the purpose of introducing cases in this course was to encourage participants to use their theoretical knowledge when analyzing mathematical errors.

Cases came from six different sources: (1) participants' errors made when solving mathematical problems at home, (2) participants' errors made when solving mathematical problems during class, (3) research papers describing quantitative studies of students' common errors, (4) research papers describing classroom interactions and a qualitative analysis of a case which involved students' mathematical errors, (5) cases taken from the participants' reports on their field observations, and (6) interviews participants conducted with high school students as they solved a mathematical problem. We investigated prospective teachers' views of how cases taken from different sources impacted on their understanding of mathematical errors.

At the end of the course, teachers were asked to rate the extent to which each case activity contributed to their understanding of error analysis and which of the six case sources they would choose to analyze if they could only choose two.

Findings indicated that participants viewed learning from their own errors as most meaningful and would rather learn from mistakes they made when solving homework assignments over learning from mistakes made during class work. Perhaps there was more anonymity in discussing mistakes made in the privacy of one's home than mistakes made in class. Regarding learning from mistakes made by others, most would choose cases based on errors reported in research studies rather than cases which came from their own observations. In addition, students preferred to learn from situations that were discussed together in class over situations that were analyzed alone.

5. Summary and Implications

In addition to the two courses described above, cases were also used as tool in three other academic courses: *Research Methods in Mathematics Education*, *Examples and Explanations in Mathematics Education*, and *Affective Aspects of Teaching and Learning Mathematics*. Findings from all courses suggested that participants felt that analyzing cases contributed to their understanding of the theories discussed in those courses, and helped them to connect what was discussed in the course to classroom events. It was also evident that the case activities promoted a positive attitude toward the practice of reading research literature.

There are several possible implications of this study for teacher education. First, not all cases have the same impact. For example, although participants agreed that it was fruitful to learn from their own mistakes, they preferred to learn from mistakes they made individually, but then discussed collectively. Thus, teacher educators might take into consideration prospective teachers' comfort zone when discussing mistakes in class as well as their need to discuss mistakes with their peers. Second, although bringing practice closer to theory is important, it does not preclude learning from previous research reports. This is in line with Tsamir (2008) who showed the effectiveness of using theories as tools in teacher education. Finally, using authentic cases taken from the prospective teachers' own reports of their observations may have limitations. In the course, *Examples and Explanations in Mathematics Education*, the authentic case analyzed in the course was not as rich in examples and non-examples as an exemplary case might have been. Perhaps a designed made-up case, would be more effective. Is authenticity a necessity? This is a question we have begun to explore. Currently, we are also exploring whether the mathematical context of a case impacts prospective teachers' analysis of affective issues. For example, if the aim of analyzing a case is to promote prospective teachers' knowledge of motivation and self-efficacy, does it make a difference if the case situation takes place in a geometry or an algebra lesson? These and other questions will be investigated as we continue to bring cases into our academic courses with the aim of not only bringing the classroom into the university, but helping prospective teachers bring theories into their classrooms.

Acknowledgment

This research was supported by the Trump Foundation grant no. 145.

References

Ball, D. L. and Forzani, F. M. (2009). The work of teaching and the challenge for teacher education, *Journal of Teacher Education*, **60**(5), 497–511.

Markovits, Z. and Even, R. (1999). The decimal point situation: A close look at the use of mathematics-classroom-situations in teacher education, *Teaching and Teacher Education*, **15**(6), 653–665.

Masingila, J. O. and Doerr, H. M. (2002). Understanding pre-service teachers' emerging practices through their analyses of a multimedia case study of practice, *Journal of Mathematics Teacher Education*, **5**(3), 235–263.

Pang, J. (2011). Case-based pedagogy for prospective teachers to learn how to teach elementary mathematics in Korea, *ZDM Mathematics Education*, **43**(6–7), 777–789.

Roesken-Winter, B. (2013). Capturing mathematics teachers' professional development in terms of beliefs, in *Proficiency and Beliefs in Learning and Teaching Mathematics* Sense Publishers, pp. 157–178.

Shulman, L. S. (1986). Those who understand: Knowledge growth in teaching, *Educational Researcher*, **15**(2), 4–14.

Tsamir, P. (2008). Using theories as tools in mathematics teacher education, in D. Tirosh and T. Wood (eds.), *The International Handbook of Mathematics Teacher Education*, Rotterdam: Sense Publishers, pp. 211–234.

Zeichner, K. (2010). Rethinking the connections between campus courses and field experiences in college-and university-based teacher education, *Journal of Teacher Education*, **61**(1–2), 89–99.

About the Authors

Pessia Tsamir is a full professor of mathematics education in the Department of Mathematics, Science and Technology Education at Tel Aviv University.

Dina Tirosh is a full professor of mathematics education in the Department of Mathematics, Science and Technology Education at Tel Aviv University.

Ruthi Barkai is a senior lecturer and a researcher in the Department of Mathematics, Science and Technology Education at Tel Aviv University, and a lecturer at the Kibbutzim Academic College of Education.

Esther Levenson is a researcher and teacher educator in the Department of Mathematics, Science and Technology Education at Tel Aviv University, and a lecturer at the Kibbutzim Academic College of Education.

Chapter 27

STEM Teaching as an Additional Profession for Scientists and Engineers: The Case of Mathematics Education Track of *Views*

Orit Hazzan* **and Iris Zodik**#

Faculty of Education in Science and Technology,
Technion — Israel Institute of Technology, Haifa, Israel
**oritha@technion.ac.il*
#*iris.zodik@gmail.com*

In this chapter, we focus on the training of prospective mathematics teachers who study at the Technion in a program called *Views*. *Views* invites Technion graduates back to the Technion Faculty of Education in Science and Technology, to study towards an additional bachelor's degree that includes a high school teaching certificate. Graduates can choose among the eight tracks of STEM education offered by the program, one of which is mathematics education. In this chapter, we introduce the *Views* program and, in particular, its mathematics education track, and present the unique characteristics of the students studying in this track. Specifically, we identify their motivational factors for joining the program as a second career, and present how they perceive their unique contribution to the education system in Israel, both as engineers and scientists and as parents.

1. Introduction

Many countries experience difficulties in attracting excellent STEM (Science, Technology, Engineering, and Mathematics) teachers, especially at the high school level. This shortage is partially explained by the fact that qualified young people, who excel in STEM, prefer to study one of the STEM subjects and work in technology and hi-tech firms as scientists and engineers rather than seek employment in the education system. This phenomenon, in turn, is partially explained by the accepted social norms in many countries, according to which teaching is a less desirable or less prestigious profession.

In response to this shortage in Israel, the Technion launched several education initiatives, one of which — the *Views*[a] program — is analyzed here. The *Views* program providing Technion graduates with an additional profession, namely high school STEM teachers. Many Technion graduates, from all scientific and engineering disciplines, ages, and professional backgrounds, are attracted to the *Views* program. In this chapter, we focus on those Technion graduates who chose to study in the mathematics education track of *Views*. The study helps understand their motivational factors for joining the program, and how they perceive their unique contribution to the education system in Israel, as more mature teachers with professional experience as engineers and scientists, as well as with more general life experience.

2. The Study Environment

2.1. *Technion's faculty of education in science and technology*

The Technion is the major supplier of scientists and engineers to the Israeli industry. The Faculty of Education in Science and Technology (henceforth: the faculty) is one of the Technion's 18 academic units. It offers eight undergraduate study tracks (mathematics, physics, biology, chemistry, computer science, environmental sciences, electrical engineering, and mechanical engineering), as well as a comprehensive graduate program. The degree bestowed by the faculty includes a high school teaching certificate for STEM subjects in one of the eight fields.

2.2. *The Views program*

Views[b] invites all Technion graduates back to the Technion's Faculty of Education in Science and Technology to study towards an additional bachelor's degree in one of the eight study tracks mentioned above.

All Technion graduates who enroll in the *Views* program receive full study scholarships from the Technion for two years, yet are *not* required

[a]The name of the *Views* program in Hebrew is Mabatim (= *Views* in English), which is the Hebrew acronym for Engineers/Scientists in Science and Technology Education.

[b]A detailed description of the *Views* program is presented in *Inside Higher Ed* (Sep. 20, 2012). The number of *Views* students is not, however, updated in the *Inside Higher Ed* paper.

Table 1. Winter 2016 Semester — Student distribution among study tracks and programs.

Track	Total	Views	Regular 4-year program
Mathematics education	69	52	17
Physics education	39	34	5
Chemistry education	46	41	5
Biology education	40	35	5
Computer science education	36	29	7
Environmental sciences education	9	9	
Electrical engineering education	25	14	11
Mechanical engineering education	34	20	14
Have not selected a track	5		5
Total	303	234	69

to commit to teaching in the education system at any time. *Views* students are required to earn 36 credits, and since the number of credits required to complete this degree is similar to that required for an MBA, the study program is organized similarly (full day or two half-days a week).

About 300 undergraduate students are currently enrolled in the faculty. Three quarters of these students are *Views* students (Table 1), of whom 60% are men and 40% women. Table 1 presents the distribution of *Views* and regular 4-year program students among the different study tracks. As can be seen, the mathematics education track is the largest track.

2.3. *The Views mathematics education track*

The mathematics education track is a high school mathematics teacher training program that includes three main tiers: disciplinary studies, education and pedagogical studies, and pedagogical-disciplinary studies.

Since its launch in October 2011, about 100 Technion graduates enrolled in *Views* program every year in all tracks. It is interesting to note that most of those who, over the years, studied or are studying in the mathematics education track are not graduates of the Faculty of Mathematics, but of another engineering department (see Table 2). Out of the 93 students, 47 were women and 46 were men. Ten jointed *Views* after they had completed a Master degree (9 students) or a Ph.D. degree (1 student) at the Technion in a topic related to their original discipline.

Table 2. Distribution of *Views* students in the mathematics education track by discipline and faculty of their original degree.

Discipline	Total in discipline	Faculty	No. students
Engineering	61	Industrial Engineering & Management	33
		Electrical Engineering	22
		Mechanical Engineering	3
		Aerospace Engineering	1
		Biomedical Engineering	1
		Electrical Engineering & Mathematics	1
Exact Sciences	27	Mathematics	12
		Computer Science	13
		Physics	1
		Physics & Mathematics	1
Life Sciences	5	Biology	2
		Chemistry	1
		Biochemistry	1
		Medical Laboratory Sciences	1
Total	93		

3. Research Framework

3.1. *Research questions*

The research questions derived from the research objective presented in the Introduction were:

- What factors motivate Technion graduates to enroll in the mathematics education track of the *Views* program?
- How do students in the *Views* mathematics education track perceive their unique contribution to mathematics education as engineers and scientists and as parents?

3.2. *Research participants*

Data was gathered from the 49 students (26 men and 23 women) who took the Algebra for Middle School course offered during the 2016 Winter semester and taught by the second author of this chapter. Of these students,

Table 3. Faculties of the students with bachelor's degree in the course.

Faculty	No. of students
Electrical engineering	14
Industrial engineering and management	12
Computer science	7
Mathematics	7
Mechanical engineering	2
Aerospace engineering	1
Physics	1
Total	44

38 were *Views* students[c]; 3 were *Views 2* students,[d] 3 were external studies students, and 5 students were studying towards their bachelor's degree in parallel to studying mathematics education. Table 3 presents the faculties in which the 44 students with their original bachelor's degree.

3.3. *Data-gathering tools*

Both quantitative and qualitative data were collected: Online surveys were completed by 31 students of the 49 students in the course.[e] The questionnaire was in Hebrew and a focus group of 5 students was conducted, which focused on the answers given in the questionnaire.

4. Data Analysis

4.1. *Motivational factors*[f]

Gero and Hazzan (2016) investigated the motivational factors among students enrolled in the *Views* program during the 2014–2016 academic years. As its theoretical framework, the study used the self-determination theory

[c]Of which 35 students were enrolled in the mathematics education track, 2 in the physics education track, and 1 in the computer science education track.

[d]In 2016, a new module of *Views* was launched, *Views 2*, which includes a master's degree in STEM education and a teaching certificate in one of the eight aforementioned disciplines.

[e]Of the students who completed the questionnaire, 13 already teach in various educational organizations.

[f]We would like to thank Dr. Ari Gero of the Technion's Faculty of Education in Science and Technology for his help in the data analysis.

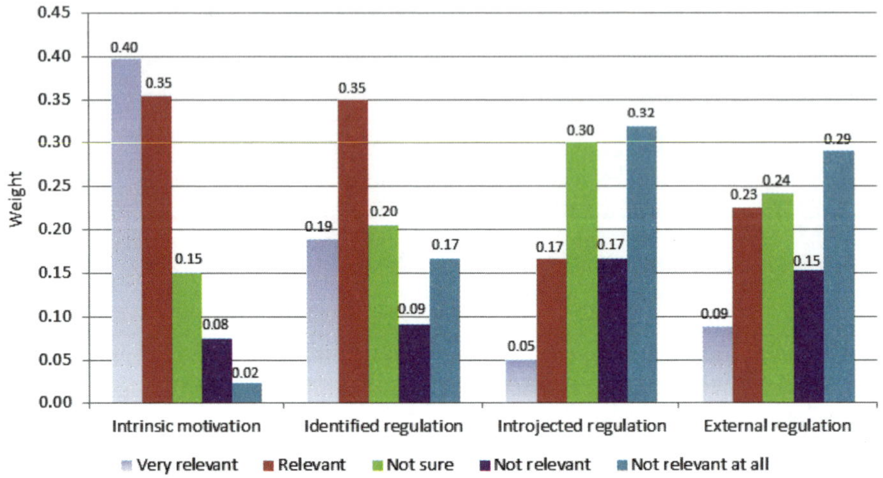

Fig. 1. Motivational factors of *Views* students for enrolling in the mathematics education track.

(Deci and Ryan, 1985), according to which the motivational factors driving an individual to invest resources in a certain behavior are located on a continuum.

As Fig. 1 indicates, *Views* students in the mathematics education track, who participated in our study, present a similar picture with respect to their motivational factors to that of the total population of *Views* students, studied by Gero and Hazzan (2016) and the *Views* students at the computer science education track studied by Hazzan and Ragonis (2014).

Specifically, the relevance of intrinsic motivation (i.e., an interest in the study of education) ranked highest, with identified regulation (i.e., future benefits the students expect to derive upon completion of the program) following closely behind. Identified regulation may be expressed by the acquisition of skills expected to promote the students' careers or help promote education in Israel.

Introjected regulation (i.e., originates from the desire to fulfill the expectations of the people who are important to the individual or from considerations of personal prestige) and external regulation (i.e., derived from the hope of receiving an (immediate) reward for the behavior, or, alternatively, from the fear of punishment) are of similar relevance and ranked at the lower end of the relevance scale, far behind intrinsic motivation and identified regulation. This ranking is partially explained by the fact that *Views* students are not required to commit to teaching after they complete the program. At the same time, however, over 50% of them do teach in various educational organizations during their studies.

4.2. *Life experience as engineers and as parents*

The diverse academic background of the students (Tables 2 and 3) has several advantages. The students, who are all Technion graduates with working experience in industry, which is characterized by high levels of uncertainty, bring the engineering perspective, as well as their life experience as parents, into the mathematics lessons. Due to space limitations, we present only a few illustrative quotes:

- Familiarity with the future job market: *Familiarity with the job market and research work at the university helps me see what work and learning processes students should develop.*
- The application of mathematics rather than its theoretical aspects: *When answering students' questions, I can give examples of practical applications in almost any area in real life.*
- Different work environment — Coping with uncertainty: *Coping with people and with complex situations.*
- Experience as parents: *Being a parent enables me to acquire skills and capabilities mainly with respect to how to approach children at any developmental stage, understand the difficulties they face, their expectations, etc.*

5. Summary and Future Research

This chapter presents the perspective of *Views* students regarding their motivational factors for joining the program, as well as their perception of their unique contribution to mathematics education — as engineers and scientists with work experience in the industry and as parents. It is important to note that two-thirds of the students in the mathematics education track are engineers with work experience in the cutting-edge Israeli hi-tech industry, who are familiar with the needs of the future job market and can use this familiarity in order to inspire their own students and increase their motivation to study advanced-level mathematics. In the future, we plan to investigate this inspiration and contribution.

References

Deci, E. L. and Ryan, R. M. (1985). *Intrinsic Motivation and Self-determination in Human Behaviour*, New York, NY: Plenum Publishing Co.

Gero, A. and Hazzan, O. (2016). Training scientists and engineers as science and engineering teachers: The motivational factors of enrollees in the *Views* programme, *World Transactions on Engineering and Technology Education*, **14**(3), 374–379.

Hazzan, O. and Ragonis, N. (2014). STEM teaching as an additional profession for scientists and engineers: The case of Computer Science education, in *Proceedings of SIGCSE 2014 — The 45th ACM Technical Symposium on Computer Science Education*, Atlanta, GA, USA, pp. 181–186.

About the Authors

Orit Hazzan is a faculty member at the Faculty of Education in Science and Technology at the Technion. Her research focuses on computer science education. In 2011–2015 Hazzan was the Faculty Dean and from 2017 she is the Technion Dean of Undergraduate Studies. See also her personal homepage on Technion website.

Iris Zodik is an associate lecturer at the Faculty of Education in Science and Technology at the Technion. She is currently leading the "Academia Classroom" project at the Technion and co-leading the Kadar Center. She has led the development of mathematics textbooks as well as a program aimed at fostering excellence of high-achieving students.

Chapter 28

Second Career Teachers Shifting from High-Tech and IDF Emeriti to Teaching Mathematics and Their Perceptions of Their New Career

Aya Steiner* and Shai Caspi#

Oranim Academic College of Education, Tivon, Israel
**ayasteiner@gmail.com*
#caspishai@gmail.com

Increasing number of individuals are entering the field of mathematics education as *second career teachers* (SCTs). Coming from successful careers in industry, accounting and in the IDF, the SCTs bring with them a range of experiences that makes their teaching qualitatively different from their first career peers. This chapter presents a survey conducted among ten second-career mathematics teachers, all are successfully teaching mathematics in secondary schools. Here we present the main perceptions and themes that came up during the interviews: (1) Inner drive for meaningful action; (2) the gap between the naive initial vision and actual job scope; (3) the unique contribution they bring to school; and (4) common perceptions regarding classroom management.

1. Introduction

The traditional concept of career has changed during the past decades. A new view of careers has emerged, transforming the occupation that an individual chooses for a livelihood from a single concept to a multiple one. One of the career opportunities mostly explored by individuals from engineering, finance, and high-tech professions is that of teaching mathematics. In recent years, the number of SCTs entering schools has grown. Following this trend, colleges have adopted teaching certification programs for those holding an academic degree, offering a fast track avenue toward certification. Though these programs are flourishing, SCTs are facing with challenges, some of which are well-known new teacher challenges, and some are inherent to the career transfer (Haim and Amdur, 2016).

2. Transition to Teaching

Those individuals who enter teacher preparation programs do so by choice. A conscious decision is made to resign from a rewarding job and sometimes a senior position, to become an educator. Therefore, these persons are, highly committed to the path they have chosen. The transition to teaching is made without gain in the financial arena. The trade-off for lower pay is the increased personal satisfaction gained by the sense of doing something meaningful.

Even though all individuals are different, SCTs possess values dependent upon their background and experiences. The collective life wisdom of SCTs enables them to approach teaching with a multitude of responses to critical teaching situations. Although they may not always see the connection between their previous careers and teaching, or may not be able to easily transfer the competencies that they bring with them to their new teaching career (Mayotte, 2003), studies indicate that they bring a diverse range of skills such as time management, work habits, coping, and problem-solving (Hunter-Johnson, 2015). They also bring communicational skills and the ability to handle multiple tasks simultaneously as well as patience and resilience in facing obstacles (Grier and Johnston, 2009).

Despite the desire to make a difference in the world, especially in the area of dealing with youth, SCTs face challenges. Some of the challenges are those that are common to all novice teachers, who may suffer from a mismatch between the ideology of the new school and their own set of beliefs. Adjusting to the organizational and social norms of the school system they need to replace previous and sometimes "ideal" concepts about teaching and learning with the dominant values of the school and classroom (McCluskey, Sim and Johnson, 2011). With habits and patterns well established, SCTs may find greater difficulties in adopting school norms.

3. Participants and Methodology

The participants of this survey are ten SCTs, which had entered teaching after having successful careers in the hi-tech industry, the financing field and in the IDF. They all teach in secondary schools with seniority range from two to seven years and were selected upon recommendations from school principals or college professors. Participants have entered the fast track teacher training program at ages ranging from 33 to 55, with a minimum of a bachelor's degree and gained full-time employment shortly after being certified.

We interviewed each participant using a prebuilt questionnaire that included questions regarding the training process, the first year at school and reflection over the educational work. Each underwent a one-hour interview that was recorded and then transcribed.

4. Themes and Perceptions

The process of identifying the themes involved inductive means of constantly comparing and analyzing the data that emerged and refining it to correspond with categories and themes. The following themes emerged: (1) inner drive for meaningful action; (2) perceptional gap between the naive initial vision and actual job scope; (3) unique contribution; and (4) classroom management.

4.1. *Inner drive for meaningful action*

When referring to the profession of teaching, all participants mentioned the word "meaningful" in the context of meaningful work. Meaningful work can be conceptualized as finding a purpose in work that is greater than the extrinsic outcomes of the work (Arnold *et al.*, 2007). In order to consider a work "meaningful" it has to have an impact on a field one considers substantial. In the absence of natural resources, the knowledge industry is considered as Israel's national resource. Hence, education in general and especially STEM is of national importance. Correspondingly, all respondents identified teaching as meaningful work, through statements that characterize education as a change-generating tool for improving the Israeli society:

> It is the most important profession. It determines the nature of the next generations. It has a strong impact on the path that we, as a society, will choose to follow.

> I think it is a mission. The field of mathematics education is critical in the national level and it is important for me to contribute within that field.

Along with choosing the right arena for impact, all participants specify their unique contribution to mathematics education. Although entering mathematics education equipped with wide mathematical knowledge, the prominent contribution they mentioned was not about mathematics itself (i.e. increasing the mathematical level, or extra-curricular mathematics), but rather giving their students the experience of learning mathematics in a joyful way:

> I wanted them to like mathematics, I wanted them to have a good experience. I wanted to reach as many students as possible, motivate the students to do mathematics by making them recognize the beauty in mathematics.

All interviewees referred to the work of teaching as weary work with no official extrinsic reword or acknowledgement from the system and related to that "inner drive" for meaningful action as essential for succeeding in this challenging system:

> You need a sense of mission, not in the fashionable manner. It has to be engraved within you, in the Zionist level, to be a teacher, to teach children.

When asked to recall the most significant experience they had since the very beginning of this career change process, all interviewees had one story in mind which heroes were themselves and one student. These stories are not stories of academic success but rather stories of caring interpersonal communication, in which their student acknowledged them for their impact. All participants recognized these acknowledgments as a compensation that helps in overcoming difficulties

> One student always sat in class, downcast eyes, not communicating with me. She got a "D" on the exam, solving only one question. I wrote to her on the exam sheet that she did an excellent job with the one question she answered and that I want her to take the exam home and complete the exam so I could give her the "A" she deserved. She did not do it, but that was the beginning of a process. Since that day, she started communicating with me and now she is a happy "A" student.

> When you teach a child, and then you succeed and then you see her smiling it is the best you can expect, and as long as you have constant dripping of instance like that you will manage all the rest.

4.2. *Perceptional gap between the naive initial vision and the teacher's actual job scope*

When taking such a life changing decision of choosing a new life course one needs to have an image of the destination. When facing reality this image might turn out to be not accurate. Two main perceptional gap themes arose. The first is about the intensity of the teacher's daily work. All participants reported that the intensity of the work was overwhelming to them. Although coming from demanding working environments, most participants testified that they are working more hours as teachers then they worked in their previous jobs. As Adam, former CEO of a big high-tech company described:

> I worked hard in the high-tech industry I worked many hours. Teaching mathematics, I work at least as intense as I worked before. Working in high-tech was a summer camp. Now, (being a mathematics teacher) I am really working.

Participants related the notion of intensity not only to the great time investment but also to the emotional load that is "an integral part of teachers' lives" (Sutton and Wheatley, 2003). This emotional work embedded in the teaching practice, came by surprise for our participant.

> Classroom management with 30 children is very difficult. It takes emotional work, inner work to stay calm. I did not foresee the main challenge would be the emotional inner work, the work of virtue.

The second theme was the portion of the educational work in mathematic teaching. Before entering school, all participants held a perception according to which the main doing of a mathematics teacher is to teach mathematics. During the first year, they found out that the main effort they need to make is in the non-disciplinary area.

> As a new teacher, I perceived my job as preparing my students for the secondary school graduation exam in mathematics. By the way, I will teach some values. Today I see myself primarily as an educator, meaningful figure to my students who inspect my behavior at every turn. I realized that my self-fulfillment is not only leading my students through the secondary school graduation exam in mathematics, but also through this major educational role.

4.3. *Unique contribution*

When asked of their unique contribution, participants did not mention professional competencies they acquired during their previous careers but indicated the entirety of their whole life experience as the source of uniqueness.

> Second career teachers bring with them life experience, perspectives, understanding of intricacies, and multidisciplinary understanding, because that is what's out there in real life, and people who worked their entire career in the school scenery might not be exposed to such intricacies. I am 45, I have kids of my own.

4.4. *Classroom management*

By classroom management, we refer to the actions teachers take to create an environment that supports and facilitates both academic and social-emotional learning (Watson and Ecken, 2003). When considering teaching challenges regarding to students the most common and basic ones are discipline issues. This great challenge of discipline might be the reason why classroom management is often thought of in terms of controlling students (McCaslin and Good, 1998). Although all interviewees acknowledged the great challenge in dealing with discipline issues, when relating to their classroom management they are not talking of belligerent actions that are meant to enforce authority. Rather they are describing actions aimed at developing caring supportive interpersonal relationships with their students, they are taking in order to establish good learning environment. Many participants are relating to students with behavioral problems as students who are experiencing difficulty.

> The main challenge is to apply regulations over an unbridled classroom. It was meeting half way. I was the CEO of a big company. People did what I said. Slowly I realized I needed to approach my students not from an authority standpoint but as a partner whose main interest is their own personal good.

> I know some teachers are very strict and intransigent, I choose other strategies, like one on one conversations, trying to figure out from where students' difficulties are coming.

5. Discussion

All participants entered teaching following deeper interests and longings for meaningful action. A desire to contribute to society by teaching youth is the impetus to make Israel a better place to live in. Not anticipating the great intensity in teaching, participants described the shocking experience they underwent during their first year at school. Although accustomed to working hard in their previous careers most participants reported that they work even harder as mathematics teachers both time and emotional wise. The transition to teaching was made without gain in the financial arena. In the absence of economic worries, participants did not consider the lower compensation associated with the new profession. The trade-off for lower pay is the personal satisfaction gained by the sense of doing something meaningful. Yet, considering the scope and intensity of the teachers' work, many used the word "insulting" regarding teachers' salary and were unable to conceive how recent college graduates get by.

Finally, many participants, with their inner drive to make a positive impact, have learnt that the ability to make a difference is a slow and arduous process. There is no comprehensive success in the classroom but rather local events of acknowledgment, which are the evidence for their success.

References

Arnold, K. A., Turner, N., Barling, J., Kelloway, E. K. and McKee, M. (2007). Transformational leadership and psychological well-being: The mediating role of meaningful work, *Journal of Occupational Health Psychology*, **12**(3), 193–203.

Grier, J. M. and Johnston, C. C. (2009). An Inquiry into the development of teacher identities in STEM career changers, *Journal of Science Teacher Education*, **20**, 57–75.

Haim, O. and Amdur, L. (2016). Teacher perspectives on their alternative fast-track induction, *Teaching Education*, 1–28.

Hunter-Johnson, Y. (2015). Demystifying the mystery of second career teachers' motivation to teach, *The Qualitative Report*, **20**, 1359–1370. Retrieved from: http://nsuworks.nova.edu/tqr/vol20/iss8/14.

Mayotte, G. A. (2003). Stepping stones to success: Previously developed career competencies and their benefits to career switchers transitioning to teaching, *Teaching and Teacher Education*, **19**, 681–695.

McCaslin, M. and Good, T. L. (1998). Moving beyond management as sheer compliance: Helping students to develop goal coordination strategies, *Educational Horizons*, **76**, 169–176.

McCluskey, K., Sim, C. and Johnson, G. (2011). Imagining a profession: A beginning teacher's story of isolation, *Teaching Education*, **22**, 79–90.

Sutton, R. E. and Wheatley, K. (2003). Teachers' emotions and teaching: A review of the literature and directions for future research, *Educational Psychology Review*, **15**(4), 327.

Watson, M. and Ecken, L. (2003). *Learning to Trust: Transforming Difficult Elementary Classrooms through Developmental Discipline*, San Francisco: Jossey-Bass.

About the Authors

Aya Steiner is a lecturer at Oranim Academic College of Education, PhD student at Haifa University working on "The development of the discourse of rational numbers" under the supervision of Prof. Anna Sfard.

Shai Caspi is the head of Mathematics Department and teacher at "Hugim" High school in Haifa. Lecturer at Oranim Academic College of Education. His PhD thesis was on "The development of algebraic discourse in junior high school" under the supervision of Prof. Anna Sfard.

Chapter 29

Club-5: Communities of Practice of High School Mathematics Teachers Who Teach Advanced-Level Mathematics

Roza Leikin*, Varda Talmon, Geula Sever, Irina Gurevitch and Revital Aisik

Department of Mathematics Education, Faculty of Education, University of Haifa, Israel

**rozal@edu.haifa.ac.il*

This chapter describes a nation-wide project — Club-5 — aimed at improving the quality of mathematics teaching in high school through management of teachers' communities of practice. Club-5 aims to create a comfort zone for project participants and enhance the inquiry components in school mathematics teaching. Mutual support, shared experiences, and professional conversations within the communities, along with implementation of challenging mathematical tasks, educational technologies, and collaborative learning in the classes, are core elements of Club-5. Special attention is paid to the professional development of community leaders.

The thought of this community thriving excites me — to think that there is another way of teaching, that it is alive and well in classrooms even today, and that the teachers it creates will go out and influence more teachers, and more, and more... I hope to see it... (Club-5 member)

1. Introduction

Raising the quality of mathematics teaching in schools is not an easy task for any educational system. The program managed by the team from the Department of Mathematical Education at the University of Haifa that we describe in this paper — Club-5 — has this ambitious goal in the context of high school advanced-level mathematics teaching. Club-5 implements an exponential model (described in Section 2) of professional development of mathematics teachers through participation in communities of practice. The core of this program is communities of practicing teachers who teach advanced-level mathematics in high school.

"Communities of practice are groups of people who share a concern or a passion for something they do and learn how to do it better as they interact regularly" (Wenger and Wenger-Trayner, 2015, p. 1). They provide spaces for critical discussion, interdisciplinarity, learner-centered thinking, and social innovation to take place. Participants can share ideas and suggestions, and engage in discourse relevant to their specific areas of academic interest. A community of practice has an identity defined by a *shared domain of interest*, *commitment to the domain,* and *shared competences*. The community members are engaged in *joint activities* and discussions, *help each other*, and *share information*. They build relationships that enable them to *learn from each other*. They have *a shared space of resources*, *skills and norms* that support and advance the level of proficiency of each member of the community (Jaworski, 2006; Krainer, 2003; Roth, 1998; Zaslavsky, Chapman and Leikin, 2003).

We report on this program at the end of its third year. The first year of the project was devoted to developing a community of community-leaders (CCL) which included 23 teachers who were considered experts in teaching advanced-level mathematics (by their colleagues, school principals and even supervisors from the Ministry of Education). During the second year of the program, community leaders (CL) in pairs (or triples) monitored 9 local communities of practice (LCPs) that included 160 teachers altogether. Fourteen additional expert teachers were chosen by CLs from within their LCPs to be trained to lead LCPs during the consequent year and to join the CCL. During the 3rd year of the project, there were 18 LCPs with about 350 high school mathematics teachers all over the country, led by our CLs.

2. Exponential Model of Development of Communities of Practice

2.1. *Core goals and elements of Club-5*

One of the main goals of Club-5 is to create a comfort zone for its members. To achieve this goal, Club-5 activities integrate mutual help, participants' sharing of positive experiences, collective solving of authentic problems related to participants' teaching practices, collaborative activities and mutual support in professional life. An additional central goal of Club-5 is advancing mathematics teachers' proficiency and developing a cohort of CLs. We employ an exponential model of professional development, and present its 3-years development schematically in Fig. 1.

CORE GOALS AND PRINCIPLES

I. ***Creating comfort zone for Mathematics teachers***: (1) Collaboration in professional development; Sharing positive experiences; (2) Solving authentic problems related to teaching practice; (3) Mutual support in professional life.

II. ***Enhancement of inquiry-based learning and advancing mathematics teachers' proficiency and educating teacher educators***: (1) Reading and analyzing research papers in mathematics education; (2) Development of a "shared terminology" and (3) Recognizing mathematical, pedagogical and educative power of tasks including but not limited to: (a) Investigation tasks with dynamic mathematics software Multiple solution tasks; (b) Cooperative learning tasks.

III. ***Implementation in LCP and in mathematics classes***

IV. ***Reflection on critical episodes from teaching practices***

YEAR 1 ACTIONS (with 23 teachers)

Developing CCL:

- Recruiting expert teachers for CCL (June 2014–September 2015)
- Course 1: Weekly meetings:
  - Integrating I, II, III, IV (November 2014–July 2015)
  - Development and implementation of activities activation of CCL by CL (February–July 2015)
- Conducting school mathematics teachers communities with implementation of Course 1 products
- Round tables at the national mathematics teachers' conference (April 2015)

Recruiting teachers for LCPs at year 2 (May – August 2015)

Annual Club-5 conference (July 2015) for CCL and future LCPs

YEAR 2 ACTIONS (with 160 teachers)

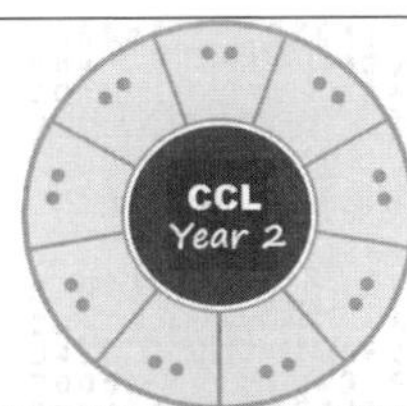

Monitoring, advancing and supporting CCL:

November 2015–July 2016

- Biweekly meetings of CCL including I, II, III, IV and reflection on monitoring LCPs
- Collaborative design of activities for LCPs
- Recruiting new CLs from LCPs (January 2016)
- Round tables conducted by CLs at the national mathematics teachers' conference (April 2016)

Developing LCPs:

- Biweekly meetings of LCPs
  - Integrating I, II, III, IV (November 2015–July 2016) through employing activities designed in CCL
  - Development and implementation of new activities by members of LCPs for LCPs and school classes (February–July 2016)
  - Reflective discussions of teaching practices

Recruiting teachers for LCPs at year 3 (May – August 2016)

Annual Club-5 conference (July 2016) for CCL and LCPs

YEAR 3 ACTIONS (with 350 teachers)

(October 2015–July 20167)

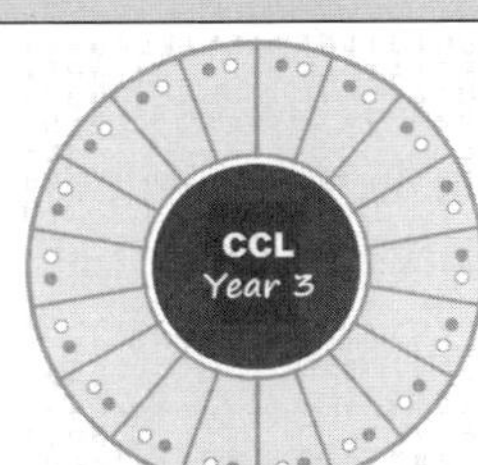

Repeating Year 2 Actions *But*

- *Triweekly* meetings in CCL and LCPs
- Design of activities *by each pair of CL for all CL*
- *Poster session* conducted by CLs at the national mathematics teachers' conference

Annual Club-5 conference (July 2017) for CCL and LCPs

Fig. 1. Exponential model of the development of communities of practice of high school teachers in Club-5.

In what follows we describe main elements of Club-5 pertaining to the development of key domains of communities of practice: shared interest and commitments, shared competences, and shared resources, skills and norms.

2.2. *Advancement of CLs' autonomy*

In October 2014, twenty-three teachers who were teaching advanced-level mathematics in high school met at the University of Haifa to join Club-5. All these teachers were experts in teaching advanced-level mathematics, and served as either mathematics department chairs or mathematics teaching supervisors. The first year in Club-5 was devoted to the development of the CCL. The shared domains of interest and commitment of CCL participants were: (a) improving teaching of advanced-level mathematics in high schools, (b) enhancement of inquiry-based learning (cf. Wells, 1999) and (c) development of the proficiency necessary for leading LPCs in Club-5.

2.2.1. *Development of CCL*

During the first year, (prospective) CLs met on a weekly basis in an academic course. The course combined reading research papers (Ball, Thames and Phelps, 2008; Fahlgren and Brunstrom, 2014; Krainer, 1993; Leikin, 2011; Leikin *et al.*, 2014; Stigler and Hiebert, 1999; Zazkis and Mamolo, 2011; Zaslavsky, Chapman and Leikin, 2003) with implementation of the research findings in practice. Reading the papers was directed at the creation of shared terminology and the development of understanding of the added value of educational research for practice. Reading and analysis of the papers was guided by the Club-5 team. Corresponding mathematical workshops were designed to illustrate the main ideas and findings presented in the papers. In this way, connections between theory and practice in mathematics education were developed. Special attention was given to inquiry-based learning, while analysis of the mathematical, pedagogical and educative power (Krainer, 1993; Zaslavsky, Chapman and Leikin, 2003) of mathematical tasks framed participants' discussion of (a) investigation tasks with dynamic mathematics software, (b) multiple solution tasks (MSTs) (see example in Fig. 2) and (c) cooperative learning tasks, those that open opportunities or even require collaboration of solvers.

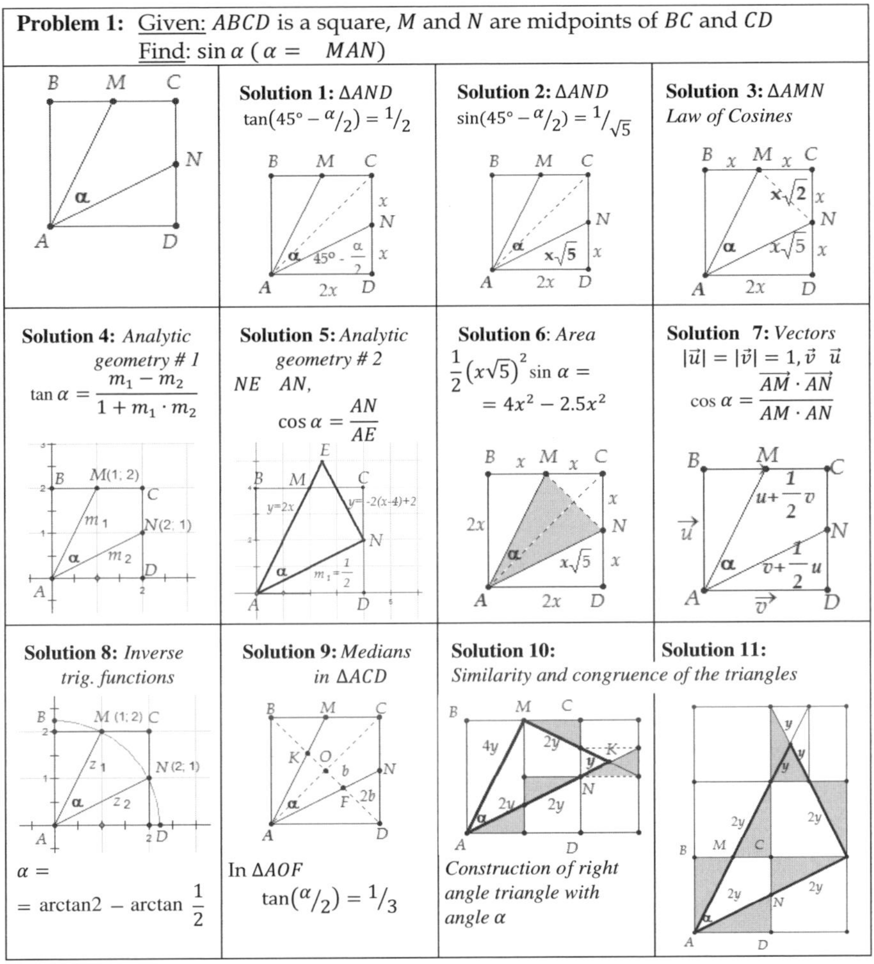

Fig. 2. Multiple solution space created by CLs collaboratively and developed in a workshop for LCPs.

At the beginning of the course mathematical workshops were designed by the Club-5 team and implemented at meetings with course participants. Later the (former) CLs developed mathematical workshops under team guidance with a gradual increase in CLs' autonomy.

Figure 2 demonstrates the collaborative solution space of one particular problem that was created by CCL members. It was used in a workshop that illustrated the connections between different mathematical topics and the variety of auxiliary constructions that constituted the mathematical power of the task. The pedagogical power of the tasks was in their embedded potential to develop mental flexibility and collaboration among students.

2.2.2. *Structure of LCP meetings*

In the middle of the first year of Club-5, CLs (in pairs) recruited participants for their local communities of practice. In collaboration with the Club-5 team, CLs developed the structure of meetings in LCPs. It was decided that each meeting of an LCP would include the three following components:

- At *sharing points* each member of the LCP has an opportunity to share either positive or problematic classroom experiences, including (a) experiences associated with the implementation of community mathematical activities from the LCP meetings, (b) teachers' ideas and initiatives, and (c) critical episodes from the classroom that require solving real-life problems related to teaching practice.
- *Warm corners* were devoted to introducing LCP members to teaching resources, such as websites, video clips, applets, games, cartoons, and mathematical puzzles to enrich the classroom experience and provide variety. Warm corners were initially prepared by the Club-5 team, later by CLs and finally by other members of the LCPs.
- *Main mathematical workshops* for teachers were directed at deepening, strengthening and enriching teaching and learning of mathematical concepts from the curriculum. These workshops were designed by CLs with the assistance of the Club-5 team.

2.2.3. *Collaboration and professional development of CLs through practice*

During the second year of the project, about 150 high school mathematics teachers were taking part in bi-weekly meetings of LCPs. CLs were involved in intensive preparation of each community meeting. All the members of the CCL worked collaboratively to prepare the meetings.

Each pair of CLs was in charge of preparing one meeting, including a workshop. The theme of the workshop was usually inspired by classroom experience, or by issues that came up within the community. First, the Club-5 team assisted the pair of CLs in developing workshops. Then the workshop was presented to all CCL members during a synchronous online meeting held once a month. Other CLs gave feedback, asked questions and suggested additional ideas. This process of collaborative creation led to the foundation of a shared space of competences, and a shared space of resources, skills and norms that included: solving problems in multiple ways, integration of technology in the teaching and learning of mathematics, integration of the history of mathematics, development of questions suited for different levels of understanding in heterogeneous classrooms, response to students' mistakes and discussion of implementation in the classroom. This shared space of principles and norms served as a framework for activities during the third year, which saw the project grow to 18 communities consisting of about 350 participating teachers. During the third year of Club-5 each pair of CLs developed a workshop for use by all CLs in their LCP (Fig. 1).

2.3. *Teachers' professional development in LCP*

The community meetings changed over the course of the year as the communities came together, with each developing its own unique character. In the beginning most of the burden of managing meetings fell on CLs, who used workshops that got other participants involved as much as possible. As each community grew more cohesive, the teachers in the communities began to express themselves and shared more and more ideas, dilemmas that came up during teaching, and mathematical knowledge. Later, as part of the community's work, teachers were asked to present the group with an example of their own method of classroom instruction.

We conclude with one teacher's feedback: *I was impressed by the depth of the discussion, by participants' readiness to share the dilemmas they faced and to share of themselves. I was no less impressed by the way you [the project team] conduct the meeting —. . . you manage to maintain the delicate balance of providing guidance that creates a framework for discussion, but allowing those present to express themselves and to be dominant. The cooperation of the participants is also inspiring — each person gets time to speak and is supported throughout by their partner.*

Acknowledgement

This project was made possible by grant no. 126 from the Trump Foundation. We would like to thank the Ministry of Education for their generous support and the University of Haifa for providing us with the necessary facilities. We would also like to thank all the Club-5 participants for their involvement and commitment to the project. Our special thanks to the members of the CCL for their enthusiastic and creative collaboration.

References

Ball, D., Thames, M. H. and Phelps, G. (2008). Content knowledge for teaching what makes it special? *Journal of Teacher Education*, **59**(5), 389–407.

Fahlgren and Brunstrom (2014). A model for task design with focus on exploration, explanation, and generalization in a dynamic geometry environment, *Technology, Knowledge and Learning*, **19**, 287–315.

Jaworski, B. (2006). Theory & practice in mathematics teaching development: Critical inquiry as a mode of learning in teaching, *Journal of Mathematics Teacher Education*, **9**(2), 187–211.

Krainer, K. (1993). Powerful tasks: A contribution to a high level of acting and reflecting in mathematics instruction, *Educational Studies in Mathematics*, **24**, 65–93.

Krainer, K. (2003). Teams, communities and networks (editorial), *Journal of Mathematics Teacher Education*, **6**, 93–105.

Leikin, R. (2011). Teaching the mathematically gifted: Featuring a teacher, *Canadian Journal of Science, Mathematics and Technology Education*, **11**, 78–89.

Leikin, R, Leikin, M., Lev, M., Paz, N. and Waisman, I. (2014). On the relationships between mathematical creativity, excellence and giftedness, in S. Oesterle and D. Allan (eds.), *Post-Proceedings of the 2013 Annual Conference of the Canadian Mathematics Education Study Group (CMESG)* CMESG/GCEDM: Burnaby, BC, pp. 3–18.

Roth, W. M. (1998). *Designing communities*. Boston: Kluwer.

Stigler, J. W. and Hiebert, J. (1999). *The Teaching Gap: Best Ideas from the World's Teachers for Improving Education in the Classroom*, New York: The Free Press.

Wells, G. (1999). *Dialogic Inquiry: Towards a Sociocultural Practice and Theory of Education*. Cambridge, UK: Cambridge University Press.

Wenger, E. and Wenger-Trayner, B. (2015). Introduction to communities of practice: A brief overview of the concept and its uses. Retrieved from: https://www.vpit.ualberta.ca/cop/doc/wenger.doc, August, 10, 2016.

Zaslavsky, O., Chapman, O. and Leikin, R. (2003). Professional development in mathematics education: Trends & tasks, in A. J. Bishop, M. A. Clements, C. Keitel, J. Kilpatrick and F. K. S. Leung (eds.), *Second International Handbook of Mathematics Education* Dordrecht, the Netherlands: Kluwer, pp. 877–915.

Zazkis, R. and Mamolo, A. (2011). Reconceptualizing knowledge at the mathematical horizon, *For the Learning of Mathematics*, **31**(2), 8–13.

About the Authors

Roza Leikin is a professor of Mathematics Education and Education of the Gifted at the Faculty of Education, University of Haifa. Her research and practice embrace mathematical creativity and ability in teachers and students. She is the head of the Department of Mathematics Education (since October 2016) and director of the RANGE Center in the University (since 2008). She served as the head of the National Advisory Committee for Mathematics Education of the Israeli Ministry of Education (2013–2017) and the President of the International Group for Mathematical Creativity and Giftedness (2012–2017). She has published more than 150 research papers.

Varda Talmon has been the head of the National Center for Secondary School Mathematics Teachers (SSTC) since 2008. She completed her PhD dissertation on Dynamic Behavior of Geometry Figures at the Department of Education at the University of Haifa. She has 25 years of experience in developing instructional activities for students and teachers of mathematics in secondary school, with a special emphasis on alternative, technology-rich teaching approaches. She teaches didactics of mathematics at the Department of Mathematical Education at the University of Haifa.

Geula Sever manages the website of the SSTC and works on the development of dynamic applets since the establishment of the center. She has 25 years of experience in teaching high school mathematics. She has an MA in Mathematics Education from the University of Haifa. She leads several key projects at the center and runs workshops for high school teachers and teacher instructors.

Irina Gurevitch Graduated with MA degree in mathematics and math education from Moscow State University. She moved with her family to Israel in January 1990. Since 1991 has been teaching mathematics at the "Reali Hebrew school" in Haifa, Israel. She has 30 years of experience in teaching high school mathematics. Irina coordinates TELEM project in the Technion since 1999. She is one of the leaders of project Club-5.

Revital Aisik completed M.A. in Mathematics Education in this University of Haifa and currently she is PhD candidate at the Department of mathematics education in this University. Her research is focused on the Development of the proficiency of community leaders in Club-5 project. She has 22 years of experience in teaching high school mathematics and she worked several years as a coordinator, guide and mentor for novice and prospective teachers. She was the member of the leading team in Club-5 during 2015-2017 school years.

Chapter 30

National Centers of Mathematics Teachers: Bridging Between Theory and Practice in Mathematics Teaching

Varda Talmon, Geula Sever, Gila Ron and Roza Leikin*

Department of Mathematics Education, Faculty of Education, University of Haifa, Haifa, Israel

**rozal@edu.haifa.ac.il*

In this chapter we describe two National Centers of Mathematics Teachers: the Primary School Teachers' Center and the Secondary School Teachers' Center. The centers are managed by a team from the Department of Mathematics Education in the University of Haifa and are monitored by the Israeli Ministry of Education. In line with their dual affiliation, the centers construct firm bridges between academic research in the field of mathematics education and pedagogical practice in schools across the country. We portray the structure and the activities which are central to the two centers and provide some examples of activities suggested by the centers.

1. Main Goals and Principles of the Centers

1.1. *Goals and principles*

The National Centers of Teachers of Mathematics play a key role in the advancement of mathematics education in Israel. The two centers — Primary School Teachers Center (PSTC) and Secondary School Teachers Center (SSTC) — work in parallel and advance corresponding areas of mathematics teaching and learning. The PSTC has operated in the University of Haifa for more than 20 years, and was initially established by Prof. Pearla Nesher as an academic unit at the Faculty of Education in the University of Haifa. The SSTC has been managed by our team for about

10 years and was preceded by the centers for mathematics teachers at the Technion and the Weitzman Institute of Science.

Since 2003, the centers have been monitored by the staff from the Department of Mathematics Education in close partnership with the Israeli Ministry of Education. They became national centers in 2007. Their work is based on the following five fundamental goals: (1) tailor activities to the goals outlined by the Ministry of Education; (2) bridge between academic research in the field of mathematics education and teachers in schools across the country; (3) work to cultivate professional leaders in mathematics education, and to support these leaders in their work with other teachers across Israel; (4) review and adapt developments and advancements in the field of mathematics education in Israel and abroad, bring them to teachers' attention and stimulate their implementation; and (5) put special emphasis on employing technological advances and informing practitioners about research in mathematics education.

1.2. *Supporting curricular changes and technological advancements in mathematics education*

In recent years mathematics education in Israel has been undergoing an upgrade that includes new primary and middle school mathematics curricula, and a new high school mathematics curriculum is in the works. The centers support the Ministry of Education in implementing curricular changes, advancing teacher proficiency and coaching teacher instructors. In order to implement curricular changes, the centers' teams, in collaboration with national teacher instructors from the Ministry of Education, provide workshops and courses for district teacher instructors who, in turn, support integration of curricular changes in schools.

In order to advance mathematics education and professional development to the 21st century, and promote meaningful, relevant, and up-to-date mathematical education in schools the centers encourage the use of technology-rich learning environments in mathematics teaching. These learning environments make use of mathematical software, dynamic and interactive applications, and electronic books. The centers train teacher instructors and leading teachers in developing technology-based materials and applications accessible to all teachers.

2. The Structure and Main Activities of the Centers

The National Centers of Teachers of Mathematics work to create, develop and cultivate professional leadership in mathematics education. The PSTC and SSTC have a similar structure that is reflected in the resources they offer. Both centers have websites that provide teachers with access to relevant sites, applications, and activities from Israel and abroad, Facebook pages with frequent updates, instructional materials, courses for teacher instructors and leading teachers and national conferences. Figure 1 depicts the main activities of the centers.

As can be seen in Fig. 1, the centers work directly with national and district instructors of teachers of mathematics who, in turn, share what they learn with a broad range of mathematics teachers throughout the country, with the help of the centers. The chief superintendents of school mathematics check that the centers' actions are in line with the goals and principles of the Ministry of Education. At the same time, the centers design instructional activities according to the needs of teachers and teacher instructors. Teachers implement the activities and provide feedback, and in response the proposed activities are transformed to better meet teachers' needs, a process which increases the feasibility of implementation of educational innovations.

Below we present a short description of some of the activities at the PSTC and SSTC. Due to the space constraints of this paper, we mainly describe activities of the SSTC.

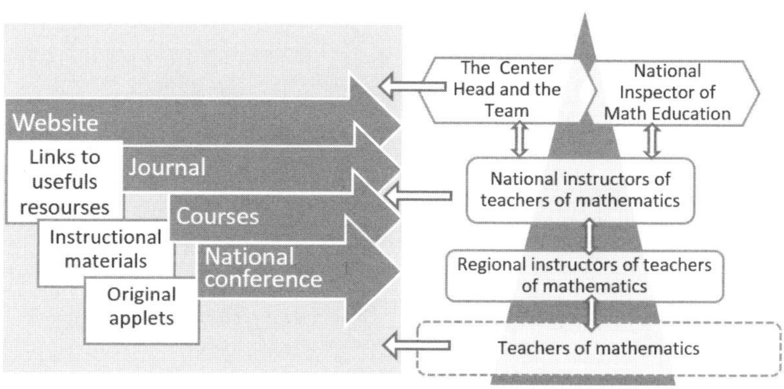

Fig. 1. The structure and main actions of the centers.

3. The Website

The website contributes to the creation and development of an active professional community that meets the needs of teachers and teacher instructors from across the country and from various sectors of society. The website also provides a gateway to mathematical, curriculum-related and pedagogical content and to updates about conferences, courses and workshops for teachers. It allows users to expand their knowledge according to their individual needs and to implement innovative didactical approaches, both in mathematics teacher education and in the classroom. The website provides teachers and teacher educators with access to innovative instructional materials developed in the center, and to original dynamic applets that accompany these materials and allow teachers to implement inquiry-based instruction.

3.1. *Dynamic applets*

Each applet includes guidelines for activities that integrate it in the classroom. The applets assist teaching instructors and leading teachers in the integration of technology in mathematics instruction. The applets have several interrelated goals: (a) using dynamic features to support definitions and proofs (see Fig. 2(a)), (b) deepening understanding of mathematical concepts and their properties through visual support (see Fig. 2(b)) with an emphasis on the common student difficulties when learning the concepts,

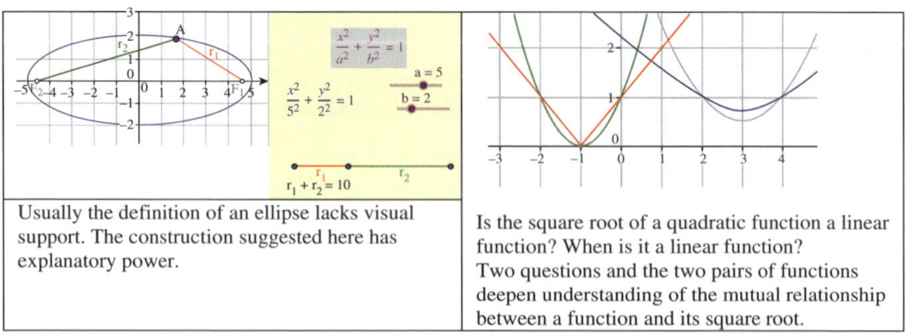

(a) (b)

Fig. 2. (a) Defining ellipse as a locus; (b) square root of a quadratic function.

(c) promoting mathematical curiosity and interest in learning mathematics by means of mathematical investigations.

3.2. *Instructional and educational videos on integrating interactive, dynamic applications*

This is an additional, unique resource accessible to teachers and teacher educators through the website. The center produces instructional videos that demonstrate mathematical concepts in various ways, with an emphasis on pedagogical and didactic aspects. The instructional videos are an integral part of activities in teacher instructors meetings. There is a special folder of educational videos for use in mathematics classrooms or in the discussion of classroom cases in a group of teachers. In parallel, the center offers teachers a course on creating their own videos.

4. Professional Journals

Each center publishes a journal (SSTC — Aleh, PSTC — Mispar Hazak). The journals are peer-reviewed and are open to publications from teachers, teacher educators and researchers in colleges and universities. There are several sections in the journals: teachers' reflections on their classroom experiences; studies conducted by mathematics education researchers; didactical and mathematical ideas for the classroom and more. For the past four years the journals have been published electronically. Thus, when possible, papers are enriched by electronic applets, short videos and other hyperlinks.

One of the latest issues featured an article by Ben-Bassat (2016), a high school mathematics teacher who teaches a class of mathematically advanced students. He suggested a "method for finding a common tangent line on two circles" that he developed following an unpredicted response from a student (see Fig. 3 for the considered cases).

The numbers of website users are huge, as shown in Table 1.

5. Conclusions

Thanks to their ties with the academic world, the national centers of mathematics teachers lead the way in promoting up-to-date educational trends and advanced technological tools and in introducing them to teachers, in developing innovative materials (such as the dynamic applications and videos

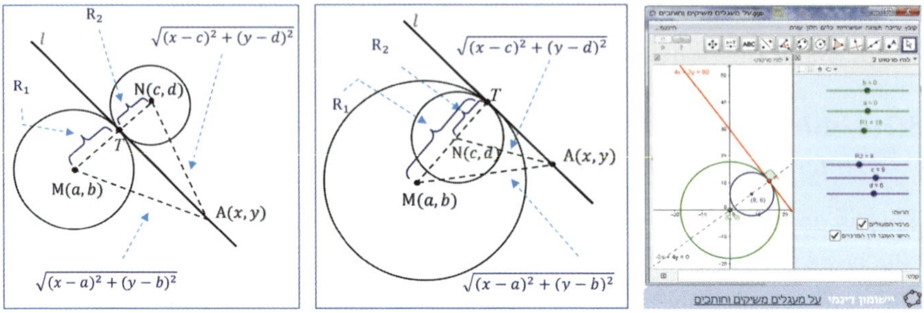

Fig. 3. Tangent line common to two tangent circles (from Ben-Bassat, 2016).

Table 1.

2015–2016 school-year	Secondary School Center	Primary School Center
Overall visits	526,470	865,640
Journals visits	40,120	47,091
Downloads of materials/applets	55,160	82,094

we have developed for the past five years), and in training leading teachers and teacher instructors in integrating the materials in the curriculum.

We believe that technology-supported, interactive, dynamic teaching promotes in-depth, connected learning and literacy, and that mathematics teachers' participation in virtual communities promotes this kind of teaching. As can be seen above, much work has already been done to develop technological tools and to establish the infrastructure for teachers' communities in different forms. It is also clear that there is a growing awareness of the importance of technology-supported teaching, particularly among district teacher instructors. We believe that these efforts should continue, and should be given further support.

Acknowledgements

The National Centers of Mathematics Teachers would like to thank academic expert staff from the Department of Mathematics Education at the University of Haifa — Dr. Dani Ben-Zvi, Dr. Irit Peled, Prof. Anna Sfard, Prof. Michal Yerushalmy, and Prof. Roza Leikin. We would also like to thank the Ministry of Education for granting our Mathematics Education Department the privilege of managing the national centers.

Reference

Ben-Bassat, A. (2016). Strategies for finding a tangent line which is common to two tangent or intersecting circles, *Aleh — The (Israeli) Secondary School Mathematics Journal*, **53**, 19–28.

About the Authors

Varda Talmon has been the head of the National Center for Secondary School Mathematics Teachers (SSTC) since 2008. She completed her PhD dissertation on Dynamic Behavior of Geometry Figures at the Department of Education at the University of Haifa. She has 25 years of experience in developing instructional activities for students and teachers of mathematics in secondary school, with a special emphasis on alternative, technology-rich teaching approaches. She teaches didactics of mathematics at the Department of Mathematical Education at the University of Haifa.

Geula Sever manages the website of the SSTC and works on the development of dynamic applets since the establishment of the center. She has 25 years of experience in teaching high school mathematics. She has an MA in Mathematics Education from the University of Haifa. She leads several key projects at the center and runs workshops for high school teachers and teacher instructors.

Gila Ron is the editor of the Aleh journal of the SSTC. She completed her PhD dissertation "Partially Correct Constructs (PaCCs) in Probability" at the University of Tel Aviv. She taught high school mathematics for over 30 years and now is a lecturer at Ohalo College. She headed the Mathematics division of the regional program for professional development for teachers and directed the development of the textbook "Learning and Teaching Analysis".

Roza Leikin is a professor of Mathematics Education and Education of the Gifted at the Faculty of Education, University of Haifa. Her research and practice embrace mathematical creativity and ability in teachers and students. She is the head of the Department of Mathematics Education (since October 2016) and director of the RANGE Center in the University (since 2008). She served as the head of the National Advisory Committee for Mathematics Education of the Israeli Ministry of Education (2013–2017) and the President of the International Group for Mathematical Creativity and Giftedness (2012–2017). She has published more than 150 research papers.

Varda, Geula and Roza manage the Club-5 project described in Chapter 29.

Chapter 31

Making Educational Excellence in Mathematics Accessible to Disadvantaged Children: The Case of Teach First Israel

Shlomit Amichai* and Saleit Ron

Teach First Israel, Lod, Israel
**AmichaiS@jdc.org*

The Teach First Israel science, technology, and mathematics track (TFI STEM) addresses the very large gaps in academic achievement between students from high and low socioeconomic backgrounds. Established in partnership with the Trump Foundation (which promotes excellence in science and mathematics teaching in Israel), the track recruits high-quality candidates with a degree in STEM subjects and a strong commitment to social engagement, gives them intensive teacher training, and places them in schools in Israel's socio-geographical periphery. These teachers receive long-term professional development and support, including individual mentoring provided by TFI STEM Leading Teachers. Interviews with teachers indicate high levels of social commitment, continued learning, and dedication to achieving excellence in their teaching and in their students' results.

1. Teach First Israel

The most recent OECD-PISA statistics (OECD, 2016) paint a disturbing picture of the Israeli education system. Overall, Israeli students perform worse than their OECD peers in reading, mathematics, and science, and gaps between the strongest and weakest in all three areas are among the largest in PISA-participating countries. These gaps are most pronounced between low-income and affluent youth, as also seen in the 2015 TIMSS reports (Israeli Ministry of Education, 2016). The community into which children are born is one of the main factors affecting their chances of reaching higher education, with students in Israel's socioeconomic and geographic periphery suffering from a lack of access to educational opportunities.

Teach First Israel (TFI, or HOTAM, Hebrew for *Impact*) believes that all children deserve the opportunity to obtain high-quality education that will enable them to choose their own futures. Provisioning skilled teachers with a strong social commitment and a belief in the ability of every child to make a real difference to educational outcomes; one large-scale study of school systems worldwide identified the three most significant factors for success as being: (1) getting the right people to become teachers; (2) developing them into effective instructors; and (3) ensuring access to the best possible instruction for every child (Barber and Mourshed, 2007).

TFI belongs to the Teach For All global network, a thriving international community of 43 partner organizations — including Teach For America (USA) and Teach First (UK) — that are committed to expanding educational opportunities in their countries. TFI was founded as a joint initiative between JDC-Israel, Israeli Ministry of Education, the Ha-Kol Hinuch Movement, and the Naomi Foundation. Shlomit Amichai is TFI's chairperson, and Asaf Banner is its CEO.

2. The TFI STEM Track

The disparities in academic success related to socioeconomic difference in Israel are very pronounced in STEM education. Schools in the socioeconomic and geographical periphery are less able to provide high-quality teachers and conducive learning environments, while also being negatively affected by lower expectations (among students, families, and staff) and a lack of visible pathways to success.

In response to these conspicuous disadvantages, TFI partnered with the Trump Foundation to create its STEM initiative in 2013. The TFI STEM track recruits and trains Israel's brightest university STEM graduates, places them in schools in the country's social and geographic periphery, and gives them long-term intensive training and support.

2.1. *Goals and theory of change*

TFI's first step in establishing the STEM track was to conduct a comprehensive study of the field of STEM teaching in Israel, of existing models for training and mentoring STEM teachers, and of the relevant activities of other organizations in the Teach For All network. A survey was conducted of TFI STEM teachers to establish their main professional needs and to collect baseline data describing the existing situation, and the findings were reviewed by a steering committee, comprising leading professionals and TFI teachers. The insights collected were used in formulating the track's theory of change, as follows:

The defined goal of the TFI STEM track is to advance students in Israel's social and geographic periphery by: (1) increasing students' belief in their ability to succeed in mathematics, sciences, and technology; (2) increasing students' motivation to excel and to choose higher-level STEM study tracks, based on an understanding of these subjects' importance; and (3) improving the achievements of students in these subjects, so that they will be able to choose any professional future they desire.

2.2. *Recruitment and placement*

Following an initial evaluation study into the retention rates of program candidates and participants, TFI has invested greatly in improving its recruitment, selection, and placement activities. Selection parameters include not only academic achievement, but also leadership capabilities, social commitment, and dedication to children's success. The painstaking selection process admits only 4% of all candidates. In order to attract and retain quality candidates, who have attractive employment alternatives, candidates have individual meetings with TFI recruitment coordinators, visit schools, observe mathematics and science lessons, and talk with their future colleagues. In the final stages of the recruitment process, TFI holds a STEM event for candidates that presents TFI's vision, social approach, and efforts to ensure that STEM teachers achieve TFI's goals. Following these changes, we have seen a marked improvement in candidate retention, and there are 170 TFI STEM teachers in schools throughout Israel's socioeconomic and geographic periphery as of 2017.

2.3. *Professional leadership development*

The professional training and support given to all TFI teachers in their first year involves a residential summer semester of intensive study, followed by studying one day a week throughout the school year, while teaching four days with support from in-school mentors and TFI tutors. In addition to this, TFI STEM teachers receive specialized courses in their subject areas, and supplementary individual support. In subsequent years, the long-term support framework includes professional, personal, and group development and mentoring (including from veteran TFI STEM Leading Teachers), with an emphasis on developing professional leadership.

Professional support for TFI STEM teachers has included in-service training programs run in cooperation with the Davidson Institute of Science Education and the Weizmann Institute of Science. In one component, provided by the Weizmann's VIDEO-LM project, teachers received instruction in methods for teaching mathematics to struggling students, and video was used to allow them to view, analyze, and discuss their own and each other's classroom teaching (Karsenty, Arcavi and Nurick, 2015).

In-service training programs are held every summer for TFI STEM Leading Teachers (who mentor new teachers), run by the Relay Graduate School of Education (NY) in partnership with the Teach For All network. The program includes components on tutoring skills and professional development, as well as components on best practices for STEM teaching.

2.4. *Impact*: *Findings from evaluation surveys and interviews*

At the end of the Relay training program described above, participating teachers in 2015 and 2016 were asked to assess its impact for them and for their future work in the classroom. In their responses, the teachers spoke about wanting to maintain a very high bar guided by their vision, in terms of both practice and discourse; to implement what they had learned, as both teachers and mentors; to engage in meta-learning and meta-thinking; and to share what they learned with their peer community of teachers. They reported gaining an appreciation for the power of shared, mutual learning, and gaining a fuller understanding of their influence as teachers in school and in society at large.

Additionally, periodic interviews are conducted with TFI STEM leading teachers, who mentor early-stage TFI STEM teachers, on the impact of the program for their teaching and mentoring. Brief summaries of the findings of interviews with four of these teachers, who participated in the

Relay training programs in 2015 and 2016, are presented below:

(i) Dr. Ornit Maimon (*Ph.D. in geophysics*; *mathematics and physics teacher*; *TFI cohort* 3; *participated in Relay in* 2015):

Ornit's 10th grade students have achieved an average score of 85–90 on internal school examinations. All her 11th grade mathematics students who took the basic level secondary school graduation exam passed, attaining an average score of 80. Following the success of two of these students, Ornit encouraged them to transfer to intermediate and advanced level tracks, where they have continued in 12th grade. In her basic-level physics class, the average grade is 90.

"The opportunity to mentor first year teachers is fascinating and instructive. I get so much from giving to them. I am reminded of the enjoyment and innocence that led me to choose to become a teacher. I want to improve as a person and a teacher; I'm constantly learning and changing. Being part of TFI gives me a supportive environment in which to learn, grow, and change."

(ii) Sarel Aiber (*B.Sc. in both computer science and psychology*; *mathematics teacher*; *TFI cohort 4*; *participated in Relay in* 2015):

In his first two years, Sarel taught mathematics in six groups: an 8th grade class of struggling students; a 9th grade basic-level class; an intermediate-level group of 10th and 11th graders; a 12th grade basic-level class; a class for struggling students in 10th and 11th grade; and a supportive class for 10th graders. In his class for struggling students in 10th and 11th grade, all students successfully passed the secondary school graduation examination. In his supportive class, the students received an average grade of 94, with four of them receiving 100. His intermediate-level class, whose initial average grade was 55, attained an average grade of 81 in the secondary school exams. He currently teaches the school's advanced-level track.

"The TFI STEM track enabled me to develop professionally quickly and in a focused manner. I'm only in my third year of teaching, and I've already accumulated experience as if I'd been teaching far longer, thanks to the support and training, and to the skills I've gained from the professional input I've had. And I also feel part of a community. I know I'm not alone. There's always someone I can turn to for help."

(iii) Yochai Peretz (*B.A. in accounting*; *studying for an M.Sc. in mathematics education*; *mathematics teacher*; *TFI cohort* 4; *participated in Relay in* 2016):

Yochai taught basic-level mathematics for grades 10–12; some of his classes are for students with particular difficulties. In his first year of teaching, his 11th graders, who were initially unable to solve even simple equations, achieved an average grade of 98.4 in the secondary school graduation examinations. In his second year, his 11th graders achieved an average grade of 97, and his 12th graders an average grade of 96.3. This year, inspired by his participation in the VIDEO-LM training program, Yochai is also running a "video club" for the mathematics teachers at his school to help improve teaching and results. He now teaches advanced-level classes, and is a member of an advanced-level mathematics teachers' community in his home town.

"When I began, the students in my class were at an extremely low academic level, with such poor mathematics skills that they couldn't even solve simple equations. Many of them were afraid of mathematics, especially the girls. The predominant phrase in the class was: 'It doesn't matter anyway; we'll never go to university.' I challenged them, I said 'Let's set a target together and see how we reach it. We can prove to those who have always said that you'll never achieve anything that they were wrong. Now's your time to show them all what you can do!'"

(iv) Adi Refaeli (*B.Sc. in mathematics*, *M.A. in educational counseling*; *mathematics teacher*; *TFI cohort* 5; *participated in Relay in* 2016):

Adi teaches a special education mathematics class. In the 10th grade, all her students achieved scores of 100 in the school's internal examinations. At the end of 11th grade, they took the second part of the basic-level secondary school graduation examination; all of them passed, some with a final grade of 100. Another of her classes took the intermediate-level exam at the end of last year, achieving an average grade of 91.2. Adi also leads a program she developed in her school, called "Magic Square," to help basic-level students progress to the intermediate-level track and improve their mathematics skills and knowledge.

"Mentoring is a role in which I get to learn from others and to think together how we can improve. It allows me to share the knowledge and experience I have acquired, while gaining feedback helping me to hone my own mathematics teaching. The role allows me to listen to their dilemmas, give advice, and share my opinions. It allows us to think together about the appropriate mathematical education for our students."

3. Insights and Challenges

Analysis of the research data as a whole reveals that TFI STEM teachers choose to engage in teaching as a means to achieving social change, and continue to work as classroom teachers for extended periods of time. Possible reasons for this include: a sophisticated selection process, which assesses suitability and commitment well; good professional support; and a broad professional horizon. Many of TFI's STEM teachers lead social and professional initiatives at their schools. They are also beginning to be active outside of their schools.

Most TFI teachers work with Welfare and Education Service (WES) classes and with students at a basic level of mathematics and science studies, in line with TFI's mission to promote instruction and learning excellence in these classes. Over the years, some of the teachers have also begun teaching higher levels and in excellence programs. This trend will be examined in more detail in the coming years. TFI teachers who have participated in professional support programs are generally satisfied with the programs offered to them. They report that the training sessions and mentoring are of high quality, and that they are able to apply what they learn in their classrooms.

The personal stories from leading TFI teachers illustrate TFI STEM teachers' many circles of influence, and the impact they have on their students, both in terms of the students' self-efficacy and their academic achievement. TFI's development team has examined and discussed the support given to TFI STEM teachers, and has used insights gained to develop parallel approaches in the humanities. Similarly, we have integrated techniques which were introduced through the STEM track — such as the use of video documentation and change-creating feedback — more widely throughout our training and support programs. The development model of the STEM Leading Teachers track has also inspired an in-service training program for all TFI Leading Teachers on the role of the "teacher-mentor."

References

Barber, M. and Mourshed, M. (2007). *How the World's Best-Performing Schools Systems Come Out on Top*, McKinsey & Company.

Israeli Ministry of Education (2016). *The 2015 TIMSS Report: An Israeli Perspective*, Ministry of Education.

Karsenty, R., Arcavi, A. and Nurick, Y. (2015). Video-based peer discussions as sources for knowledge growth of secondary teachers, in K. Krainer and N. Vondrová (eds.), *Proceedings of the 9th Congress of the European Society for Research in Mathematics Education*, Prague: ERME, pp. 2825–2832.

OECD (2016). *PISA 2015 Results (Volume I): Excellence and Equity in Education*, Paris: OECD Publishing.

About the Authors

Shlomit Amichai served two stints as the director-general of Israel's Ministry of Education, and was a divisional director of JDC-Israel. She is the chairperson of Teach First Israel, and was instrumental in its founding. Shlomit holds a bachelor's degree in Israeli history and political science from the Hebrew University of Jerusalem, and a master's degree in educational administration from Tel Aviv University.

Saleit Ron is the STEM coordinator at Teach First Israel. She holds a doctorate in science education from the Technion — Israel Institute of Technology, and a master's degree in genetics from the Hebrew University of Jerusalem. Saleit has over twenty years of experience in science education as a teacher, pre-service and in-service teacher trainer, and faculty member.

Chapter 32

Teacher Residency Programs for Secondary School Mathematics: Different Models, Similar Challenges

Talli Nachlieli[*,], Tili Wagner[#,##], Dalia Imanuel-Noy[#,‡‡], Ruthi Barkai[‡,§] and Atara Shriki[¶,||]**

[*]*Levinsky Academic College of Education, Tel Aviv, Israel*
[#]*Beit Berl Academic College of Education, Kfar Saba, Israel*
[‡]*Kibbutzim Academic College of Education, Tel Aviv, Israel*
[§]*Department of Mathematics, Science and Technology, School of Education, Tel Aviv University, Tel Aviv, Israel*
[¶]*Oranim Academic College of Education, Tivon, Israel*
[||]*Kesher-Cham Technion, Haifa, Israel*
[**]*tallin@levinsky.beit*
[##]*tiliw@beitberl*
[‡‡]*imanuelda@beitberl*

This chapter describes four secondary school mathematics teacher-preparation practice-based programs established in Israel at four colleges of education during 2013–2014 with the support of the Trump Foundation. The chapter focuses on the clinical aspects of the programs and ends with a discussion of the programs' strengths and the challenges faced by program developers and graduates.

1. Introduction

According to Darling-Hammond, "*[i]f the political will and educational conditions for strengthening teaching are substantially absent, I do not believe it is an overstatement to say we will see in our lifetimes the modern-day equivalent of the fall of Rome.... colleges of teacher education have a major responsibility for which path the nation travels*" (2010, p. 35).

This statement stems from and reflects a profound understanding of the importance of quality teaching for society.

In 2011, the Trump Foundation was established to help the Israeli education system curb the deterioration in secondary school students' achievements in mathematics and science and to bring about a significant improvement. Realizing that excellent teachers are what make a difference, the foundation chose to invest in teachers and teaching. This investment yielded various projects that focus on teacher education. One of those projects included the launching of several practice-based clinical teacher programs for secondary school mathematics and science teachers. In our capacity as the heads of the first four such programs, we were given the opportunity to plan and develop our dream programs.

In this chapter we describe four (out of a current total of eleven) teacher preparatory programs for secondary school mathematics and science teachers that are supported by the Trump Foundation: *Delta* at Levinsky Academic College of Education, *Excellence in Teaching* at Oranim Academic College of Education, *Mathematics Above and Beyond* at Beit Berl Academic College of Education, and the 700 *Program* at Kibbutzim Academic College of Education. All four programs aim at increasing the number of students that elect to study advanced-level mathematics by educating student teachers (STs) to become excellent teachers. The programs share several characteristics, including three primary ones: (1) they recruit top candidates; (2) they are designed as "clinical and practice-based" programs; and (3) they include a two-year supplementary induction and retention program.

In this chapter we focus on the programs' clinical aspects. Specifically, we address the following questions: What are the clinical components of the programs and how are these reflected and represented in the programs? What are the programs' strengths? What are the challenges in developing and maintaining these strengths?

2. Secondary School Mathematics in Israel

In Israel, primary school comprises grades 1–6, middle school comprises grades 7–9, and high school comprises grades 10–12. At the end of the 11th and 12th grades, students take secondary school exams in mathematics at either a basic, intermediate, or advanced level. In most high schools, each of the three tracks is taught separately. The main goal required of the preparation programs described in this chapter was to prepare secondary school teachers to help increase the number of students studying advanced-level high school mathematics.

3. Problem Description

As one of the world's leading countries in the medical, biotech and high-tech sectors, Israel has been labeled a "Start-Up Nation" (Ben-David, 2013; Senor and Singer, 2010). Yet the number of high school students who take the advanced-level secondary school exam in mathematics is declining. In 2006 almost 13,000 students (about 14% of 12th grade graduates) took the advanced-level exam. Since then, this number has decreased steadily to only 8869 (less than 11%) in 2012. This decline raised concerns among the country's leaders regarding its potentially negative impact on Israel's technological and economic future.

In 2011, the Trump Foundation (http://www.trump.org.il/?lang=en) began working in Israel with the stated goal of increasing the number of high school graduates who study advanced-level mathematics and science. Correspondingly, in 2013 the Israeli Ministry of Education set a high priority goal: to double this number for mathematics within five years (http://www.5p2.org.il/english/). Achieving this goal necessitated training a large number of mathematics teachers qualified to teach advanced-level mathematics. These efforts have already had an impact: While the number of high school graduates who studied advanced-level mathematics decreased to around 9000 in 2013, since then it has been steadily increasing, reaching 12,800 in 2016 (http://www.trump.org.il/scoreboard/?lang=en).

4. Clinical Aspects of Teacher Preparation Programs

The Trump Foundation's main requirement of its sponsored programs was that they would be practice-based clinical programs. In this chapter we refer to the programs' clinical aspects. Specifically, we consider the following three clinical dimensions of teaching programs (Grossman *et al.*, 2009): (a) Representations of practice: records of classroom instruction such as videos, student work, and lesson plans (e.g., Lampert and Ball, 1998). Discussions and encounters around such representations may be central in helping students learn to notice and learn from practice as well as in linking theory and research to enhance the alignment between academic course work and hands-on teaching. (b) Decomposition of practice: components of the complex profession of teaching. The mere identification of such components (such as focusing on aspects of lesson planning) ensures that STs' field experiences enhance their awareness of the constituent aspects of teaching. (c) Approximation of practice: practices that approximate teaching, such as simulations of teaching and teaching in a classroom with a collaborating teacher. Such approximations can provide STs with opportunities to experiment with new skills and approaches and enable them to receive more support and feedback than actual practice in the field allows. Approximations may require more elaborated versions of practice than what novices will enact later in their teaching and may also provide opportunities to experience instructive failure.

5. The Stories of Our Programs

5.1. *Israeli background*

According to the Council for Higher Education (CHE), teacher-preparation programs at colleges of education for academic prospective middle school mathematics teachers (bachelor's degree at least) should include 26 one-semester courses in mathematics, 12 courses in education (e.g., sociology of education), four courses in mathematics pedagogy, and one school-day a week. Students complete the program in one or two years, depending on the number of credits allocated for their previous studies. Our programs were planned based on these requirements.

5.2. *Four practice-based clinical programs*: *A beginning*

In the summer of 2013 the Trump Foundation asked the Levinsky Academic College of Education and Oranim Academic College of Education to propose a unique training program for excellent candidates holding academic degrees in mathematics or a mathematics-intensive discipline who wished to change their career and become secondary school mathematics teachers. The requirement was to plan a practice-based program that would be highly prestigious and involve rigorous recruiting and screening. Another requirement was to design a program that could be maintained by the college after the foundation's three years of financial and professional support came to an end. This was an opportunity to rethink the attributes of an optimal teacher training program, to closely examine the components of existing programs and to consider how to improve, modify, and extend these components. We sought to plan practice-based clinical programs that would be consistent with this objective and would consider the inherent limitations of the current preparation program.

In September 2013 the first two programs were instituted: the *Delta* program at Levinsky Academic College of Education, led by Talli Nachlieli (mathematics) and Yael Kashtan (biology), and the *Excellence in Teaching* program at Oranim Academic College of Education, led by Atara Shriki (mathematics). In August 2014 two additional programs were added: the *700 Program* at the Kibbutzim Academic College of Education, led by Ruthi Barkai (mathematics) and Ilana Avisar (physics), and *Mathematics Above and Beyond* at Beit Berl College, led by Tili Wagner.

The directors of these programs met regularly to discuss program standards and principles. In September 2014 the program directors traveled to the US to meet with the professional staff of various residency programs (The Woodrow Wilson National Fellowship Foundation, TNTP Teaching Fellows, Boston Residency Program, New Visions for Public Schools, UTR United) and learn about different processes and aspects of practice-based residency programs.

5.3. *Clinical aspects of the programs*

The central part of the preparation programs is the day participants spend in the schools. The STs in our programs spend one day a week at school with a pedagogical instructor from their college, who is responsible for generating learning opportunities for the STs. Groups of two to four students are assigned to a cooperating teacher from the school and spend the entire day with this teacher. The students observe lessons, work with individual students and groups in class, and teach the whole class. Sometimes the entire group (up to 10 STs) observes a lesson together taught by a cooperating teacher or by one of the students and then discusses their observations with the instructor. In each program, at least one of the pedagogical courses is taught at the school, either by the instructor or by the cooperating teacher. This allows the cooperating teachers to participate and strengthens the connections between practice and theory. During those lessons the group usually analyzes specific cases that occurred during the school day. Besides the school experience, the students also take 12 courses in education and pedagogy. These courses are designed to be closely related to the STs' school experience, either by referring to parts of videotaped lessons from the school day or by connecting the school experience to theories studied at the college.

We now consider the clinical components of our programs according to three categories: representations of practice, decomposition of practice and approximations of practice (Imanuel-Noy and Wagner, 2016).

Representations of practice: We focus on three main features: (1) enhancing STs' mathematics discourse; (2) linking theory and practice; (3) analyzing STs' videotaped lessons.

Mathematics discourse is a central component of all the programs. To help enhance their discourse, the STs take advanced mathematics courses either during the summer prior to their school practice (the *Mathematics Above and Beyond* and the *Excellence in Teaching* programs) or during the school year. They also take pedagogical courses that focus on different aspects of mathematics discourse (as in the *Delta* and 700 programs). Mathematics discourse is a major component that pedagogical instructors focus on when analyzing the lessons that the STs observed or taught during their day at school.

Theory and practice are linked through specially designed courses in education and pedagogy and by relating practice experiences to theoretical aspects. One such example is the modular field-based themes that have replaced the traditional introductory course structure in *Mathematics Above and Beyond*. The themes emerge from the practical world of teachers. For instance, the *psychology* module focuses on issues such as development of mathematical thinking, mathematics anxiety and enhancement of self-efficacy.

In some of the programs (*Delta* and 700 *Program*), students videotape their lessons and choose a 5–10-minute segment to watch and analyze with the other students in the group. As Grossman *et al.* (2009) suggest, "in videos of practitioners at work, the interactive features of the practice may be visible, but the professional reasoning underlying the practitioners' actions may be invisible" (p. 2066). For this purpose, lessons that students observe are analyzed and reflected upon in a group led by the pedagogical instructor.

Decomposition of practice: The complexity of the work of a teacher is manifested in intensive school practice and in different courses. These experiences provide STs with opportunities to learn about the complexity of their role, their decision-making processes and the establishment of their professional identity. In pedagogy and education courses, students learn about specific teaching practices. For example, in the *mathematics discourse* course in the *Delta* program, STs learn to lead productive mathematics discussions. In *Teaching for Excellence*, STs have access to *RAMZOR*, an online environment that includes a large repository of lesson plans written by teachers (see Chapter 34). The STs discuss different lesson plans and add their own lesson plans to that environment. By focusing on planning and on written plans, STs have the opportunity to learn about the importance of extensive planning, of different lesson components and of teachers' decision-making. In the didactic seminar in the *Mathematics Above and Beyond* program, STs discuss the various roles of teachers, including class manager and organizer, diagnostician, evaluator and provider of feedback, and developer of a positive social climate. During this workshop, the STs discuss dilemmas and problems from the complex work of the teacher.

While observing lessons, students are often asked to focus on specific components of practice, such as students' or teachers' questions, teachers' reactions to students' questions or difficulties, and teachers' mediation while students work alone or in groups. Each of these components is then discussed in a post-lesson reflective discussion and during a course on analyzing practice that takes place at the school. In this course, they study theory in the context of students' observations and teaching.

Specific components are also referred to in other education courses. In the 700 *Program*, for example, students are asked to cite cases from their school experience when referring to ideas and theories of teaching and learning. Moreover, the instructors use the cases brought up by the students to refer to different theoretical ideas. That is, these courses both shape and reflect the STs' practice.

Approximation of practice: Our STs encounter various approximations of practice, including classroom teaching (in the presence of a teacher-mentor and a pedagogical instructor); designing apps and parts of lessons in different courses (e.g., using technology in teaching); analyzing classroom situations (e.g., in classroom management courses); using short clips of STs' videotaped lessons to closely analyze teaching practices, learning processes, and teacher decision-making.

In *Mathematics Above and Beyond*, approximations of practice are also reflected in group assignments that the STs carry out using the Project-Based

Learning (PBL) method. As the common goal of the program is to encourage students to choose to study advanced-level mathematics, the STs are divided into investigation groups. Each group researches one issue emerging from the practical context that is expressed in the theoretical context. Examples of the issues considered include cooperative learning, flipped-classroom, use of software such as Geogebra, filming of clips of foundation classes, and learning from error analysis. At a later stage, the products are processed and analyzed together with the STs.

A different type of approximation of teaching is found in the *Excellence in Teaching* program. The use of a repository of lesson plans and the focus on the planning part of the lesson are more elaborate than usual. This approximation of the practice of teaching allows STs to observe the complexity as well as the affordances of intense planning.

6. Looking Ahead: Strengths and Challenges

The STs studying in these programs have been carefully selected. The main strength of the programs lies in creating and nurturing learning environments at the school and at the college in which the groups think and work together. The STs form a community of learners. After completing the program, the participants continue to participate in intern workshops during their first year of teaching and to meet with former graduates for professional development days and social evenings. In addition, a tailor-made support program is available for any graduate who seeks such support. We believe that such an environment is important to enhance the STs' learning during the preparation stages as well as later when they become teachers. Even after completing the program, they still have a safe group to think and connect with. The clinical aspects of the program contribute to the development of a learning community that includes the academic staff, the cooperative teachers and the STs. This community is the driving force of the program. Its partners are devoted to the program goals, share a common discourse and support one another.

Nevertheless, the creation of such a community is not consistent with the current situation in teacher education, where students study in large groups and pedagogical instructors visit a school only once every two or three weeks. Due to the constraints of large classes, providing courses at the schools is problematic if not impossible. Moreover, connecting education and pedagogy courses to STs' practice is more complicated.

Another challenge is related to the need to cope with discrepancies between the organizational culture of the college and that of the school and to establish a climate of acceptance among the school staff.

One final challenge concerns the entry of our graduates into the school system and our collaboration with the Ministry of Education. The main dispute concerns human resources, and specifically the number of years of seniority the Ministry of Education is willing to recognize from our graduates' previous jobs and the number of years of teaching experience required before they are assigned to teach advanced-level mathematics in schools, which is what they have been trained to do.

References

Ben-David, D. (2013). Labor Productivity in Israel, *State of the Nation Report*: *Society, Economy and Policy in Israel*, 95–115.

Darling-Hammond, L. (2010). Teacher education and the American future, *Journal of Teacher Education*, **61**(1–2), 35–47.

Grossman, P., Compton, C., Igra, D., Ronfeldt, M., Shahan, E. and Williamson, P. (2009). Teaching practice: A cross-professional perspective, *Teachers College Record*, **111**(9), 2055–2100.

Imanuel-Noy, D. and Wagner, T. (2016). Unpacking the clinical and participatory dimensions of the trump math-teacher-residency-program, *Australian Journal of Teacher Education*, **41**(7), 88–109.

Lampert, M. and Ball, D. L. (1998). *Teaching, Multimedia, and Mathematics*: *Investigations of Real Practice. The Practitioner Inquiry Series*, New York: Teachers College Press.

Senor, D. and Singer, S. (2010). What next for the start-up nation? *The Wilson Quarterly* (1976–), **34**(3), 62–66.

About the Authors

Talli Nachlieli is a Mathematics Educator and Researcher at Levinsky Academic College of Education. Talli is the head of the *Delta* teacher preparation residency program. Her research interests include teacher preparation and education, and mathematics classroom discourse.

Tili Wagner is an assistant professor and head of the Mathematics and Physics Residency teacher education programs at Beit Berl Academic College of Education. Her major fields of research and expertise are science education (development of scientific literacy) and teacher education (unique teacher education models).

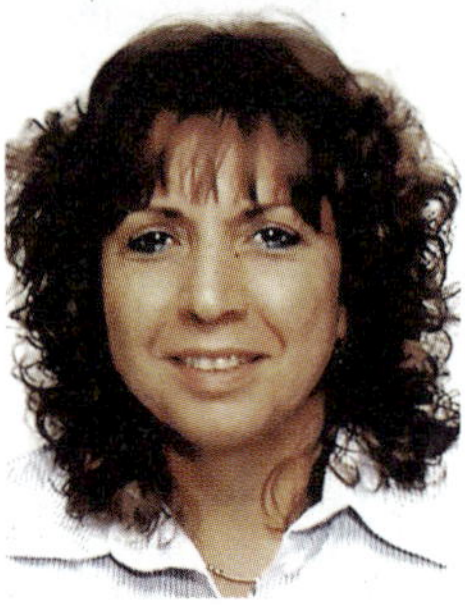

Dalia Imanuel-Noy is a senior lecturer at Beit Berl Academic College of Education, head of the Curriculum & Assessment of MEd program and director of the Academy-Classroom Partnership Program of the Israeli Ministry of Education. Her professional expertise includes curriculum development, assessment and evaluation, educational reforms, teacher education in PDS programs, and induction of beginning teachers.

Ruthi Barkai is a mathematics educator and researcher at Kibbutzim Academic College of Education. She is the head of the teacher preparation residency program. Ruthi is also a researcher and teacher educator in the Department of Mathematics, Science and Technology Education at Tel Aviv University.

Atara Shriki is an associate professor who teaches mathematics and mathematical didactics at Oranim Academic College of Education. She also works as a researcher at the Samuel Neamann Institute for National Policy Research and at Kesher-Cham — R&D center for mathematics education at Technion — Israel Institute of Technology.

Chapter 33

Enhancing Mathematics Teachers' Knowledge and Reflection Through Peer-Discussions of Videotaped Lessons: A Pioneer Program in Israel

Abraham Arcavi* **and Ronnie Karsenty**#

Department of Science Teaching,
Weizmann Institute of Science,
Rehovot, Israel
**abraham.arcavi@weizmann.ac.il*
#ronnie.karsenty@weizmann.ac.il

VIDEO-LM is a professional development project for secondary school mathematics teachers, designed for enhancing mathematical knowledge and supporting reflection on practice. Using six "viewing lenses", teachers watch authentic mathematics lessons of unknown colleagues and collectively reflect on the teaching practices observed. We describe the rationale, tools, and implementation of the project.

1. Background

Since the early 1960s, in-service teacher education has been a central concern in Israel. However, its aims, formats, and perspectives have evolved and changed over the years. In the 1970s, when the Israeli Center for Science & Technology Education (MALAM) was established as a joint effort of several universities and the Israeli Ministry of Education, in-service teacher programs became widespread as an integral part of teachers' work. These programs consisted mostly of short-term courses devoted both to enhancing teachers' subject matter knowledge and to supporting the implementation of new curricula. In the 1990s, with the launch of the comprehensive

"Tomorrow 98" reform in mathematics and science education, a strong emphasis was placed on the professional development of teachers. The modular and episodic teacher courses, common in the 1980s, turned into courses with comprehensive goals: not only to support teachers in adopting new curricula into their classrooms but also to assist them in becoming professional, well-informed and active practitioners. Teachers were introduced to research-based results about students' learning difficulties and were encouraged to reflect on their practice and on their decision-making. Thus, teachers began to realize the power of teacher professional communities as a source for advancing and improving their profession.

Technological advancements played a significant role in broadening the goals of professional development. Video recording of classrooms, already in use in the 1970s, became widely accessible since the 1990s. In this chapter, we briefly review the uses of video for teacher professional development. We then describe the conception, development, implementation, and evaluation of an ambitious large-scale pioneer program, named VIDEO-LM (Viewing, Investigating, and Discussing Environments of Learning Mathematics, or ADASHA in Hebrew), designed for using videotaped authentic mathematics lessons as a means to advance teachers' mathematical knowledge for teaching (Ball, Thames and Phelps, 2008) and their reflective skills.

2. Video-Based Professional Development

Videotapes are a powerful record of classroom teaching, allowing for:

(a) access to more classroom events (especially fleeting moments) than those which can be captured by an on-site observation;
(b) correcting or complementing "faulty memories" of events which do not reliably reflect what happened;
(c) repeated replaying of the same event with different foci of attention;
(d) discussions about intriguing occurrences.

Nowadays, the low cost of portable easy-to-use digitized video recording devices, coupled with accessible means of editing and exchanging clips, increase the attractiveness of this technology. Two of the objectives described in a recent comprehensive review of video uses in teacher education (Gaudin and Chaliès, 2015) are: (1) learning how to interpret and reflect, and (2) building knowledge on "what to do". Additionally, this review inspects the nature of different teachers' activities as they watch a classroom video and the kinds of videos used (e.g., one's own teaching, a lesson by a peer, or a lesson by an unknown teacher). It reports on the effects of video watching on teachers' motivation, cognition and classroom practice, concluding that video is a unique and powerful tool.

3. VIDEO-LM: Principles and Tools

Among the many possible uses of video pointed out in the literature (e.g., Sherin, 2004; Gaudin and Chaliès, 2015), we decided to focus VIDEO-LM on strengthening teachers' mathematical knowledge for teaching and supporting reflection on the practice of teaching. Following Schoenfeld (2010), we assumed that teacher actions and decisions are rooted in (1) their knowledge resources; (2) the global or local goals they are trying (explicitly or implicitly) to achieve; and (3) their orientations and perspectives on mathematics, its teaching, and its learning. Since this triad (resources, goals, and orientations) is assumed to shape how teachers teach (and therefore influence what students learn), professional development may profit by explicitly supporting teachers' awareness to, and reflection on, precisely their resources, goals, and orientations in and around practice. Moreover, the teaching of mathematics requires a specialized kind of knowledge which includes familiarity with students' ways of reasoning; fluency in providing appropriate examples; connections to previous and future mathematical ideas; and alternative ways of accessing new ideas and concepts (Ball, Thames and Phelps, 2008). Therefore, professional development programs can benefit from explicitly addressing the development and revision of this knowledge.

Following various definitions of reflection (e.g., Dewey, 1910; Schön, 1983; Clarke, 2000; Ricks, 2011), and in order to refine our goals, we characterized teacher reflection on their practice as the detailed, analytical and careful re-inspection of "what was done" in order to attempt to explicate intentions, plans, actions, and utterances, and to consider alternative decisions and their possible implementation.

Watching videotapes which capture and record the complex reality of classroom teaching may lead to conversations about aspects of that complexity. Thus, we aimed for a video-based tool that would (a) direct the observation of a lesson; (b) support teachers' reflection on their own resources, goals, and orientations; (c) bring to the fore issues related to mathematical knowledge for teaching; and (d) provide a common language to refer, organize, and make sense of what was seen. We decided to base the design of such a tool on the following principles:

- *The teacher is at the center*. Teachers' reflection should be centered on teacher actions, decisions, utterances, goals, and beliefs, and include the enactment of student-centered teaching approaches.
- *Best practices rather than best practice*. Since there exist *many* possible ways of "good" teaching, lessons selected to be observed and discussed are not a model to imitate, but rather an opportunity to launch relevant discussions on aspects of the teaching practice.
- *Shunning evaluation*. We consider evaluative/judgmental comments on a lesson as not conducive to reflection. Thus evaluative talk needs to be redirected into "issues to think of" emerging from the practice.
- *Whole lessons*. In many cases, the context of a whole lesson enables following the development of a topic from opening to closure, depicting a "story" rather than an isolated event.
- *Unknown teachers*. Discussions in which the filmed teacher is not present enable a free discussion, in which hypothetical interpretations of what is watched can be raised, expanding the reflective space.
- *Mathematics at the core*. Discussions should concentrate on the characteristics of teaching mathematics rather than on generic issues (e.g., general classroom management, body language).

The tool we designed consists of six "viewing lenses" that focus the observations and the discussions, based on the above principles. In the following, we describe the lenses and instances of their use.

Mathematical and meta-mathematical ideas. Under this lens, reflection refers to making explicit the mathematics around a certain lesson. The teaching of a mathematical topic may imply a choice of which central ideas to stress, as well as a choice of meta-mathematical ideas (e.g., providing a counter example to disprove an assertion). In many of our VIDEO-LM workshops, we often witnessed that eliciting these ideas before watching a lesson was a new exercise to many participants. Moreover, after observing the lesson, teachers were sometimes surprised to see that the filmed teacher selected ideas different than those they have mentioned (Nurick, 2015; Karsenty and Arcavi, 2017). We also witnessed examples of discussion and subsequent refinement of mathematical definitions (e.g., reconsidering the definition of inflection points, after watching an 11th-grade calculus lesson).

Goals. When designing and implementing a lesson, as well as when taking a "local" action, teachers have goals, or in other words, they have implicit or explicit reasons for what they choose to do. Seldom, teachers have opportunities to discuss these reasons. In using this lens, teachers are asked the following: "Assuming the filmed teachers acted in the best interest of their students, why did they do what they did?" Since the lesson is taught by an unknown teacher, there is no access to direct questioning, and the exercise becomes one of hypothetical attribution. This reflective exercise usually yields surprises: even a single action by a teacher during a lesson can have various possible goals (Karsenty and Arcavi, 2017). The repeated use of this lens is likely to become an internalized tool; some teachers report they have systematically started to ask themselves about the reasons for many of their own actions in class.

Tasks. Usually learning goals are accomplished through engaging students in tasks and problems. This lens allows for reflection on how the task (as analyzed and solved by professional development participants before watching the lesson) unfolds in a lesson, sometimes undertaking different goals than those planned. Tasks can also elicit discussions about potential student difficulties, and sometimes even motivate teachers to try tasks they never saw before. Also, as a consequence of watching a certain lesson, teachers dare to try tasks they previously considered too challenging for their own class, and reflect on what an appropriate challenge for one's own students is (Karsenty and Arcavi, 2017).

Interactions. Using this lens implies noticing and reflecting on: the questions teachers ask during a lesson and when they ask them; the questions students ask, as well as students' solution suggestions, and how these are taken or ignored by the teacher (and why); who receives permission to talk; who is leading the construction of knowledge; which questions are pursued within a class discussion, and the like.

Dilemmas. This lens exercises the spotting of a teacher dilemma, its articulation and the discussion with peers about alternatives to resolve it. A dilemma is posed by situations in which there are many plausible and sensible alternatives to go about, and the reflection consists of weighing carefully the advantages and disadvantages of each. Using this lens provides teachers with reflective capacities to cope with future dilemmas, thoughtfully resolve it, and explain decisions.

Beliefs. The questions leading to reflection with this lens are posed as follows: "On the basis of the observations, what may be the teacher's vision of the nature of mathematics? How does the teacher perceive his/her role in the lesson? How does the teacher see the role of the students in the lesson? What is the teacher's perspective on teaching, and on learning?" As with goals and dilemmas, this lens also invites participants to infer and to attribute beliefs to the filmed teacher, basing such attributions on "evidence" from the lesson. This showed to be the most intricate lens, and it takes most teachers quite some time to get used to it and to feel comfortable discussing it. However, quotes from participants such as "and what about me, what do I believe in?" indicate that "beliefs" can also become a powerful introspective tool.

4. Implementation and Upscale

A course in the VIDEO-LM program consists of a series of sessions of collectively viewing and discussing videotaped lessons. There are several "watching modes": sometimes guided conversations with teachers may precede the screening of a lesson, other times a session may open with watching the lesson. Facilitators may decide when and how to introduce the viewing lenses. The same lesson can be watched first with no directing questions, and then a second time, having subgroups use just one lens each, and then the whole group reconvenes to present, compare, and contrast their observations, leading to reflections on aspects that went unnoticed in the first watching. Facilitators may decide which mode to adopt and adapt or alternatively to propose their own, depending on the character of the lesson to watch, and on the target audience.

To date, about 800 teachers from all over Israel have participated in 30-hour courses, either in regional teacher centers ("Psagot") or within their school departments. The courses are conducted by about 25 facilitators who, after participating themselves in a VIDEO-LM course, took a second course specially designed to prepare them as facilitators.

A website (adasha.weizmann.ac.il in Hebrew) with over 60 videotaped lessons and their corresponding watching guidelines and complementary materials is accessible to teachers and teacher educators. Course facilitators can find suggestions for conducting sessions with teachers as well as related mathematical materials. As of the end of 2017, the website had over 2000 registered users, and the numbers are rising.

The ongoing analyses we are conducting on the ways teachers interact and reflect during these courses indicate that teachers enrich their mathematical knowledge for teaching and adopt the lens terminology for describing and communicating with peers about their practice. Some teachers indicated that the lenses guide them in making changes in their practices (Nurick, 2015; Schwarts, 2016; Karsenty and Arcavi, 2017).

5. Concluding Remarks

In line with the main conclusions in the literature (e.g., Gaudin and Chaliès, 2015), our experience with VIDEO-LM confirms that this powerful technology can exert a positive and long-lasting influence on teacher reflection and learning. However, as repeatedly shown in both research and practice, viewing videotaped lessons per se does not ensure teacher learning; a clearly delineated objective should be established and a robust toolkit should be developed (in our case the six lenses), at the service of pursuing that objective. Moreover, a skillful facilitation is essential: facilitators should be capable of listening and containing teachers' reactions, and of subtlety re-directing unproductive discussions.

The isolation of the professional teacher, with all its negative consequences, dissipates when a video provides a communal space in which colleagues share concerns, doubts, thoughts and proposals, and produce a fresh spectrum of views about classroom complexities.

A long-term goal of VIDEO-LM is to have as many teachers as possible using video as a handy tool that produces conversational pieces with which to enrich professional exchanges, using a language that supports communication and reflection. We launched a seemingly successful first stage, yet the challenge of creating communities which adopt these tools in their daily practices lies ahead. This is not an impossible goal, as we learn from strong existence proofs provided in some other countries (e.g., Fernandez and Yoshida, 2004).

References

Ball, D. L., Thames, M. H. and Phelps, G. (2008). Content knowledge for teaching: What makes it special? *Journal of Teacher Education*, **59**(5), 389–407.

Clarke, D. (2000). Time to reflect, *Journal of Mathematics Teacher Education*, **3**(3), 201–203.

Dewey, J. (1910). *How We Think*, Lexington: D.C. Health & Company.

Fernandez, C. and Yoshida, M. (2004). *Lesson Study: A Japanese Approach to Improving Mathematics Teaching and Learning*, Mahwah, NJ: Erlbaum.

Gaudin, C. and Chaliès, S. (2015). Video viewing in teacher education and professional development: A literature review, *Educational Research Review*, **16**, 41–67.

Karsenty, R. and Arcavi, A. (2017). Mathematics, lenses and videotapes: A framework and a language for developing reflective practices of teaching, *Journal of Mathematics Teacher Education*, **20**(5), 433–455.

Nurick, Y. (2015). *The Crystallization of Mathematical Knowledge for Teaching of High School Teachers in Video-Based Peer Discussions*, Unpublished Master's thesis, Weizmann Institute of Science (in Hebrew).

Ricks, T. E. (2011). Process reflection during Japanese lesson study experiences by prospective secondary mathematics teachers, *Journal of Mathematics Teacher Education*, **14**(4), 251–267.

Schoenfeld, A. H. (2010). *How We Think: A Theory of Goal-Oriented Decision Making and its Educational Applications*, New York: Routledge.

Schön, D. A. (1983). *The Reflective Practitioner: How Professionals Think in Action*, New York: Basic Books.

Schwarts, G. (2016). *Characterizing Video-Based Peer Discussions of Israeli Mathematics Teachers Watching a Japanese Lesson*, Unpublished Master's thesis, Weizmann Institute of Science (in Hebrew).

Sherin, M. G. (2004). New perspectives on the role of video in teacher education, in J. Brophy (ed.), *Using Video in Teacher Education: Advances in Research on Teaching*, Vol. 10, Oxford: Elsevier Press, pp. 1–27.

About the Authors

Abraham Arcavi and **Ronnie Karsenty** are faculty members of the Department of Science Teaching at the Weizmann Institute of Science. Abraham's main interests are research and development in the areas of student and teacher learning. Ronnie's main interests are assessment and the professional development of mathematics teachers and mentors.

Chapter 34

Mathematics Teachers Documenting, Sharing, and Improving Their Work Using a Newly Developed Software

Nitsa Movshovitz-Hadar

Kesher-Cham, R&D Center for Reviving Mathematics Education
Technion — Israel Institute of Technology, Haifa, Israel
nitsa@technion.ac.il

The increasing use of Wikipedia, social networks like Facebook, and databases with search engines such as Google, inspired us to consider a suitable computer-based platform for preserving, accumulating, and continuously improving mathematics teachers' practice, through collaborative efforts, aiming at empowering the mathematics teachers' community of practice. This chapter describes a continuing effort invested in the software development and in its assimilation.

1. Introduction: A Neglected Key-Factor in the Teaching Profession

It is well known that teachers need to possess a wide range of skills and various types of knowledge (Shriki and Lavy, 2012). But above all, they must be able to integrate these skills and knowledge and translate them into practice. In fact, designing lesson plans — (LPs) and planning longer periods of teaching, as well as evaluation tools, are at the heart of the teaching profession. However, in many cases teachers prepare their LPs "in mind", and the preparation in writing of a detailed LP is considered to be an "unnecessary burden" required only in preservice education. In addition, most of the teachers operate in isolation, each producing his/her own plans and evaluation tools, seldom sharing them with colleagues. This means that teachers constantly "re-invent the wheel", as all teachers follow the mandatory school mathematics curricula. Moreover, they do not get feedback that could improve these products. As a result, at the individual level, drawing conclusions towards the next time the same topic is taught is limited to memorized facts, and at the community level there is a lack of sharing practical knowledge with colleagues (Movshovitz-Hadar, Shriki and Zohar, 2014).

2. A Few Prohibiting Conditions

The common high school mathematics teaching workload requires a typical mathematics teacher in Israel to prepare weekly for about 20 lessons of 45 minutes each, which amount to about 600 lessons a year. Not any two of them necessarily focused on different topics, but even if the teacher re-teaches the same topic in two classes, it is usually not a mere repetition as adapting previous implementation plan to the state of the current class must take place. Moreover, being a professional mathematics teacher implies the competence to prepare about 450 *different* lessons for middle school grades (3 years $\times$ 30 weeks $\times$ 4 or 5 periods a week), plus about 1050 *different* ones for the various strands and levels at the 3 years of high school. Preservice education can barely provide for this demand. Clearly, teachers' continuous professional development from the moment they graduate is central to enhancing students' learning. As indicated in the OECD (2011) report: "Learning that improves individual competencies, and collaboration among teachers to produce better instruction in the classroom, must go hand-in-hand...".

3. Our Initiative and Steps Taken Towards its Realization

The rapid development of the internet and particularly the increasing use of Wikipedia, social networks like Facebook and databases with search engines such as Google inspired us to consider a suitable web-platform for preserving, accumulating, and continuously improving mathematics teachers' practice, through collaborative efforts at lesson planning, planning longer periods of teaching and designing items for evaluation.

We carried out three related preliminary studies (Segal, Shriki and Movshovitz-Hadar, 2016), each aimed at examining some deliberation about shaping up our initiative. These studies involved the development of a tailor-made computer software we named RAMZOR (Which in Hebrew means: Stoplight), as metaphorically, this software is supposed to allow ordinary mathematics teacher to:

- STOP to explore and consider other colleagues' ideas;
- PREPARE for teaching a particular class under consideration;
- GO in the classroom, implement and possibly reflect upon the implementation of the plan, or share it with others.

Following these preliminary studies, during 2014–2017 a team of four experts conducted a 3-year intervention (supported by Trump Foundation grant no. 120) in 20 high schools in the Northern Region of Israel based upon three central pillars:

(i) experienced teachers' mentoring of novice teachers;
(ii) assimilation of RAMZOR software;
(iii) professional development workshops for mentors and mentees. (Movshovitz-Hadar *et al.*, 2017).

As of September 2017 we have been in the midst of yet another intervention and implementation program — aimed at supporting the evolution of collaborative team work in 13 urban high schools in the city of Rishon LeTsiyon. Assimilating RAMZOR software is one of the pillars of this program too.

It should be noted that control over the contents which teachers upload is kept to a minimum. Only grammatical errors and mathematics errors are corrected by the project team. The rest is left to interactive dialogue between the authors and the community, very much like the way Wiki models operate.

Our long term goal is to establish from within the mathematics teachers' community of practice an interactive repository of lesson plans, longer teaching units designs, and evaluation tools which suit the mandatory curricula for all levels. The repository will be accessible to all members of this community for employing it, improving its parts and updating them according to current changes in the mandatory curriculum. The ultimate goal of Project RAMZOR is to upgrade the daily teaching of mathematics towards improving students' achievements and ability, in conceptual understanding, problem-solving, higher-order mathematical thinking and application of mathematics to other areas by creating a collaborative educational site for the Israeli Mathematics Education System.

4. Features of RAMZOR Platform

RAMZOR from "Omnisol Information Systems LTD", is a "structured wiki" platform — a wiki which manages structured objects. Its "wiki" capabilities include "using MS Word" for text editing, discussion page, version control, personal user pages, alerts and more, while its "structured" capabilities include advanced search capabilities, monitoring and permissions management. RAMZOR allows collaborative construction of teaching and learning materials of various types; it allows fast and convenient detection of desirable learning and teaching materials; and it also allows and encourages professional interaction among teachers.

Needless to say there is no one single way to teach any subject and the more alternatives the better. We now possess a beta version of RAMZOR platform with 1000 lesson plans. Some parts of it are accessible through RAMZOR website (http://ramzor.sni.technion.ac.il).

5. Future Steps

Ramzor software was originated to serve the needs of high school mathematics teachers, however in some future point it will be extended to elementary school mathematics, and possibly serve not only teachers in service, but also prospective teachers who study at teachers colleges.

In the last year we were approcahed by Computer Science teachers who teach mathematics as well, and asked to adapt the system to the teaching of Computer Science. We also had a few biology and chemistry teachers who asked about a similar possibility. Yet another future direction is making the interactive features of RAMZOR software available on the website so that the software itself will remain "behind the scene".

Acknowledgments

Project RAMZOR is conducted at Samuel Neamann Institute for Policy Studies in collaboration with Kesher-Cham, at Technion — Israel Institute of Technology. Atara Shriki, Varda Zigerson, Ruti Segal, Naomi Buchnik, Tamara Avissar, and Orna Imber take part in it. Ramzor Team appreciates the support of Trump Foundation Grant no. 120.

References

Movshovitz-Hadar, N., Shriki, A. and Zohar, O. (2014). Collaborative structuring of the pedagogical content knowledge accumulating within mathematics teachers' community of practice, Paper presented (in Hebrew) at a symposium, in *The second Jerusalem Conference on Research in Mathematics Education* (JCRME2), Jerusalem College of Technology, Israel.

Movshovitz-Hadar, N., Shriki, A., Segal, R. and Zigerson, V. (2017). *Ramzor to the North Final Report.* Technion, Israel: Samuel Neamann institute for policy studies. https://www.neaman.org.il/Ramzor-to-the-North-publication, (in Hebrew).

OECD (2011). "Building a High-Quality Teaching Profession Lessons from around The world", Background Report for the International Summit on the Teaching Profession. http://asiasociety.org/files/lwtw-teachersummit.pdf.

Segal, R., Shriki, A. and Movshovitz-Hadar, N. (2016). Facilitating mathematics teachers' sharing of lesson plans, Presented by R. Segal and published, in Csikos, C., Rausch, A., & Szitanyi, J. (eds.), *Proceedings of the 40th Conference of PME — the International Group for the Psychology of Mathematics Education*, Vol. 4, Szeged, Hungary, Aug. 3–7, 2016, pp. 171–178.

Shriki, A. and Lavy, I. (2012). Perceptions of Israeli mathematics teachers regarding their professional development needs, *Professional Development in Education*, **38**(3), 411–433.

Shriki, A. and Movshovitz-Hadar, N. (2011). Nurturing a community of practice through a collaborative design of lesson plans on Wiki system, *Interdisciplinary Journal of E-Learning and Learning Objects*, **7**, 339–357. http://www.ijello.org/Volume7/IJELLOv7p339-357Shriki768.pdf.

About the Author

Nitsa Movshovitz-Hadar is an Emerita Professor of mathematics education at Technion — Israel Institute of Technology. She is a Senior Researcher at Samuel Neamann Institute for national policy studies and the Head of Kesher-Cham — R&D center for reviving mathematics education in Israel, which she established in 1986. See also her homepage on Technion website.

Part V

Issues and Innovations of General Interest Related to Mathematics Education in Israel

Chapter 35

Creating a Modern Hebrew Language for Mathematics

Leo Corry

Cohn Institute for History and Philosophy of Science
Tel Aviv University, Israel
corry@post.tau.ac.il

The roots of the modern mathematical Hebrew lexicon are found in the Haskalah movement at the end of the 19th century. The next important stages are found in the context of the efforts of Eliezer Ben Yehuda and the institutions he developed for fostering modern Hebrew in general in Palestine. Of particular importance was the contribution of Avraham Baruch Rosenstein in Tel Aviv.

1. Introduction

The story of the teaching of mathematics in Israel, both before and after the creation of the State, cannot be told without telling the story of the development of a modern Hebrew lexicon for mathematics. In this chapter I indicate some of the main milestones in this intriguing story.[a]

2. "Haskalah" and the Modern Scientific Hebrew Lexicon

One of the most important cultural phenomena in the Jewish world before the Zionist revival of the Hebrew language was the European Jewish Enlightenment, or "Haskalah," beginning in the late 18th century. Inspired

[a] A more detailed version of this story appears in Corry and Schappacher, (2010, pp. 449–457). I am indebted to Norbert Schappacher for our joint work in that article and for allowing me to use here the relevant parts, which I had written in the original version. I also thank Michael Fried for his help in preparing the final form of the current version.

by classical enlightenment values and motivated by the wish to achieve a fuller integration of Jews into European society, the Enlightened Jews, or "Maskilim," promoted the study of secular topics among Jews (Feiner, 2004). They advocated a revival of the Hebrew language and of Hebrew literature.

The revival of Hebrew in the context of the "Haskalah" can be divided loosely into two stages. In the first stage, from about 1780 to 1855, the "Maskilim" aimed specifically at developing a language that would be as close as possible to the original, Biblical Hebrew. In the second stage, roughly between 1855 and 1880, the new Hebrew literature in Europe shifted to broader topics, including scientific topics, and with that came a freer use of post-Biblical Hebrew and neologisms.

Among the Maskilim who deserve particular attention in this story there stands out the remarkable figure of Chaim Selig Slonimski (1810–1904). He was a distinguished Talmudist proficient in several European languages as well as in Latin. He was well-versed in the science of his time and well-connected with the international community. Slonimski was among the first (if not the very first) to write books on science for a broad Jewish audience, focusing especially on astronomy, physics and mathematics. In 1862, he established in Warsaw the widely-read Hebrew journal, *Ha-Zefirah*, specializing in scientific topics. (Soffer, 2007).

Slonimski's books illustrate the difficulties in creating an adequate modern mathematical lexicon in Hebrew, difficulties connected with appropriate terminology, symbols, and the right-to-left text direction of Hebrew. Indeed, in the introduction to his 1865 *Sefer Yesodei Hachkmat Hashiur* (ספר יסודי חכמת השיעור — "A Book of Basic Mathematics"), Slonimski discusses these issues explicitly and tells the reader how he often had to adopt foreign terms, that is, terms in "the language of the nations." Nevertheless, like his fellow *Maskilim* writing other kinds of texts, Slonimski actually took great pains to use Hebrew words wherever possible in his mathematical book — sometimes modifying the accepted sense of terms — rather than transliterating foreign words. Only in a few cases did he follow the terminology used by "the learned of all nations". Thus, for instance, while "arithmetic" appeared as "*hokhmat hashiur*" (חוכמת השיעור — roughly: "the learning of ratios"), and "geometry" as "*medida*" (מדידה — "measurement"), "algebra" was simply transliterated. Sometimes,

Table 1. Slonimski's suggestions for a mathematical lexicon.

Numerator (of a fraction)	מונה	"*moneh*": counter (Slonimski adds in "parentheses צעהלער, i.e., a transliteration of the German "Zähler")
Denominator (of a fraction)	איכה	"*eikhah*": an archaic Biblical term for "how" (Slonimski adds in parentheses נענער, i.e., "Nenner")
Power (of a number)	מדרגה	"*madregah*": stair
Exponent	רכס	"*rechess*": crest
Edge (of a polyhedron)	צלע	"*tzelah*": rib

he added a German translation. As in the case of "Differential-Rechnung", for which he used "*heshbon pishionot*" (חשבון פשיונות). "Heshbon" was and remains the accepted term for calculus, while the Talmudic term "*pishion*" (פשיון) is completely out of use nowadays. Table 1 presents examples for translations suggested by Slonimski for elementary mathematical concepts.

As for the thorny issue of mathematical symbolism for Hebrew texts, Slonimski's own choice was *not* the one that was eventually adopted. In Slonimski's approach, the quadratic formula would appear as follows (read from right to left, of course):

$$. \; \overline{({}^{2}ב\tfrac{1}{4} - א\tfrac{1}{2})}\sqrt{\;} \pm א\tfrac{1}{2} \; = \; ק$$

3. Mathematical Lexicons in the Land of Israel

The most important figure associated with the revival of the spoken Hebrew language in Palestine, Eliezer ben-Yehuda (1858–1922), emerged from the same kind of *Haskalah* background as Slonimski. He was born in the town of Luzhky, near Vilna, and moved to Jerusalem in 1881. Almost immediately, he started working towards establishing Hebrew as the everyday means of communication for Jews coming to Palestine from all regions of the world. He regarded this common language as the key ingredient for the national revival of the Jewish people. At the turn of the 20th century he established the *Vaad Halashon Haivry* (ועד הלשון העברי — Committee for the Hebrew Language), which would become the most prominent vehicle for

Table 2. The *Vaad*'s suggestions for a mathematical lexicon.

Numerator (of a fraction)	כמה	"*kamah*": literally "how many", also "quantifier"
Denominator (of a fraction)	מנה	"*manah*": portion or ratio, but also related with "counting" (למנות)
Power (of a number)	מדרגה	"*madregah*": stair
Exponent	רכס	"*rechess*": crest
Edge (of a polyhedron)	חוד/פה	"*peh*": mouth; or "*khod*": cusp

the realization of his plans. In 1953 the *Vaad* became the National Academy of the Hebrew Language.

The *Vaad* was directed by Hebrew teachers intent on "adapting the Hebrew language for use as a spoken language in all aspects of life: at home, at school, in the public sphere, in commerce, in industry, in art, and in science." As part of their efforts, they periodically published lists of new terms to be used variously. In 1913, they published one such list containing mathematical terms as shown in Table 2.

Besides the *Vaad*, various primary and secondary schoolteachers' unions took upon themselves the task of turning Hebrew into the main and only teaching language in schools of the *Yishuv* (the term used to refer to the Jewish residents of Palestine before the establishment of the State of Israel). The question as to who was better qualified to develop Hebrew lexicons occasionally gave rise to fierce theoretical discussions, as well as open power struggles, between these teachers' associations and the *Vaad* (Efrati, 2004, pp. 49–80).

4. Language Wars in Palestine

The story of the establishment of a modern Hebrew lexicon cannot be told without mentioning the so-called *Language War* in Palestine which broke out in 1913. Jewish educational institutions had been established since the end of the 19th century not only by Zionist initiatives, but also as part of the efforts of European Jewish philanthropic organizations such as the French *Alliance Israélite Universelle* and the German *Hilfsverein der Deutschen Juden* (עזרה).

In 1907, the director of the *Hilfsverein*, Paul Nathan (1857–1927), initiated the establishment of an institution of higher learning for engineering in Haifa, the *Technikum* — which eventually became the *Technion* — as well as a *Realgymnasium* working close to it — which eventually became the Hebrew Reali High School of Haifa. Against the background of a technological gap increasingly felt at the time in the Ottoman Empire, Nathan assumed that the Jewish population of Palestine, if properly trained in the German technological tradition, would be a welcomed contribution to the development plans then encouraged by the Ottoman government, which, in turn, would encourage Jewish immigration into Palestine.

Nathan's initiative was welcomed at the *Yishuv*, but in 1913, just as the first buildings were being constructed, the Board of Governors of the *Hilfsverein* in Berlin passed a decision to the effect that all teaching in the *Technikum* would be in German. The decision was based on the assumption that no adequate Hebrew books existed for advanced technological teaching so that German books would be used anyway. The decision, however, triggered furious opposition from the local Jewish population. Teachers of the Jerusalem Teacher's Seminar (created and supported by the *Hilfsverein*) renounced their posts, while students went on strike. This debate sparked intense emotions. In an open letter signed by Ben-Yehuda and some representatives of the teachers' associations, the decision was described as "...an open attack against the soul of the Hebrew nation and we consider it to be a national disaster," and the position that Hebrew must be the language of all schools in the Land of Israel was reasserted trenchantly (Efrati, 2004, p. 147).

The "war" ended victoriously for the Hebrew camp in February 1914. The *Hilfsverein* agreed that teaching at the *Technikum* would be in Hebrew, and that all teachers not yet fluent in Hebrew will be compelled to learn the language within four years. By the time the first courses were taught there in 1924, the British Mandate had passed a decision (in 1922) recognizing Hebrew as one of its official languages, alongside English and Arabic, consummating the victory in the "language war."

5. Rosenstein

If we need to single out the person whose influence was the most decisive in establishing the Hebrew mathematical lexicon in use up to this day, that person would be Avraham Baruch Rosenstein (1881–1950), a renowned mathematics teacher at the emblematic Gymnasia Herzliya in Tel Aviv (Razi-Stein, 1991). The Gymnasia was established in 1905 as the first high school to conduct its teaching entirely in the Hebrew language.

Rosenstein joined the staff of the Gymnasia in 1911 shortly after completing a dissertation in mathematics in Vienna. He immediately undertook the task of creating an appropriate lexicon for his teaching, and, more importantly, of writing the necessary textbooks in all fields of high school mathematics and physics. Over the years, he authored or co-authored twenty-five books that continued to be used across the country until the 1950s. As an enthusiastic former student put it, thanks to the efforts of Rosenstein, high school students at Gymnasia Herzliya "spoke algebra, geometry, trigonometry, logarithms and differential calculus as if these disciplines had been created and initially taught in Hebrew" (Ben Yehuda and Ofek, 1971, pp. 83–84).

In his quest for a modern mathematical lexicon, Rosenstein went back to ancient Hebrew sources, including Avraham Bar Hiyya (1070–1136) and Levi ben Gerson (1288–1344), both of whom wrote treatises on arithmetic and geometry. He also consulted more recent Hebrew scientific authors such as Slonimski. As early as 1912, he was teaching summer seminars on the Hebrew mathematical lexicon to teachers from all around the *Yishuv*. The participants of his seminars decided to write to the *Vaad* demanding that they define their mathematical lexicon based on the suggestions of the teachers' union (which were actually Rosenstein's) (Efrati, 2004, p. 71). The list published by the *Vaad* in 1913 from which we quoted above shows that this demand was not met. Some examples of Rosenstein's Lexicon are shown in Table 3.

It is important to stress that, while rejecting the specific mathematical choices of the *Vaad*, the neologisms introduced by Rosenstein did often follow some new kinds of general declensions, introduced into Hebrew by the *Vaad* with the explicit aim of allowing the consistent production of a broad range of neologisms based on existing roots. Thus, for instance, Rosenstein's "chezkah" was an adaptation of the root (חזק) to one such new type of declension.

No less important than the lexicon itself in shaping the new Hebrew mathematical language was Rosenstein's decision to use Latin characters for

Table 3. Rosenstein's suggestions for a mathematical lexicon.

Numerator (of a fraction)	מונה	"*moneh*": counter, like Slonimski
Denominator (of a fraction)	מכנה	"*mekhaneh*": namer
Power (of a number)	חזקה	"*chezkah*": a neologism derived from the root חזק (strong or powerful)
Exponent	מעריך	"*ma'arikh*": a neologism derived from the root ערך (value)
Edge (of a polyhedron)	מקצוע	"*miktzoah*": a forgotten Biblical term, whose original meaning was "corner"

all mathematical and physical equations and formulae, rather than Hebrew ones as Slonimski had done previously. Rosenstein wrote them from left to right, as in any European text, and this is the way that equations continue to be written in Hebrew scientific texts.

Rosenstein's influence was also decisive concerning the way in which the modern Hebrew mathematical lexicon of advanced academic research took its shape. This was mainly through the mediation of two Gymnasia *alumni*, Jacob Levitzky (1904–1956) and Binyamin Amira (1896–1968). Both were among the young faculty of the Institute of Mathematics at the Hebrew University in its early days. Both earned their Ph.D. in Göttingen, where they met Edmund Landau (1877–1938), who was to become the first Professor of Mathematics in Jerusalem in 1927–1928. Landau's teaching in Jerusalem and several texts that he wrote at the time adopted Rosenstein's lexicon and style, and so did all other faculty members at the time.

References

Ben Yehuda, B. and Ofek, U. (eds.) (1971). *The Story of Gymnasia Herzliya*, Tel Avi: Gymnasia Herzliya Publications (in Hebrew).

Corry, L. and Schappacher, N. (2010). Zionist internationalism through number theory: Edmund Landau at the opening of the Hebrew University in 1925, *Science in Context*, **23**(4), 427–471.

Efrati, N. (2004). *From a Language of Individuals to a Nation's Language. Speaking Hebrew in the Land of Israel 1881–1922*, Jerusalem: National Academy for the Hebrew Language (in Hebrew).

Feiner, S. (2004). *Haskalah and History: The Emergence of a Modern Jewish Historical Consciousness*, Oxford: Littman Library of Jewish Civilization.

Razi-Stein, R. (1991). First in mathematics, *Aleh*, **9**, 20–24.

Rinot, M. (1972). *The "Ezrah" Association for the German Jews. A Chapter in the History of Hebrew Education in Palestine*, Jerusalem: The Hebrew University (in Hebrew).
Soffer, O. (2007). *"There is No Place for Pilpul!" HaTzfira Journal and the Modernization of the Sociopolitical Discourse*, Jerusalem: The Bialik Institute (in Hebrew).

About the Author

Leo Corry is the Bert and Barbara Cohn Professor for History and Philosophy of Science at Tel Aviv University. Currently he serves as Dean of the Lester and Sally Entin Faculty of Humanities.

Chapter 36

Technology for Mathematics Education: The Case of CET

Sara Hershkovitz*, Osnat Efrat and Maureen Hoch

Center for Educational Technology (CET), Tel Aviv, Israel
**sarah@cet.ac.il*

Technological environments have become an integral part of the learning-teaching process. We discuss here the advantages of such environments — full courses and content repositories — and present examples from CET's repository of interactive books, courses, and activities.

1. Introduction

In recent years, different countries have published national strategic plans for integrating technology into the education system (Anderson, 2010; ATCS, 2010; Bellanca and Brandt, 2010; Christensen and Horn, 2008). These plans include development of technological environments that support teaching, learning, and assessment at all levels of education. The integration of digital technology into the mathematics classroom is an ongoing process (Labbord and Straber, 2010, p. 122) which follows this international tendency. It seems that the process of integrating new technologies into learning and instruction is inevitable.

The Center for Educational Technology (CET) is a non-governmental and non-profit organization in Israel. CET's educational world-view, from its inception in the early 1970s, was to promote learning via technology and to adapt learning to the variation among students, believing that the best learning takes place when it is adapted to the abilities of individual students and their own learning styles. Max Rowe, one of CET's founders, expressed this in his book (Rowe, 1985) which presented the importance of equal educational opportunities for all. He argued that every student, irrespective of his/her individual skills, is entitled to a fair chance to develop within the limits of their intellectual ability (p. 17), and he pledged to develop technology to this aim. The mathematics department within CET develops textbooks (printed and digital) and technological platforms in line with the mathematics curriculum for K-12 and with special regards to the needs of teachers.

2. Technology for Mathematics Education

The main goals in mathematics education are defined by NCTM (1991) and Kilpatrick, Swafford, and Findell (2001) as follows: (1) conceptual understanding — comprehension of mathematical concepts, operations, and relations; (2) procedural fluency — skill in carrying out procedures flexibly, accurately, efficiently, and appropriately; (3) productive disposition — habitual inclination to see mathematics as sensible, useful, and worthwhile, coupled with a belief in diligence and one's own efficacy. The Common Core (2010) added: (4) problem solving and modeling; (5) generalization, reasoning, and proof; (6) using representations; (7) communication; and (8) connections.

The integration of technology into the mathematics classroom takes into account all the above-mentioned goals and reveals new interactive possibilities, as well as new pedagogical opportunities (Pierce and Stacey, 2010) for designing mathematics lessons. Some of the technological possibilities are: (1) representations and computerized laboratories designed especially for each mathematical context; (2) tools which enable students to explore mathematical ideas dynamically in topics such as geometry, algebra, and statistics; (3) different types of feedback such as "feed-forward" (Ferrell and Gray, 2016) that gives students hints for further learning, "intellectual mirror" (Schwartz, 1989) that reflects the learner's conjecture, and feedback that tells if answers are correct; (4) using real and updated data for problem solving in authentic situations; (5) communication and collaboration — between students, between students and teachers, and between students and experts.

3. Technological Environments

Technology is an integral part of the learning-teaching process. We are witness to a varied technological environment that supports the general aims of mathematics education. There are different models of technology integration (blended learning) from the model that combines technological activities within traditional teaching (the best of both worlds) to virtual courses which are almost entirely technology based. In virtual courses we also find different learning models for individual student learning, interspersed with teacher intervention, and student–student discourse (Horn and Staker, 2015, pp. 67–81).

In this chapter we refer to digital books and courses (linear environments — see Chapter 15 of this book for a description of the Virtual High School), and repositories of technological activities for learning and assessment (modular environments), which blend with conventional instruction (CETa,[a] 2006). The modular repositories enable teachers to create a sequence of activities individually matched to their students' needs, for deeper learning and more efficient assessment; and to choose the technological tools according to their particular teaching style and curriculum requirements. See, for example, The Mathematics Challenge[b] which provides all the curricular material for three years of advanced-level high school mathematics in PowerPoint presentations, video clips, and interactive questionnaires.

We present here an example from the primary school repository — the "Birthday Cake Riddle"[c] (Fig. 1). This problem provides students with an

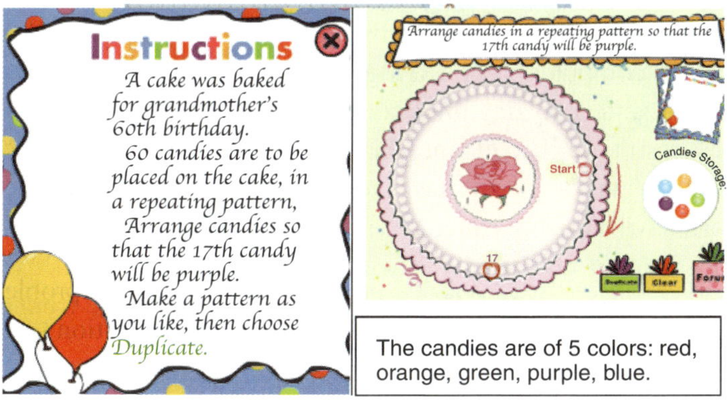

Fig. 1. The Birthday Cake Riddle.

[a]CET's technological environments are used in 90% of schools in Israel.
[b]http://ebagcourses.cet.ac.il/%D7%94%D7%90%D7%AA%D7%92%D7%A85.
[c]http://ofek.cet.ac.il/units/en/math/Unit97/act1.aspx.

interactive tool for solving it. They can explore the phenomena and receive feed-forward reactions which enhance generalization, that is important for extending or expanding their range of reasoning beyond the case considered (Harel and Tall, 1991; Carpenter and Franke, 2001). This process characterizes the activity of professional mathematicians (Kaput, 1999).

The computer can also help the learner to improve mental models and to enhance performance on more complex cognitive tasks (Norman, 1993). The use of computers can also help learners who are not able to cope with a regular curriculum to find new motivation and challenge by learning how to cope with computer-based tasks.

4. Assessment and Evaluation

The computer becomes an important part of evaluation, presenting learners with interactive tasks which enable them to demonstrate practical skills. Other advantages of these interactive tasks are: (1) they can be divided into levels, e.g., "novice — apprentice — expert"; (2) they can reveal a student's "submerged talent" not apparent in the classroom; (3) they do not require above-average curriculum knowledge (Pead, 2012); (4) they reduce language demands, thus helping learners who are verbally challenged (Quellmalz *et al.*, 2012).

The use of interactive evaluation tasks also fits the challenges mentioned by Pead (2012): (1) requesting students to give a written explanation in the form of a description or a justification; (2) challenging students, e.g., by a "Beat the computer" task; (3) logging the interactions between student and computer, and analyzing this data; (4) heuristic inference — based only on the best results of the student (this does not lend itself to all tasks); (5) extension problems — further tasks to demonstrate understanding.

Linear environments (interactive textbooks and courses) enable systematic instruction and supply a wide range of interactive activities for learning and assessment [CETb,[d] 2008]. Here is an example from a digital geometry textbook for 5th grade (Fig. 2).[e]

This task is integrated in a laboratory for investigating area and perimeter of parallelograms. The laboratory enables locking (keeping constant) various components of the parallelogram to investigate the effect on its

[d]CET's digital textbooks are used in 60% of schools in Israel.
[e]http://www.school.kotar.co.il/KotarApp/Viewer.aspx?nBookID=72611806#174.3959.4.default.

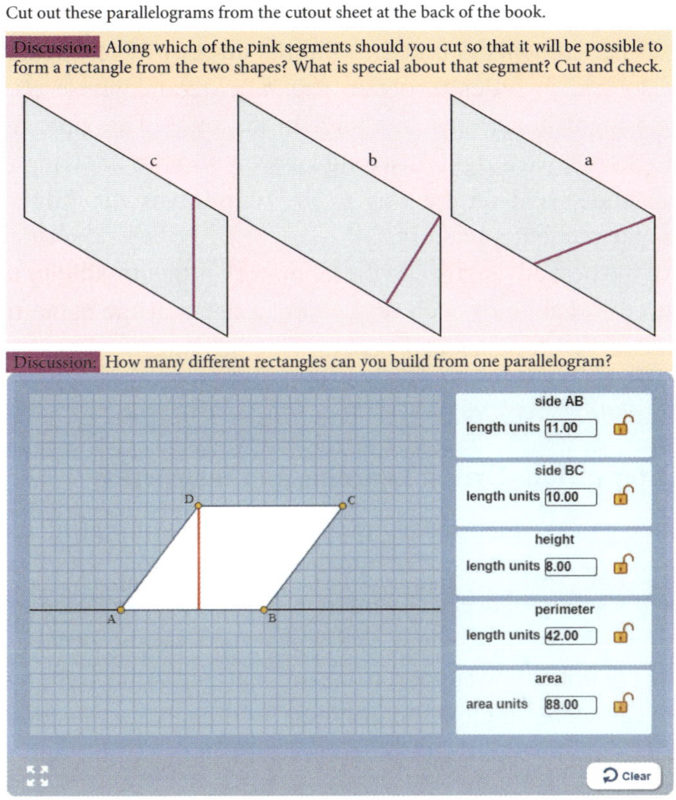

Fig. 2. Investigating parallelograms.

other components. For example, locking two adjacent sides and dragging the vertices will cause the area to change but the perimeter will remain constant (although it is not locked). In this task students are asked how many different rectangles can be drawn whose area are equal to that of the given parallelogram.

The integration of such interactive laboratories into the digital textbook broadens the potential of exploring mathematical phenomena and also the potential for formative assessment.

The added value of using mathematics laboratories for tasks integrated in digital textbooks has many aspects. For example, the learning process becomes observable. Often teachers are unable to follow students' reasoning processes, even when they have requested them to explain their methods. They can see the final answer but not how it was reached. To illustrate this we will use the example of the parallelogram laboratory. Let us assume that the teacher asks her students to find the minimal number of components which should be locked in order to obtain a parallelogram whose components will stay constant under any dragging. In the case of a paper and pencil test, we might receive a right or wrong answer, without knowing the stages the student underwent on the way to the written answer. The laboratory documents the student's process.

One of the best advantages of the laboratory is the possibility to observe the measure of change of each of the components at the same time. This cannot be done with paper and pencil. This advantage allows teachers to ask questions which demand various skills and different levels of thinking.

5. Next Steps: Data-Driven Learning and Instruction

CET is moving to personalized learning (adaptive learning) which we define according to Friend *et al.* (2017): "Personalized learning tailors learning to each student's strength, needs and interest. Students have 'voice and choice' in determining what, how, when, and where the learning occurs. Teachers provide the flexibility and support to ensure mastery of the highest standards possible". The technological environment will enable a dynamic analysis of a student's preferences and outcomes will indicate the optimum next learning step.

The personalized learning system will allow teachers to adapt their teaching to the most precise possible approach for the student or interest group; and it will allow the development teams to assess the effectiveness and the contribution of content units to the learning process and to change them where needed. In addition, the adaptive system will allow the collection of data on the content, on learning pathways, and on difficulties at the individual student level, at the whole-class level, and at a national, systemic level. All of these can provide tools to the teacher for constant monitoring, for planning how each lesson will be taught, and for long term planning.

References

Anderson, J. (2010). *ICT Transforming Education, A Regional Guide*, Published by UNESCO Bangkok, Available: http://unesdoc.unesco.org/images/0018/001892/189216e.pdf.

ATCS (2010). *Assessment and Teaching of 21st Century Skills*, Available: http://www.atc21s.org/.

Bellanca, J. and Brandt, R. (2010). *21st Century Skills — Rethinking How Students Learn*, Canada: Solution Tree Press.

Carpenter, T. and Franke, M. (2001). Developing algebraic reasoning in elementary school: Generalization and proof, in H. Chick, K. Stacey and J. Vincent (eds.), *Proceedings of the 12th ICMI Study Conf: The Future of the Teaching and Learning of Algebra*, Melbourne, Australia: The University of Melbourne, pp. 155–162.

CETa (2006 — ongoing development). "OFEK — Interactive activities for learning and assessment in mathematics for 1st–9th grades": http://ebag.cet.ac.il/%D7%9E%D7%AA%D7%9E%D7%98%D7%99%D7%A7%D7%94/.

CETb (2008 — ongoing development). "KOTAR — Digital textbook in mathematics for 1st–9th grades": http://www.school.kotar.co.il/KotarApp/BrowseBooks.aspx?ExpandNodeID=2330$2342&lang=he,en.

Christensen, C. M. and Horn, C. M. (2008). How do we transform our schools? *Educ Next*: http://educationnext.org/how-do-we-transform-our-schools/.

Common Core State Standards Initiative (2010). Mathematics curriculum standards, http://www.corestandards.org/Math.

Ferrell, G. and Gray, L. (2016). *Feedback and Feed Forward — Using Technology to Support Students' Progression Over Time*, Updated: 7 March 2016: https://www.jisc.ac.uk/guides/feedback-and-feed-forward.

Friend, B., Patrick, S., Schneider, C. and Vander Ark, T. (2017). *What's Possible with Personalized Learning?* Vienna, VA: International Association for K-12 Online Learning (iNACOL).

Harel, G. and Tall, D. (1991). The general, the abstract, and the generic, *For the Learning of Mathematics*, **11**, 38–42.

Horn, M. B. and Staker, H. (2015). *Blended: Using Disruptive Innovation to Improve Schools*, Jossey-Bass, A Wiley Brand.

Kaput, J. (1999). Teaching and learning a new algebra with understanding, in E. Fennema and T. Romberg (eds.), *Mathematical Classrooms that Promote Understanding*, Mahwah, NJ: Erlbaum, pp. 133–155.

Kilpatrick, J., Swafford, J. and Findell, B. (2001). *Adding It Up: Helping Children Learn Mathematics*, Washington, DC: National Academy Press.

Labbord, C. and Straber, R. (2010). Place and use of new technology in the teaching of mathematics: ICMI activities in the past 25 years, *ZDM, International Journal of Mathematics Education*, **42**(7), 121–133.

NCTM — National Council of Teachers of Mathematics (1991). *Professional Standards for Teaching Mathematics*, Reston, VA.

Norman, D. A. (1993). *Things That Make Us Smart*. MA: Reading, Addison-Wesley.

Pead, D. (2012) World class tests: Summative assessment of problem-solving sing technology, *Educational Designer*, **2**(5). Retrieved from: http://www.educationaldesigner.org/ed/volume2/issue5/article18.

Pierce, R. and Stacey, K. (2010). Mapping pedagogical opportunities provided by mathematics analysis software, *International Journal of Computers for Mathematical Learning*, **15**(1), 1–20.

Quellmalz, E., Timms, M., Buckley, B., Davenport, J., Loveland, M. and Silberglitt, M. (2012). 21st century dynamic assessment, in M. Mayrath, J. Clarke-Midura and D. Robinson (eds.), *Technology-Based Assessments for 21st Century Skills: Theoretical and Practical Implications from Modern Research*, Charlotte, NC: Information Age Publishers, pp. 55–90.

Rowe, M. (1985). *From a School System to a Learning System in Israel*, Max Rowe Publishing, http://www.kotar.co.il/KotarApp/Viewer.aspx?nBookID=94613335 (in Hebrew).

Schwartz, J. (1989). Intellectual mirrors: A step in the direction of making schools knowledge-making places, *Harvard Educational Review*, **59**(1), 51–62.

About the Authors

Sara Hershkovitz, PhD, is head of Mathematics Department, Center for Educational Technology. She is a Professor at Shaanan Religious Academic College of Education, Israel.

Osnat Efrat, MA, is head of Primary School Development, at Mathematics Department, Center for Educational Technology.

Maureen Hoch, PhD, is a curriculum developer, at Mathematics Department, Center for Educational Technology.

Chapter 37

Challenges Teachers Face When Designing Their Resources: The Case of Technology-Based Formative Assessment

Michal Yerushalmy* and Rachel Hess-Green#

School of Education, University of Haifa, Haifa, Israel

**michalyr@edu.haifa.ac.il*

#rachely.hg@gmail.com

Teachers face challenges when conducting lessons built on student-designed solution strategies. The work we describe here is part of an effort to explore how technological support for classroom formative assessment may play a role in attempts to reform mathematics instruction. We focus on the challenges involved in teachers' attempts to design and experience a new genre of formative assessment with an automated formative assessment platform in mathematics (STEP).

1. Introduction

An important aim of pedagogical tasks is to grant opportunities for students to explore for the purpose of formulating ideas. The tasks, often referred to as rich "expert tasks," require experimentation to help make sense of problems, and the effective use of strategies, concepts, and skills. They often involve construction of examples (Burkhardt and Swan 2013). Task design involves extensive use of visual means and interactive linked representations (interactive diagrams, or IDs), to support students in their work with mathematical objects and processes (Yerushalmy, 2015). IDs are a primary resource in designing such problems.

Studies of learning and teaching algebra and functions curriculum with the Visual Mathematics (*LIROT MATEMATICA*, in Hebrew) illustrate the challenges involved in scaling up the implementation of a technology-based, guided-inquiry curriculum (e.g., Yerushalmy, 2013). The rationale for designing the STEP — an automated online formative assessment platform (Olsher, Yerushalmy and Chazan, 2016) — was that technology-based guided inquiry requires constant feedback for teachers regarding the actual implementation of the designed learning process, to support further design and explorative teaching and learning. Design based on interactive technology for assessment raises issues related both to technology and to representation and analysis of the automatically checked solutions. In this paper we follow the challenges faced by a teachers' educator designing the main components of the interactive assessment process.

2. Automated Rich Assessment

New technologies give rise to the learning of skills, such as problem solving, creativity, critical thinking, and risk-taking, but e-assessment has focused primarily on testing factual knowledge (Scalise and Gifford, 2006). The STEP platform uses technological means to support teachers in routinely identifying patterns and characteristics of their students' thinking, which are apparent in the students' submissions of problem-solving work. STEP was developed at the University of Haifa (Olsher, Yerushalmy and Chazan, 2016). The assessment environment is built upon three elements: (a) tasks that allow using interactive resources to experiment with linked multiple mathematical representations that reflect and mirror the student's actions in various ways; (b) each task requires answers provided through dynamic interaction, which illustrate the solution process (mainly by construction of supporting or contradicting examples); and (c) tasks (and collections of related tasks–the assessment) are characterized mathematically. The submitted answers can be identified and analyzed automatically according to the characteristics identified. We adopted Black and Wiliam's (1998) conceptualization of formative assessment as "all those activities undertaken by teachers, and/or by their students, which provide information to be used as feedback to modify the teaching and learning activities in which they are engaged" (1998, pp. 7–8). Nagari-Haddif and Yerushalmy (2016), and Olsher and Yerushalmy (2017) used this framework to design calculus and geometry assessments, and studied responses of Israeli high school students and teachers, automatically collected and analyzed with the STEP platform.

3. Observations Based on Working With Teachers as Designers

Design of instructional material, including that of formative assessment material, is assumed to be a routine part of a teacher's preparation for teaching. To study possible implementation methods of the STEP platform in the educational system, we established a framework for a professional development project. The first stage of the scaling-up project involved teachers who joined the development team and were trained to be STEP coaches. Each trainee was experienced in teaching with rich tasks and technology, and a few were teacher educators. But they were only starting to think about formative assessment, about the use of technology for this purpose, and about designing assessment with STEP. As part of their training, they worked with a small team of high-school mathematics teachers to develop diagnostic student assignments in response to their practical needs, and test them in the classroom. The environment enabled these teachers to provide feedback and suggestions to the development team for improving the technological tools and the formats they support. The two main design challenges are (a) fitting the tool to curriculum content the teacher wants to assess, and (b) phrasing the task in a way that provides important pedagogical and mathematical assessment criteria and information about students' work.

The coaches-in-training (the trainees) participated in group meetings where the developers introduced the principles of design and the tools used to implement them with existing STEP tasks, and the trainees presented the tasks and experiments they designed with their students and teams (a team consisted of one trainee, the teachers mentored by the trainee, and the teachers' students). The data collected and accumulated by STEP were open for all participants to view and analyze, triggering critical analysis and iterations of revisions.

Below we follow the process that one trainee, Omar, went through, designing an activity aimed to teach an advanced topic of high school geometry. Omar's recollection from his own teaching and from his mentoring of teachers, was that students do not easily solve problems that require the use of theorems and properties concerning angle bisectors in triangles. He reported that angle bisectors were considered mainly in the form of the main angle bisector in isosceles triangles. According to Omar, students had difficulty recognizing and using the bisector property of dividing the opposite

Table 1. Three stages of evolving design of the angle bisector task.

<table>
<tr><th></th><th>First version</th><th>Second version</th><th>Third version</th></tr>
<tr><td>The task</td><td>Given: In triangle ABC, point D is on AB. CD is the angle bisector. The length of CB is equal to 10 cm. Task: Adjust the dynamic figure and submit three versions of it that approximately describe the given.</td><td colspan="2">Given: In triangle ABC, point D is on AB. Task: Insert in the appropriate boxes the lengths of BC, AC, BD, and AD, such that CD divides angle C into two equal parts. Submit three figures that are different from one another.</td></tr>
<tr><td>Construction of examples by:</td><td>Dragging the triangle</td><td>Entering numeric values</td><td>Entering numeric values</td></tr>
<tr><td>The given example</td><td>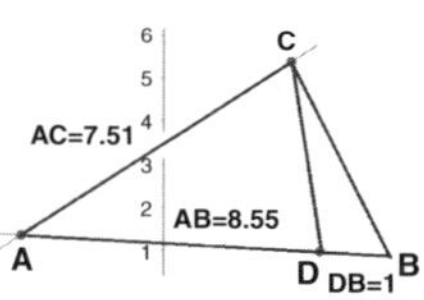
</td><td>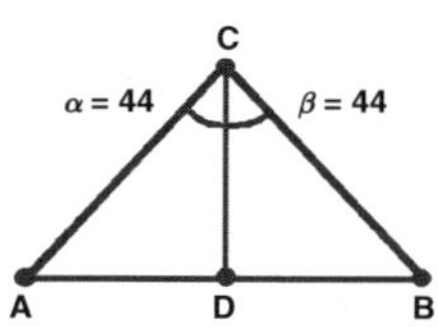
</td><td>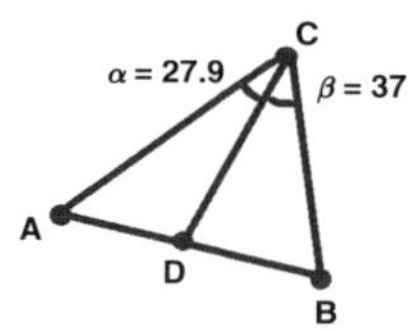
</td></tr>
</table>

side in the ratio of the two other sides ("angle bisector theorem"), which is often required in proving tasks. Omar designed a task to support students' construction of angle bisectors in triangles, hoping to find a better way of teaching the theorem by challenging students with a construction problem in a dynamic geometry environment (DGE) rather than by traditional paper-and-pencil means. He expected STEP to help him analyze submissions and diagnose students' perceptions and difficulties, which would in turn improve his ability to support the students. We describe the design process using the documentation framework (Gueudet and Trouche, 2009), in which teachers engage with resources in a "documentational genesis". The documentation process, which involved individual and team work, included three discussions that led to three versions of the task (Table 1). The analysis of this process reveals important challenges involved in the design of tasks. As part of this process, they create "a document", describing the mechanism by which the resources in question influence the teaching of the task, and how modifying the teacher's resources affects this process.

The challenge of "fair judgment"

Initially, the task asked students to adjust the dynamic figure so that it approximately describes the given, and to submit three examples of it. Omar introduced the term "approximately" because he valued construction by experimentation with dragging; he assumed that it would lead students to pay attention to the approximate lengths of the sides and the segments, to develop awareness of specific ratios between sides and segments, and to develop conjectures based on these views. He also hoped that the variety of figures of triangles, which can then be assessed by STEP, would help him discuss related conjectures and eventually state the theorem. Group members who worked with the task raised the mechanical difficulty they faced in dragging point D to form an angle bisector. Some suggested determining the location of D by indirect manipulation (for example, using a slider) or direct numeric input of the length of the sides of the triangle. Others used the collected examples to demonstrate the difficulty in judging the correctness of the submission, and the pedagogical implications of the decision that the teacher must make if accepting mathematically incorrect (or inaccurate) answers as correct submissions. Omar decided to reformulate the instruction, asking students to construct the segment by specifying numeric values that can support conclusions based on accurate measures. This may sound like a discussion of technical details of the applet design, but in practice it raised the central issue of assessment of open rich problems, or of "fair judgment," in other words, the automatic non-interpretive evaluation of the "truth" of constructions that "approximately describe the given" (Table 1, first version). Group members suggested that there was a large gap between the experimentation mode being offered and the demand to state a specific numeric proportion, and that the submissions would not help the teacher follow the students' thinking. Therefore, Omar decided to introduce another change, and to refer to a specific given example (2nd version).

The challenge of "argumentation by example"

Feeling comfortable with this design, the group turned its attention to the instructional situation that Omar would face in the classroom. A principle of task design, whereby answers are based on examples, dictates special ways of regarding proofs and refutations. There is an enormous difference in using examples to prove or to disprove a proposition. A finite set of examples is not sufficient to prove a proposition, but a single example is sufficient to disprove it. This issue prompted a discussion about the logical aspects of proof and refutation tasks. There was agreement that examples can help generalize phenomena or formulate a conjecture, but cannot replace a proof. This discussion led to the analysis of the instrumentation process students would have to go through when arriving at a set of examples that are assessed as correct by the system. Would students perceive this as "proving" the theorem empirically, which in turn might cause them to question the need for a proof? Omar restated his belief that the task that leads students to generating correct examples should also lead to a general conjecture, but it should not replace a proof. Omar realized, however, that his instructional routines related to proving would have to change, and that the challenge may be greater for experienced teachers, who would have to change their established routines. The need to rethink familiar teaching practices is central to the guided inquiry approach, and has been studied extensively in connection with the use of technology.

4. Discussion

The instance of a design process by a team of educators, which we analyzed through the documentational lens, illustrates the challenges of the technological opportunities that were designed into STEP. Chazan, Herbst and Clark, (2016) have conceptualized teaching as negotiating a variety of obligations. The tasks were part of the institutional agenda, and decisions were made based on individuals' pedagogical needs; nevertheless, it is the teachers' obligation to the mathematics that comes into focus. The responsibility that the team took upon itself to design automatic formative assessment tasks, as well as principles for their design, prompted them to probe the discipline.

The intervention described here focuses on technological resources as an important instrument of change. The challenges are not new to the study of problem solving within technologically-supported guided-inquiry settings, but they have not yet been discussed as challenges of assessment. Our goal was to show the opportunities for supporting teaching with tools for formative assessment, based on the assumption that examples generated by IDs can prove useful in revealing how students understand the function that examples fulfill in mathematical reasoning. IDs can also drive professional discussions about the integrity of assessment.

Omar was enthusiastic about guiding inquiry within a DGE, yet all three versions of assessment he helped design were enhancements to help students arrive at the exact formula of a given theorem. From our point of view, the mission of the DGE in such assessment is to provide sufficiently rich submission spaces, which, when analyzed, can characterize the students' problem-solving processes. We may have been too ambitious in assuming that teaching practices that involve experimentation and construction within a DGE would naturally carry over to assessment. This should serve as a reminder that new technologies require significant shifts in approaches to teaching, even changes to the mathematical content at stake. Attempting to create new platforms of innovative assessment to affect the way we teach entails challenges in elaborating and supporting productive instrumentalization of the new tool.

Acknowledgment

This research was supported by the Israel Science Foundation (grant 522/13) and the Trump Foundation (grant 191).

References

Black, P. and William, D. (1998). Assessment and classroom learning, *Assessment in Education: Principles, Policy & Practice*, **5**(1), 7–74.

Burkhardt, H. and Swan, M. (2013). Task design for systemic improvement: Principles and frameworks, *Task Design in Mathematics Education*, 431–439.

Chazan, D., Herbst, P. G. and Clark, L. M. (2016). Research on the teaching of mathematics: A call to theorize the role of society and schooling in mathematics instruction, in D. H. Gitomer and C. A. Bell (eds.), *Handbook of Research on Teaching*, pp. 1039–1097.

Gueudet, G. and Trouche, L. (2009). Towards new documentation systems for mathematics teachers? *Educational Studies in Mathematics*, **71**(3), 199–218.

Nagari-Haddif, G. and Yerushalmy, M. (2016). Construction e-tasks: Design considerations, in *Proceedings of the 13th International Congress on Mathematical Education* (*ICME 13*), Hamburg, Germany.

Olsher, S. and Yerushalmy, M. (2017). What can we learn about mathematics multiple choice problems from attached supporting examples? *CERME 10*, Dublin.

Olsher, S., Yerushalmy, M. and Chazan, D. (2016). How might the use of technology in formative assessment support changes in mathematics teaching? *For the Learning of Mathematics*, **36**(3), 11–18.

Scalise, K. and Gifford, B. (2006). Computer-based assessment in e-learning: A framework for constructing "intermediate constraint" questions and tasks for technology platforms, *The Journal of Technology, Learning and Assessment*, **4**(6).

Yerushalmy, M. (2013). Designing for inquiry in school mathematics, *Educational Designer*, **2**(6).

Yerushalmy, M. (2015). E-textbooks for mathematical guided inquiry: Design of tasks and task sequences, in *Task Design In Mathematics Education*, Springer International Publishing, pp. 229–247.

About the Authors

Michal Yerushalmy studies mathematical learning and teaching with technology.

Rachel Hess-Green's post-doctoral research focuses on the professional development of teachers using the STEP platform.

Chapter 38

The Commognitive Theory Framework: From Theory to Implementation

Einat Heyd-Metzuyanim[*,‡] **and Michal Tabach**[#,§]

**Technion — Israel Institute of Technology, Haifa, Israel*

#Tel Aviv University, Tel Aviv, Israel

‡einathm@technion.ac.il

§tabachm@post.tau.ac.il

This chapter describes several projects deriving from Sfard's Commognitive Theory. After providing a brief outline of its major tenets and concepts, we describe implications of this theoretical framework in four areas of practice: preservice teacher preparation, in-service professional development, introduction of mathematical texts to high school students and diagnosis of mathematical learning difficulties. We end with a discussion of the affordances and challenges of linking the Commognitive Theory with practice.

1. Introduction

The activities described in this chapter began almost 20 years ago, when Anna Sfard first published a paper describing what she called the Commognitive Theory (Sfard, 1998). Though this is a highly theoretical framework, several practical implications have emerged in recent years in the Israeli educational context. Our aim is to describe them.

2. The Commognitive Theory

The Commognitive Theory belongs to a socio-cultural theory that derives from the writings of Vygotsky, Lave, Wenger and other scholars, all of whom advocated the view that learning entails becoming a participant in a historically established activity (Sfard, 2008). The main tenet of the theory is that thinking can be regarded as a form of intra-personal communication that qualitatively does not differ from interpersonal communication. Sfard (1991) began by identifying the cognitive operation of *reification*, which enables learners to think about processes as if they were also their own products. For example, a function can be discussed both as a process (e.g., matching values in a table) and as an end result of this process (as in a function that has certain properties). Reification is indispensable in learning mathematics, since mathematical objects such as numbers must be reified to be dealt with in more sophisticated discourse. In 1998, Sfard pointed out that this reification is also prevalent in *research* on learning, where reification expresses itself in the consideration of learning as the act of adding some new entities known as *concepts*, *mental schemes* or *knowledge* to the assortment of those the learner already possesses. Such reification conceals the process of learning and limit researchers' ability to understand the mechanisms underlying these processes. The framework offers an alternative way of communicating about learning by proposing a set of conceptual tools for considering learning mainly in terms of processes.

A central concept of the Commognitive Theory that captures learning processes is that of *routines*. Routines are repetitive patterns of communication that produce certain *endorsed narratives*. For example, counting is a routine that produces a narrative about the number of objects in a certain situation. Routines often make use of *visual mediators* such as digits, algebraic notations and graphs. Sfard and Lavie (2005) categorized routines into several types, two of which are *rituals* and *explorations*. Ritual routines are characterized by rigid rule-following in which correctness is determined by an external authority. They often deal with mathematical signifiers detached from their meaning. Explorations, on the other hand, are flexible, often deal with mathematical objects and are justified by mathematical means.

The first consolidated edition of the Commognitive Theory appeared in two publications. A short version was presented in Sfard (2007), where theoretical claims were derived from findings on the learning of negative numbers and of basic geometry; a fuller and more elaborate exposition of the theory appeared in the book titled *Thinking as Communicating* (Sfard, 2008).

Although the focus during the first two decades of theoretical development was on research-based theory building, empirical findings and theoretical ideas ripened into several practical projects. Among the theoretical ideas most relevant to this chapter is that of the ritual vs. explorative instruction. Ritual participation has been defined (Heyd-Metzuyanim, 2015; Sfard and Lavie, 2005) as participation for the sake of pleasing or connecting with others, whereas explorative participation is geared towards constructing mathematical narratives about the world for their own sake. According to the Commognitive Theory, ritual participation is often a necessary stage for participating in a new discourse. Yet given the prevalence of ritual participation in classrooms and even in preservice courses (e.g., Heyd-Metzuyanim, Tabach and Nachlieli, 2016) many of the current activities inspired by the theory are focused on enhancing explorations and minimizing ritual participation. The rest of this chapter describes some of these activities.

3. Projects Inspired by the Commognitive Theory

3.1. *Preservice teacher preparation*

Studies of preservice students' participation in mathematics courses at the college level based on the Commognitive Theory (Heyd-Metzuyanim, Tabach and Nachlieli, 2016; Tabach and Nachlieli, 2015) have shown that students often participate in these courses ritually. Based on these studies, Nachlieli together with Yafim Katz set out to design a problem-solving course specifically aimed at engaging students in explorative participation. To examine the success of implementing this course, they video-recorded and studied the lessons using analytical tools based on the Commognitive Theory. Their analysis (Nachlieli and Katz, 2017) revealed that their careful thinking about how to engage students in explorative participation had paid off. Indeed, students in the course gradually became accustomed to thinking about problems in multiple ways and enjoyed constructing mathematical narratives for their own sake, not just to "solve the problem and move on".

The Commognitive Theory lens has had a major impact on the M.Ed. program at the same college of education, and particularly on an M.Ed. seminar for mathematics in-service teachers. The seminar involved designing, implementing and reflecting upon the participants' own teaching.

The participants reflected by videotaping, transcribing and analyzing their own instruction in accordance with the Commognitive Theory framework. According to the participants' self-reports and the instructors' observations, these experiences changed how the participants taught and looked at their own practice (Nachlieli, personal communication, 26/4/2017).

3.2. *In-service mathematics teachers' professional development*

Commognitive Theory has also had an impact on in-service professional development. Here, the aim has been to enhance explorative participation of primary and middle school students by providing professional development for in-service teachers. The training focuses on designing tasks that afford multiple realizations of mathematical objects and on orchestrating productive discourse around students' solutions to these tasks (Weingarden, Heyd-Metzuyanim and Nachlieli, 2017). In this context, Commognitive Theory has also been used to examine teachers' discourse on instruction, including their professional identity and mathematical discourse (Shabtay and Heyd-Metzuyanim, 2017). This research is intended to help modify professional development programs such as the "Five Practices for Orchestrating Productive Discussions" (Smith and Stein, 2011) and Accountable Talk (Resnick, Michaels and O'Connor, 2010) by adapting them to the specific needs of teachers in the Israeli context.

3.3. *Incorporating text-based activities in high school curriculum*

Another effort at engaging students in explorative mathematical learning was pursued by Avital Elbaum-Cohen (2016) in her attempts to teach high school students to read mathematical texts. Commognitive theory contributes to this attempt to broaden the repertoire of mathematics teaching and learning both by providing appropriate and useful language for reporting on the research and by motivating students to engage with communicational activities. In her first attempts, Elbaum-Cohen taught small groups comprising advanced-level mathematics students. She found that although reading mathematical texts was designed to engage students with mathematics in an explorative manner, the students' conduct within the new learning environment had to be established ritually. Currently Elbaum-Cohen is training several groups of practicing high school teachers to incorporate texts in their advanced-level program. Furthermore, Elbaum-Cohen is currently writing a teachers' guide on these issues. Finally, the committee for the new advanced-level high school mathematics curriculum is presently considering the incorporation of reading mathematical texts.

3.4. *Assessing mathematical learning difficulties*

Ben-Yehuda (2003) developed a unique tool for assessing mathematical difficulties based on discursive ideas. The study assessed middle school students (12–13 years old), though the tool was later also adapted for primary school students. The main innovation of the tool was in focusing on students' communication and talk while solving mathematical problems as an end in itself, rather than viewing this talk as a window to some inner mental deficits. Moreover, a later publication (Ben-Yehuda *et al.*, 2005) focusing on older students (18 years old) showed how behaviors often diagnosed as "learning disabilities" may be produced in repetitive forms of talk that turn students into ritual rule followers rather than explorative problem solvers. For the past 15 years, Ben-Yehuda has been teaching this form of diagnosis at a teachers' college, thus making a long-lasting impact on diagnosing mathematical learning difficulties in Israel.

4. Discussion — The Relation Between Commognitive Theory and Practice

As can be seen in the projects described above, the Commognitive Theory has already had a major impact on education in Israel. We believe this is because viewing learning as a community-based process guides researchers influenced by Commognitive Theory to design contexts where optimal opportunities for explorative participation can occur and provides them with the tools for examining the success of their efforts.

Nevertheless, applying Commognitive Theory to educational practice is not without its challenges, first and foremost, because of the specificity and complexity of its concepts. Commognitive Theory defines and employs a series of concepts (e.g., "routines," "narratives," and "mathematical objects") that are mostly unfamiliar to practitioners (e.g., teachers and teacher educators). Therefore, there is always tension between researchers' epistemological stance and their ability to communicate this epistemology explicitly to participants. Researchers in the above described projects handled this tension in different ways. For example, in her work with pre-service and in-service teachers within college courses, Nachlieli (personal communication, 24/4/2017) was able to teach the relevant concepts and engage teachers with analysis that was actually based on Commognitive Theory (e.g., analyzing classroom transcriptions according to routines and endorsed narratives). In their in-service professional development program, however, Nachlieli, Heyd-Metzuyanim and Weingarden (e.g., Weingarden, Heyd-Metzuyanim and Nachlieli, 2017) found it more difficult to engage teachers with such an analysis. Indeed, the researchers suspected that teachers would be reluctant to engage with too much "theory". They therefore chose to keep the Commognitive Theory-based tools in the background, using them as guides for designing the professional development in ways that would enhance teachers' explorative participation in mathematical discussions and assist them in affording such experiences to their students. For example, during the sessions and in the classroom tasks the researchers stressed the importance of objectification through exposure to different realizations of mathematical objects. Elbaum-Cohen (2016) made a similar choice. In her work, she does not talk about Commognitive Theory concepts with teachers, and certainly not with high school students. Still, the Commognitive Theory strongly guides her thinking about the learning experiences that she strives to afford her students and about modifications needed in the program. Likewise, Ben-Yehuda (personal communication, 25/4/2017) does not teach her future didactic diagnosticians the concepts of Commognition. Still, her way of viewing assessment as a communicational process is epistemically rooted in Commognitive Theory. Moreover, there are good reasons to believe that her students will look at learning difficulties through such a processual lens.

5. Postscript

Almost all research related to the Commognitive Theory prior to 2016 was published in English, a second and sometimes even a third language for most Israeli students. This has made the theory less accessible to the

Israeli audience even though most of the research takes place in Israel. A recent special issue of the Hebrew journal *Research in Mathematics Education* (Nachlieli and Tabach, 2016) represents an important step in reaching the Israeli audience. It includes a collection of studies carried out in Israel deriving from the Commognitive Theory. We believe this will encourage even more practical applications of this theory in Israel in the future.

References

Ben-Yehuda, M. (2003). *Using Discourse Analysis to Investigate Arithmetical Thinking Processes of Students with Learning Difficulties* (*Unpublished Dissertation*), Haifa: Haifa University.

Ben-Yehuda, M., Lavy, I., Linchevsky, L. and Sfard, A. (2005). Doing wrong with words: What bars students' access to arithmetical discourses, *Journal for Research in Mathematics Education*, **36**(3), 176–247.

Elbaum-Cohen, A. (2016). Kriat textim mathematiyim betikhon (reading mathematical texts in high-school), *Mehkar VeIyun BeHinuch Matemati*, **4**, http://shaanan.ac.il/?page_id=24077.

Heyd-Metzuyanim, E. (2015). Vicious cycles of identifying and mathematizing — A case study of the development of mathematical failure, *Journal of the Learning Sciences*, **24**(4), 504–549.

Heyd-Metzuyanim, E., Tabach, M. and Nachlieli, T. (2016). Opportunities for learning given to prospective mathematics teachers: Between ritual and explorative instruction, *Journal of Mathematics Teacher Education*, **19**(6), 547–574.

Nachlieli, T. and Katz, Y. (2017). Ritual vs. Explorative classroom participation of pre-service elementary school mathematics teachers, *The Tenth Congress of the European Society for Research in Mathematics Education*.

Nachlieli, T. and Tabach, M. (eds.) (2016). *Mehkar VeIyun BeHinuch Matemati*.

Resnick, L. B., Michaels, S. and O'Connor, M. C. (2010). How (well structured) talk builds the mind, in D. D. Preiss and R. I. Sternberg (eds.), *Innovations in Educational Psychology*, Springer Publishing Company, pp. 163–194.

Sfard, A. (1991). On the dual nature of mathematical conceptions: Reflections on processes and objects as different sides of the same coin, *Educational Studies in Mathematics*, **22**(1), 1–36.

Sfard, A. (1998). On two metaphors for learning and the dangers of choosing just one, *Educational Researcher*, **27**(4), 4–13.

Sfard, A. (2007). When the rules of discourse change, but nobody tells you: Making sense of mathematics learning from a commognitive standpoint, *Journal of the Learning Sciences*, **16**(4), 565–613.

Sfard, A. (2008). *Thinking as Communicating*, New York: Cambridge University Press.

Sfard, A. and Lavie, I. (2005). Why cannot children see as the same what grown-ups cannot see as different? — Early numerical thinking revisited, *Cognition and Instruction*, **23**(2), 237–309.

Shabtay, G. and Heyd-Metzuyanim, E. (2017). Teachers' discourse on students' conceptual understanding and struggle, in B. Kaur, W. K. Ho, T. L. Toh and B. H. Choy (eds.), *Proceedings of the 41th Conference of the International Group for the Psychology of Mathematics Education (Vol. 4)* (pp. 177–184). Singapore: PME.

Smith, M. S. and Stein, M. K. (2011). *5 Practices for Orchestrating Productive Mathematics Discussions*, Reston: VA: National Council of Teachers of Mathematics.

Tabach, M. and Nachlieli, T. (2015). Classroom engagement towards using definitions for developing mathematical objects: The case of function, *Educational Studies in Mathematics*, **90**(2), 163–187.

Weingarden, M., Heyd-Metzuyanim, E. and Nachlieli, T. (2017). The realization tree assessment tool: Assessing the exposure to mathematical objects during a lesson, in *The Tenth Congress of the European Society for Research in Mathematics Education*.

About the Authors

Einat Heyd-Metzuyanim is a senior lecturer at the Faculty of Education in Science and Technology, Technion.

Michal Tabach is currently head of the Department of Education in Mathematics, Science and Technology, at the School of Education, Tel Aviv University.

Chapter 39

The Story of Mathematics Education in the Kibbutz Movement

Ilana Lavy

The Yezreel Valley Academic College of Education, Yezreel Valley, Israel
ilanal@yvc.ac.il

The chapter presents the story of mathematics education in the Kibbutz Movement. It starts with a description of the kibbutz way of life and its unique educational system. Then the evolution of the education system up to the 1940s follows, focusing on mathematics education for different age groups. Subsequent sections focus on mathematics education in various levels (pre-school, primary, middle, and high school) in the Kibbutz Movement from the 1940s until today, with emphasis on the meaningful changes that led to increasing number of students who study intermediate and advanced-level mathematics and pass the secondary school graduation exams in these levels.

1. Kibbutz as a Way of Life

At the beginning of the 20th century, a new form of settlement, called "kibbutz", was created in Israel. This way of life was developed by young men and women who left their comfortable lives in Europe, came to work as farmers, and settled down in what would become the State of Israel. In fact, the kibbutz was a combination of the Zionist dream and socialist life. Since then, more than 250 kibbutzim[a] have been established. They can be found across Israel and constitute about 2% of the total population.

The kibbutz is based on principles of social equality among all its members. The kibbutz function as an autonomous unit and has its own

[a]The word "Kibbutzim" is plural for "Kibbutz" in Hebrew and is used in this chapter as such (rather than Kibbutzs).

management and service systems such as an education system, health clinic, laundry service, dining room, social activities, gardening and so on. Despite the autonomous nature of the kibbutz life, to many kibbutz members it was important to be involved and to play an active role in shaping the newly established state in 1948, in light of political and security processes during those years. Some members joined the country's leadership in the struggle to achieve the national goals of the Jewish people and shape society and the State of Israel (Yoffe, 1992).

In recent years, farming has been supplanted, partly, by other economic branches, including industrial plants and high tech enterprises. In the past two decades, most kibbutzim have been privatized and changes have been made in the communal lifestyle. Nevertheless, kibbutz members still share many responsibilities.

2. The Kibbutz Movement Education Systems — Early Years

The education system of the kibbutz was entirely separated from the general education system in the country — from nursery to the end of high school. Moreover, these settings differed among kibbutzim based on the kibbutz's political affiliation and ideology. In fact, there are three main Kibbutz Movements: The United Kibbutz Movement (Hatakam) which is largely aligned with the Israeli Labor Party and its predecessors; The Hashomer Hatzair Movement which is a Socialist-Zionist youth movement, aligned with the Marxist-Zionist political party Mapam; and the Religious Kibbutz Movement which is aligned with the mainstream religious political parties. Kibbutzim that belong to Hashomer Hatzair Movement developed an education system in which students lived, learned and worked in a boarding school (Golan, 1961). The school was located in one kibbutz and students came from several kibbutzim nearby. Students spent the whole week with the educational staff, which included teachers, social instructors and nannies. Only in the weekends could they meet their families.

There were minor differences among the education schools of the kibbutzim (not including the religious ones). Teachers accompanied the students throughout the day in both academic and social activities, and had a major impact on shaping their educational and moral values. The education system in kibbutzim that belonged to the United Kibbutz Movement functioned differently. The schools were day schools, much like city schools. However, students had a complementary education system, after school hours. In this system, students performed their educational duties, worked in the kibbutz, and had free time for social activities. It allowed their parents to dedicate their time and attention to their work.

The kibbutz education system had two conflicting goals: On one hand, focusing on the individual and providing him with a humanistic, Jewish and Zionist education that would enable personal growth and fulfillment. On the other hand, focusing on building expectations from, and pressures on, individuals to remain in the kibbutz (Yoffe, 1992).

3. The Kibbutz Education System Until 1940

As was previously mentioned, the first members of the kibbutz were young people. Consequently, educational issues started to arise only when the first children were born. These issues continued emerging as the children matured. At every stage of the children's growth, the kibbutz had to pave the way and provide solutions, which would address the needs of each age group. These solutions had to fit the ideology of the kibbutz and other external circumstances such as financial means and political propensities (Porat, 1987). The basic educational ideology put the responsibility of education of the youth in the hands of the educational staff, which included teachers, nannies, and social instructors. Parents had a secondary role, at best.

The education system included a local primary school, children and youth houses, an informal social education, and a labor system in which the students worked before or after school hours. It provided students with the autonomy to manage their own lives. Each group of children had its own committees and functionaries and ran itself (a kibbutz within the kibbutz). The relationships between the educational staff and the children were based on equality and lack of formality, which sometimes caused conflicts in cases requiring disciplinary measures (Porat, 1987). As to high schools, they were autonomous and were completely independent from other municipal schools. Due to the small number of children, several kibbutzim shared a regional high school.

The learning in schools was based on general themes taken from the kibbutz life and was based on self-learning of the students. They worked in small groups on worksheets. Only mathematics and Bible classes were taught in the afternoons, separately from the other subjects. The teaching of mathematics was utilized to learn topics related to the kibbutz life: economic calculations, work lists, measuring areas, and estimates of distances. The learning method was based on class discussions and independent learning of the students. Since the students had to work after school, they were not given homework.

There were slight differences between the kibbutzim schools at that time. For example, the teachers of Kibbutz Ein Harod attached great importance to homework, but conditions were difficult (Porat, 1987). Students used to do their homework in the communal houses, which did not provide a peaceful environment to support this activity. High school students studied in the afternoon and worked before noon. This situation was changed at the beginning of 1940. From this year and on, all students up to 11th grade studied five days a week before noon and one day was dedicated to work. Twelfth grade students had to work five days and studied one day.

4. Mathematics Education in the Kibbutz Preschools

Nursery schools and preschools in the kibbutzim underwent significant changes over the years. These changes influenced the nature of the educational activities. In the early years, most kibbutzim had communal children houses in which the children actually lived. Between 0–6 years of age, all activities, from bathing, eating, and sleeping to educational activities were performed in these houses by nannies who received special professional guidance. The educational staff was in charge of a small number of children, and was responsible for all the activities that took place in the houses. During those years, these communal houses were only for kibbutz children, in small multi-age groups. Infants acquired basic mathematical concepts while playing in the yard outside the communal houses, when nannies introduced concepts related to counting and to basic geometric shapes. They also encouraged the infants to classify toys by types, colors or shapes while playing in the yard. Even social activities carried out inside the communal houses involved learning of basic mathematical concepts such as counting and identification of geometrical shapes. The small number of infants per nanny allowed nannies to give their full attention to the infants' needs. That made the preschool kibbutz education system one of the best one can find.

In most kibbutzim today, the communal house is no longer the actual children's home in which most of their life activities take place. Today, these houses are mostly a social and cultural club during the morning hours (for preschoolers) or during the afternoon (for school age children) (Dror, 1997).

5. Mathematics Education in the Kibbutz Primary Schools

Mathematics was taught similarly to other education systems in the country. To prepare kibbutz members for teaching according to the kibbutz spirit, prospective teachers studied the art of teaching in three colleges dedicated to this purpose: Oranim Academic College of Education near Haifa, Beit Berl Academic College of Education near Kfar Saba, and Kibbutzim Academic College of Education in Tel Aviv. Didactical materials included all the mandatory mathematical concepts for each grade level, but the word problems involved stories taken from the kibbutz life. One of these specialists was Miriam Nussdorf (1907–1974), who wrote the book "Methods of Mathematics" (1959) — a methodological guidebook for teaching mathematics in grades 1–4.

The teachers did not adhere strictly to the mathematics school curriculum. Mathematics, as well as other subjects, was based on self-study practice that was done with work sheets. The number of students in classes was relatively small compared to numbers in equivalent classes in city schools, which allowed teachers to pay more attention to each individual student and provide them with the maximum help (Lavi, 1990).

Since the 1980s, kibbutz primary schools grew, opened their gates to children from nearby communities, and mathematics teaching has undergone a significant change. The number of students per class grew, and many teachers were not kibbutz members. The growing number of students per class enabled grouping students in mathematics according to their needs and abilities. Today, most primary schools in kibbutzim are regional schools and although the number of students in each school grew dramatically, they are still smaller in comparison to urban primary schools.

6. Mathematics Education in the Kibbutz Middle and High Schools

Most graduates of kibbutzim schools completed their high school education without taking the secondary school graduation exams, because the aim of their studies was learning for the sake of knowing (Yoffe, 1992). Rather than successfully passing the exams which is necessary for gaining entry to universities and colleges.

In fact, official studies in the kibbutz high schools, until the 1960s, ended in the 11th grade. In the 12th grade, the youth worked in the kibbutz until drafted by the army. Only a few students decided to take the secondary school graduation exams and had to study for them after school hours, without teacher guidance. Since there was no external motivation for mathematics studies (e.g., passing the graduation exam) there was no need to follow the school curriculum strictly and meet deadlines. As a result, the vast majority of students studied basic-level mathematics and very few studied at the intermediate or advanced levels. From the 1970s and onward, there was a growing demand to learn similarly to city students. This demand caused teachers to teach according to the national school mathematics curriculum and its timetables. From a situation (in the 1970s) in which only one or two students took the matriculation exams, students' motivation to excel increased over the years and most of these school graduates gained access to academic studies at universities and colleges. For example, between the years 2013–2015, the following percentages of kibbutz students took each secondary school exam: basic-level — 58%, intermediate level — 30%, and advanced level — 12%.

The privatization of the kibbutzim, which led to differential salaries based on occupation, also served as a catalyst for parents to encourage their children to excel in studies so that they will be able to acquire a good profession.

As for the educational staff, most mathematics teachers in recent years are hired ones, not Kibbutz Members, and as a result, their teaching style is not influenced by the principles and life style of the kibbutz. The class atmosphere changed over the years as well: while 30 years ago, students were driven only by internal motivation (i.e., interest, curiosity, challenge), today they are driven by external motivations as well (e.g., passing the secondary school graduation exams, pleasing their parents by receiving good grade reports).

References

Dror, Y. (1997). *Kibbutz Education in its Environment*, Tel Aviv: 'Ramot' Tel Aviv University Press (in Hebrew).
Golan, S. (1961). *Collective Education in the Kibbutz*, Tel Aviv: Education Department of the Kibbutz-Artzi, Hashomer Hatzair (in Hebrew).
Lavi, Z. (ed.) (1990). *Kibbutz Members Study Kibbutz Children*, New York: Greenwood Press.
Nussdorf, M. (1959). *Methods of Mathematics*, Beit-Berl: Beit Berl Academic College of Education (in Hebrew).
Porat, R. (1987). *Ha-yachid Ve-hayachad: Toldot Beit Ha-sefer Ha-kibbutzi*, Tel Aviv: Ha-kibbutz Ha-meuhad. (Together but on our own: The creation of schools in the Kibbutz Movement) (in Hebrew).
Yoffe, M. (1992). *What Will be the Future Direction of the Kibbutz's High-School in Israel?* Unpublished master's thesis, Hebrew Union College, Los Angeles.

About the Author

Ilana Lavy is a Kibbutz member of Ein Harod, and a mathematics educator holding a position of associate professor at Yezreel Valley Academic College of Education. Her main research interests are mathematics teachers' professional development, and students' acquisition and understanding of concepts in mathematics and computer science. She has published over 100 papers and research reports (some in Hebrew).

Chapter 40

Math Literacy in Israel: What Does PISA Tell Us That We Did Not Know Before?

Zemira R. Mevarech

School of Education, Bar-Ilan University, Ramat-Gan, Israel
David Yellin Academic College of Education, Jerusalem, Israel
mevarz@mail.biu.ac.il
zemiram@dyellin.ac.il

This chapter focuses on Israeli students' performance in mathematical literacy as assessed by PISA. It discusses the PISA definition of mathematical literacy, and describes how it is assessed by PISA. The chapter then reviews the results of PISA assessment, showing low performance on all four mathematical content areas (quantity, space and shapes, change and relationships, and uncertainty and data), extremely large variations between and within schools, and difficulties in students' ability to handle certain mathematical processes. The general picture was not new to the Israel people, but the detailed analyses, the depth of the variations, and the very fact that Israel lagged so much behind the OECD average were not anticipated.

1. What is Math Literacy? The PISA Approach

The dictionary definition of "literacy" is "the ability to read and write" (Oxford Dictionary, 1963). Yet, the Programme for International Student Assessment (PISA) enlarged "literacy" concept much beyond the "traditional" approach described above. Looking on literacy skills from the world of tomorrow perspectives, PISA distinguished between three types of literacy: reading, mathematics, and science. Hereafter, the chapter refers only to mathematics literacy, defined by PISA as:

> "An individuals" capacity to formulate, employ, and interpret mathematics in a variety of contents. It includes reasoning, understanding, and using mathematics

concepts, procedures, facts, and tools to describe, explain and predict phenomena. It assists individuals in recognizing the role that mathematics plays in the world and to make well founded judgments and decisions needed by constructive, engaged and reflective citizen" (PISA, 2012, p. 1).

The major emphasis of PISA literacy is on the **use** of mathematical knowledge and skills rather than just memorizing mathematical facts and routine problem solving. In particular, PISA focuses "not only whether students can reproduce knowledge, but also whether they can extrapolate from what they have learned and apply their knowledge in new situations. It emphasizes the mastery of processes, the understanding of concepts and the ability to function in various types of situations" (PISA, 2012, p. 1).

2. How is Mathematical Literacy Assessed by PISA?

PISA research is organized in three-year cycles. Each cycle assesses all three literacy domains: reading, mathematics, and science, but in each cycle one domain is more dominant than the others. For example, in 2015 the main domain was science literacy, in 2012 mathematical literacy, and in 2009 reading literacy. The main domain covers about two-thirds of the test items, whereas the other two domains include about one-third of the items.

As in reading and science, three broad dimensions are used for the transformation of mathematical literacy into an assessment: contents, processes, and contexts or situations. The *content* refers to the "big ideas" in four areas: quantity, space and shapes, change and relationships, and uncertainty. The *processes* are organized in three "competency clusters": (a) *formulating* situations mathematically, (b) *employing* mathematical concepts, facts, procedures and reasoning, and (c) *interpreting*, applying, and evaluating mathematical outcomes (referred to in an abbreviated form as *formulate*, *employ*, and *interpret*). The *contexts*/situations represent various kinds of problems encountered in real-life, from those affecting the students directly to scientific problems of more general interest. PISA further classified the situations into personal, occupational, societal, and scientific (PISA, 2012).

The mathematical literacy test includes multiple-choice and open ended items (about one-third and two-thirds of the items, respectively). All items represent authentic situations, and many include diagrams and verbal information. The multiple-choice items were scored 1 or 0 (correct or incorrect, respectively). For more complex items, the scoring includes, full or partial credits. Very detailed rubrics were provided to ensure high reliability.

In order to analyze trends, the scores are calibrated as follows: the range is 200–800, mean OECD score is 500, and standard deviation (SD) is equal to 100. The scores are clustered into six proficiency levels, the highest level includes scores above 669, the lowest between 358 and 420; Scores below 358 are labeled "below level 1".

PISA tests assess students' literacy at the age of 15–16 (usually, 10th graders). In addition to the literacy assessment, students answer a questionnaire that focuses on the students themselves, their homes, and school learning experiences. Principals are also asked to fulfill a questionnaire that covers the school system and the learning environment. The battery includes, therefore, mathematical literacy test to complete in two hours, a student questionnaire to complete in half an hour, and a principal questionnaire to complete in 20 minutes. This chapter refers only to students' mathematical literacy performance.

Israel participated in all PISA cycles from 2002 on, with the exception of the 2003 cycle. The findings reported here are all taken from the PISA (2016) report (the cycle itself was implemented in 2015). Detailed analyses referring to the contents and processes were reviewed from the PISA (2012) report, in which mathematics was the main domain.

3. Israel Mean Scores and Ranking: The Big Surprise

Israel was ranked at the bottom third of all participating countries with a mean score of 467, about one-third standard deviation below OECD mean score. PISA indicates a continual increase in the mean scores from about 454 in 2006 to 467 in 2015 (as indicated Israel did not participate in PISA 2003).

Detailed analyses of the contents and processes subscales indicated a mean score of 462 on "change and relationships", 465 on "uncertainty and data", 449 on "space and shapes", and 480 on "quantity". Interestingly, while the difference between the content subscales was about 1/3 SD, the mean scores on the three processes subscales was quite the same. On **formulate** Israel mean score was 462, 35 points below the OECD mean score;

on **employ** Israel mean scored was 465, 27 points below the OECD mean score; and on **interpret** Israel mean score was 469, 36 points below the OECD mean score.

What did Israel learn from these findings? In 2003, when PISA findings were first published, Israel was highly surprised. The Israel newspapers' titles asked again and again "how come Israel, the "start-up nation", scored so low on mathematical literacy?" "Will Israel lose its priority in computer and natural/exact sciences in the coming years?" "Why did it happen?" These issues were raised in the Knesset, Israel Parliament, by parents, teachers, and the public in large. Yet, no clear answers were provided. Some people "blamed" the teachers, others the curriculum that is very traditional, and still others the students themselves who were not motivated to answer the test items because it did not affect their scores on the report cards (the tests are anonymous). Similarly to the weather, people talked about PISA findings, but not much has been done in order to improve Israel mathematical literacy.

Even though Israel participates in PISA research for more than a decade and in other international studies (e.g., TIMSS) since the sixties, it seems that on all cycles people were surprised again and again to find out how low Israel performance was. Probably, this is due to the large gap between PISA results and the facts that Israel scientific accomplishments are very high and is one of the worldwide leaders in the digital era.

The low performance of Israel students on the various contents and processes subscales challenged mathematics educators and researchers. Israel school mathematics curricula emphasize "quantities" at all levels of education which indeed turned out to be the best subdomain performance, only 5 points below the OECD mean score. In contrast, the Israel school mathematics curricula put less emphasis on "space and shapes" on which the performance was the lowest: 51 points below OECD mean score. These findings raise a general question: to what extent a country has to change its curriculum in order to be in accord with the "big ideas" of the OECD committee? A similar question relates also to the three processes specified by PISA: formulate, employ and interpret. On all three subscales, Israel scored about 30 points below the OECD mean score.

Finally, a lot of research has been published on the effects of metacognitive pedagogies on mathematics achievement and reasoning (e.g., Mevarech and Kramarski, 2014). Yet, there is scarce research on the effects of these methods on mathematical literacy and in particular on the "interpretation and reflection" process as described by PISA. This issue merits future research.

4. Extremely High Variations Between and Within Israel Schools: Was it Expected?

The low mean scores went hand in hand by extremely high variation. In all PISA cycles the variation in Israel was either the largest or second largest among all participating countries. Large variation was found not only between schools, but also within schools.

The large variation was reflected also in the very large gap between the high-achievers (levels five and six) and the low-achievers (level 1 and below). While only 9% of the Israeli students scored on the top five and six levels, 32% of the students scored on the lowest levels (below level two). The proficiency finding is even more alarming considering the differences between the Arabic and Jewish sectors (see below). The proficiency level of the Jewish sector is slightly lower than the OECD one, but among the Arabic students only 1% of the students were rated at the top 5–6 levels and about 64% were rated below level 2.

The large variation was found not only between schools, but also within schools. A large difference between schools was expected since Israel education system is divided into two big sectors and several smaller subsectors. The two big sectors are: Hebrew speaking schools (Jewish schools) and Arabic speaking schools. Within the Jewish schools there are three subsectors: Secular, Orthodox, and ultra-Orthodox schools. Within the Arabic sector there are also three subsectors: Muslims, Druze, and Bedouins. For years, it has been known that there is a large gap between the Hebrew and Arabic speaking schools. Nevertheless, a difference of almost one standard deviation between the two sectors has to ring a warning bell.

Another finding that Israel was not aware of is the large within-school variation. Given the fact that Israel education system includes so many subsectors, one would expect a large variation between schools, but a small variation within schools due to the fact that the schools are more homogenous in terms of student population. However, that is not the case. In many Israel schools, students are grouped by ability, or by the levels to which they are prepared for the secondary school graduation exam. This causes a high within-school variation.

5. Comparison of Mathematical Literacy among Israel Educational Sectors

Although the Israel education system is divided into a large number of subsectors, PISA sample size did not allow analyses of the ultra-Orthodox boys and the subsectors within the Arabic-speaking sector. Therefore, the following analyses refer to the main subsectors within the Jewish schools: secular, Orthodox, and ultra-Orthodox girls. The analyses are based on the 2012 cycle.

The findings indicate that students in the secular schools scored higher than the students in the Orthodox schools, who in turn scored higher than the ultra-Orthodox girl schools (Mean scores = 500, 488 and 465, respectively). Many people predicted it because the ultra-Orthodox subsector emphasizes reading literacy (of mainly religious texts), but much less math literacy. Yet, a comparison of the girls mean scores within these three subsectors showed much smaller differences of about a quarter of a SD (Mean scores = 489, 478, 465 for the secular, Orthodox, and ultra-Orthodox, respectively). Since the ultra-Orthodox educational system is a "terra incognita", the relative small difference between the girls' subsectors was surprising. Several explanations were provided, each merits further research. First, the core curriculum of the ultra-Orthodox schools is reading comprehension which is a prerequisite for answering PISA tests that include a lot of verbal information. Second, while the secular and Orthodox students disrespect anonymous tests, the ultra-Orthodox girls who are generally more obedient were driven to do their best on the tests regardless the fact that the tests are anonymous. Finally, Pygmalion effect worked for the ultra-Orthodox girls who are educated from a very early age to be the main employee in the family and therefore had to excel in education.

6. No Significant Gender Differences: The Good News

Israel was one of the countries in which the gender difference was not significant (boys outperformed girls by 8 points, but this difference is statistically not significant). Further analyses indicate that the largest gender differences were obtained on the "interpret" process and the smallest and insignificant on "employ". Interestingly, The Arabic girls outperformed the Arabic boys by 12 points, whereas in the Jewish sector, the boys outperformed the girls by only 4 points.

Gender-differences in mathematics education have been largely explored in the last decades (e.g., Else-Quest, Hyde and Linn, 2010). Many of the studies showed that boys outperformed girls, mainly in high school and higher education. Until lately, Israel was not exceptional. Given these findings, Israel Ministries of Education have put a lot of efforts in order to eliminate, or at least minimize, the gender differences. The insignificant gender differences found in 2016 may serve as a springboard to encourage girls to specialize in mathematics, a key factor for being constructive, engaged, and reflective citizens.

7. Mathematical Literacy and Student Family Background: Tell Me Who You are and I'll Tell You Your Achievement Score

PISA developed a special index for assessing students Economic, Social and Cultural Status (ESCS), or in short "family background". Students reported about parents' education and occupation, excess to educational, cultural and economic resources, and the existing of other educational resources in home. The socio-economic background was divided into three categories: low (bottom 30%), middle (40%) and high (up 30%).

In Israel, the differences between highest and lowest SES students were particularly large, almost one standard deviation (Mean scores = 510 and 423, respectively). Surprisingly, the SES gap within the Arabic sector was much smaller than within the Jewish sector (Mean gap = 32 and 84 points, respectively).

The effect of family background on students' achievement is well known in the educational research. Children from low SES families start school with many more difficulties compared to their counterparts from higher SES families, and the disadvantages that started at kindergarten, increases over the years. Research pinpoints many reasons for explaining this phenomenon: the lower SES families are occupied by "survival" issues and therefore they are less involved in their children education, the level of the lower SES schools is poor, and the students themselves are less motivated to engage in education. Yet, several countries were able to eliminate the SES gap; unfortunately, Israel is not one of them. A gap of almost one standard deviation between the highest and lowest SES students is much beyond what was anticipated.

8. Summary and Concluding Remarks

"Mirror, mirror on the wall, who is finest of them all?" PISA mirror provides a clear picture: Israel's mean score on mathematical literacy is low, the variation is extremely large, the percentage of low-achieving students is very high and the percentage of high-achieving students is relatively low.

What is the role of PISA mirror? Does it have to tell the world "who is finest of the all"? Or, is it a diagnostic system of the participating education systems? With regard to Israel, although the diagnostic information that PISA provided was generally known, nevertheless it was unexpected to find out how severe the level of mathematical literacy was. PISA shows Israel how much it lagged below the OECD mean score and how deep are the variations in mathematical literacy. Maybe the big surprise is that over more than a decade, the annoying findings remain quite the same.

References

Else-Quest, N. M., Hyde, J. S. and Linn, M. C. (2010). Cross national patterns of gender differences in mathematics: A Meta-Analysis, *Psychological Bulletin*, **135**, 103–127.

Mevarech, Z. R. and Kramarski, B. (2014). *Critical Maths for Innovative Societies: The Role of Metacognitive Pedagogies*, Paris: OECD.

Oxford Dictionary (1963). Oxford University Press.

PISA (2012). *PISA 2012 Results: What Students Know and Can Do: Student Performance in Mathematics, Reading and Science*, Paris: OECD.

PISA (2016). *PISA 2016 Results: Excellence and Equity in Education*, Vol. 1, Paris: OECD.

About the Author

Zemira R. Mevarech is professor (emerita) of education, currently the president of David Yellin Academic College of Education. She was the Head of the School of Education, the Dean of the Social Sciences Faculty, and the Vice Rector at Bar-Ilan University. Prof. Mevarech was the Chief Scientist of Israeli Ministry of Education. Her research focuses on mathematics education, meta-cognition, teacher education and professional development. Her most recent book "Critical Maths for Innovative Societies: The Role of Meta cognitive Pedagogies" was published by the OECD and was translated into Japanese and Spanish. Prof. Mevarech holds a PhD from the University of Chicago.

Chapter 41

Kidumatica: A Multicultural Fulcrum for Leveraging the Israeli Periphery's Untapped Reservoir of Mathematical Excellence and Talent Toward Academic and Professional Success

Miriam (Miri) Amit

Ben-Gurion University of the Negev, Beer-Sheva, Israel
amit@bgu.ac.il

"Kidumatica" is an extra-curricular program that, for 20 years, has worked to identify and develop the potential of mathematically talented students in Israel's geographic and socio-economical periphery. It is built on a cyclic model: its **sound theoretical foundation** shapes the **implementation of Kidumatica experience**, which in turn is a platform for generating **new research**. Every year, approximately 550 students come to the Ben-Gurion University campus to participate in the program, some for up to 5 years. Kidumatica demonstrates how, through a specially designed structure and tailor-made materials and methods, we can help talented students break through social barriers and attain great academic and professional heights.

1. Kidumatica: Tapping into the Nation's Talent Pool

The economic, scientific, and technological future of the State of Israel depends largely on identifying and cultivating a new generation of talented youth who will lead the country in facing the technological world of tomorrow and achieving economic independence. Because it is not rich in natural resources, it is doubly important for Israel to find ways to harness the full potential of its citizens. Both social justice and national expediency demand that this include the residents of Israel's geographical and socio-economical periphery. Discovering, developing and implementing their intelligence and talent is the personal prerogative and the national duty of every single citizen. In this chapter, we present "Kidumatica — The Mathematics Youth Club for Excellence and Creativity" as a model that turns the goal of unlocking the potential in Israel's periphery into a reality.

In Hebrew, "Kidumatica" is a play on words, combining "kidum", meaning promotion and development, with "matematica" — mathematics. It reflects the twin pillars that define the program: one social and one scientific. The social pillar emphasizes the social justice of providing equal opportunities for everyone to realize their potential and achieve personal success. The scientific pillar consists of research into the cognitive and affective aspects of the learning experience, particularly mathematics.

Kidumatica was designed to identify and promote the mathematical abilities, skills and creativity of talented students from diverse cultural backgrounds, all of whom live in a relatively underprivileged area of Israel's Southern Region, the Negev. Mathematics was recognized as the most appropriate means achieving this goal for several reasons. First, because it is an indicator and predictor of ability and success. Second, mathematics is applicable in other areas and to everyday life. Third, success in mathematics confers prestige in today's society. Finally, mathematics is an important key for anyone wishing to participate in Israel's global role as a "start-up nation".

In the nearly 20 years since its inception, thousands of Kidumatica graduates have gone on to higher education in prestigious faculties, followed by lucrative careers in a variety of mathematically rich domains, where residents of the southern periphery are significantly under-represented. The project and its success stand as an ongoing testament to the possibility of traversing the cultural, ethnic, linguistic and socio-economic barriers that so often prevent bright and deserving young people from realizing their full potential.

2. Kidumatica — Facts and Figures

Founded in 1998 by the author of this paper, Kidumatica operates out of Ben-Gurion University of the Negev. Students in the Negev tend to score relatively low on national exams, and many resources have been invested in the struggling students in an effort to bring them up to the national average. In this context, however, the talented students (of which, it turns out, there are many!) have not been allocated any special attention, and as a result have rather fallen through the cracks. Kidumatica was created to address this gap — to seek out the more talented students, develop their potential and expand their horizons.

Every year, the program brings together an average of 550 students, aged 10–17, who study in Kidumatica for 3–6 years. The students come from 100 different schools in 35 communities throughout the Negev. While many students are recommended for the program by their schools, students are also encouraged to apply directly. Apart from cognitive empowerment, one of Kidumatica's central goals is fostering a culture of tolerance and cooperation by bringing together students from different cultural, religious and ethnic backgrounds in an atmosphere of camaraderie, teamwork and mutual support. It has drawn students from 26 different counties of origin, many of whom still speak one of 14 different languages at home. Approximately 15% of the students are from Bedouin tribes living in the Negev Desert. Another phenomenon that sets Kidumatica apart is that 40% of its participants are girls — a number almost unheard of in mathematically-oriented enrichment programs! Throughout the program's existence, its students have consistently won prestigious awards in national and international mathematics competitions. The term "Kidumatica" has become a byword for mathematical excellence, earning a place at the top of Israel's mathematical pyramid and marking a major achievement for the Negev's morale and reputation.

Kidumatica relies on a three-part cyclic continual structure: (a) the scientific research upon which the Kidumatica model is based; (b) the "Kidumatica experience" — the model's development and implementation (the main focus of this chapter); (c) new research derived from Kidumatica's activity, which serves both to improve the program's implementation and to contribute new theoretical and practical knowledge to the field of excellence education.

3. The Theoretical Basis for Kidumatica

The brief description provided here does not represent the full extent of Kidumatica's theoretical basis, but is rather a summary of the theories whose impact on the program was most direct and profound. These include cognitive theories like Krutetskii's (1976) description of the cognitive characteristics and abilities of gifted students, Renzulli's (2002) extended "Three Ring" conception of giftedness as a combination of above average ability, creativity and commitment, and Guilford's (1975) theories about creativity. These theories are central to the identification of new students for Kidumatica, and serve as guidelines for determining the qualities Kidumatica strives to develop in its students. Pedagogically, Kidumatica relies heavily on theories expressed in Sriraman (2008), with special emphasis on the role of challenging problem solving as a central teaching strategy. Finally, Kidumatica is based on research into the affective aspects of learning, and the importance of fostering students' determination, ambition, motivation, confidence and willingness to face challenges. The development and implementation of Kidumatica is based on numerous additional sources, which address issues like representations, argumentation, classroom management and performance assessment, but due to the limited scope of this paper, they will not be detailed here (for more, see Amit, 2010).

4. Implementation: The Kidumatica Experience

In this section we describe how Kidumatica's theoretical foundations are implemented in all its aspects, including identification, structure, content, pedagogy and assessment.

4.1. *Equal-opportunity identification*

Kidumatica's entry exams are carefully designed to bypass any reliance on culture, or on previous academic experience. Applicants are asked to sit two separate exams, spaced one week apart. This serves a dual purpose: first, it increases the exams' credibility; second, it decreases the potential impact of anxiety and novelty on the students' performance. Language assistance is given to non-Hebrew speakers (immigrants, Arabs, etc.).

The exams are "potential testing"; they rely as little as possible on knowledge acquired in school to minimize any bias against applicants from less prosperous schools. The test consists primarily of open-ended, non-routine questions that allow for a variety of creative solutions and require persistence and focus to solve. Here, for instance, are two examples that test (respectively) the number sense and the visualization skills of ten-year-old applicants (prealgebra):

> 1. Who Am I? I am a two-digit number.
> If you double me and subtract 1, you will get me in reverse order.
> Who am I? (Explain all your steps and justify your answer.)
> 2. Draw two triangles, the overlapping area of which is a triangle.
> Is it possible to draw two triangles with an overlapping area that forms a pentagon? If yes, draw them. Can you form other shapes? Show us.

Scoring of these problems is based not only on the final product, but also on the process, argumentation, quality and creativity of solutions.

4.2. *Structural elements — how Kidumatica works*

Structure and location: Kidumatica workshops take place twice a week for four hours in the afternoons, on the Ben-Gurion University campus. They are conducted in small groups, divided by age and achievement level. Because of its location inside the university, the program offers these students — and often their parents — their first glimpse of a university environment, and an opportunity to gain a familiarity with what might otherwise have seemed a strange and inaccessible place.

Pedagogical approach: Kidumatica's pedagogical paradigm utilizes problem based learning (PBL) and an interactive approach that encourages collaborative and experiential learning. Its classes include almost no frontal teaching; instead, students are given a problem and work with the instructor towards its solution, acquiring and developing mathematical knowledge as they go. The skills and tools they acquire focus on logical thinking, quantitative understanding, spatial vision, and creative use of mathematical, scientific and technological tools to complete the challenge.

Mathematical content — areas of cultivation: Kidumatica is meant to enrich its students, broaden and deepen their understanding; hence, there is no acceleration in formal school curriculum. Instead, a "tailor made" curriculum was developed, based on mathematical areas of knowledge that are not taught in school, to provide students with an alternate perspective on mathematics that "revitalizes" it as a field of study. This unique curriculum includes topics like: logic, critical thinking, optimization, linear algebra, number theory, sets and patterns, non-routine problem solving, probabilistic thinking, history of mathematics, mathematics in art and music, modeling of real life situations and more.

Unique assessment method that prompts learning: There are no tests in Kidumatica. Students are assessed through a specially designed "rating system". Each student starts with 1000 "rating points", and can earn more based on parameters like active participation in the group, giving an exceptional answer or asking an exceptional question, solving homework or classwork creatively, winning competitions, etc. Every week, the names of the highest rated students are displayed on a bulletin board, and annual prizes are given to those rated highest. The rating system is designed to increase motivation and provide positive reinforcement for a wide range of qualities, such as perseverance, teamwork and creativity, so that everyone — not just the academically strongest students — has a chance to shine.

Activity days and field trips: Every five weeks Kidumatica holds special activity days. These include scientific guest lectures, competitions and the publication of a mathematical newspaper. Students are also taken on field trips to national research facilities and science museums. Mathematical festivals take place twice a year and a special open day for parents is conducted to give them the "taste of Kidumatica".

Instructors: The Kidumatica instructors are usually graduate students, PhDs and experts in specific mathematical fields. Each instructor teaches one or two topics in their area of interest or research. The instructors develop their own teaching materials, and teach every group, adapting the material to their age and ability. The material is taught in the form of "mini-courses" or "workshops" approximately 4–5 meetings long.

Mentors: Because the vast majority of Kidumatica students come from homes that can offer them limited academic support in mathematics, the program provides each group with a **mentor**. These are university students from mathematically rich fields, often Kidumatica graduates themselves. They are "older siblings" to their students, who help them cognitively and emotionally, and follow their attendance and progress.

5. Kidumatica as a Site for Developing New Educational Research and Practice

While Kidumatica has made tremendous strides in promoting excellence and equity, it has also become a fruitful field for research. The research conducted in Kidumatica can be roughly divided into two types. The first consists of examinations of Kidumatica itself — its structure, identification process, instructors and mentors, interaction with schools and parents, characteristics of success, etc. These serve as a means of identifying the project's strengths, as well as areas that are in need of further development.

The main body of research conducted in Kidumatica, however, is made up of studies devoted to designing, implementing, researching and developing new and appropriate research-based learning materials in innovative content areas, and teaching approaches for mathematically talented students. Doctoral studies have used Kidumatica as a basis for research on topics such as the cultivation of higher order thinking, critical thinking, explicit teaching, strategic approaches to non-routine problem solving, advancing creativity through mathematical modeling of real-life situations and more. The fruits of their research are then translated into new materials and methods for use in Kidumatica.

6. Kidumatica's Legacy — Proving that it is Possible

For nearly 20 years, the success of Kidumatica's formula has been exemplified by the consistently large demand for participation in the program, and by the outstanding achievements of its students. By far the most striking of these is the phenomenal integration of Kidumatica's graduates into institutes of higher learning, elite cyber-oriented military units and prestigious, high paying jobs. Their success serves as constructive proof that it is possible — given the right model — to break through substantial social barriers. In helping students overcome these barriers, Kidumatica fulfills what should be a national obligation to give each and every individual student the opportunity to compete and cooperate with their fellow citizens as equals. In doing so, it also promotes the interests of the nation as a whole by helping to channel the bright and creative young minds of the Negev into the national talent pool.

References

Amit, M. (2010). Gifted students' representation: Creative utilization of knowledge, flexible acclimatization of thoughts and motivation for exhaustive solutions, *Mediterranean Journal for Research in Mathematics Education*, **9**(1), 135–162.

Guilford, J. P. (1975). Creativity a quarter century of progress, in I. A. Taylor and J. W. Getzels (eds.), *Perspective in Creativity*, Chicago: Aldine Pub., pp. 37–60.

Krutetskii, V. A. (1976). *The Psychology of Mathematics Ability in School Children*, Chicago: University of Chicago Press.

Renzulli, J. S. (2002). Expanding the conception of giftedness to include co-cognitive traits and to promote social capita, *Phi Delta Kappan*, **84**(1), 33–58.

Sriraman, B. (2008). *Creativity, Giftedness, and Talent Development in Mathematics*, Charlotte, NC: IAP — Information Age.

About the Author

Miriam (Miri) Amit served as chief superintendent of school mathematics in the Ministry of Education and then as a Department Chair and Dean at Ben-Gurion University (BGU). She got her MSc at the Technion and her PhD at BGU. Her interest areas are: excellence and equity, ethnomathematics, creativity and problem solving, assessment and educational policy.

Chapter 42

Opening the Door to Ethnomathematics in Israel

Ada Katsap

Kaye Academic College of Education, Beer-Sheva, Israel
adak@013.net

The chapter describes ideas examining the interweaving of ethnomathematical approach into preservice mathematics teachers education and mathematics school curriculum in Israel. Appraised are the advantages ethnomathematics offers teachers in accommodation of pedagogical skills for effective coping with the cultural diversity of the classroom population, and understanding the perspective of multicultural mathematics with its pedagogical implications.

1. Ethnomathematics

In an era of social networks that transformed the world into a global village, can cultural uniqueness still exist in science and mathematics? Most people, including that part of the public actually engaged in mathematics, are surprised by the mere posing of the question. However, when forced, they might recall the contributions of ancient Egypt and Greece and that of the Chinese, Indian and Arab mathematicians, as well as famous European mathematicians in modern era of mathematics, which brought glory to their respective countries. The Jewish people might list names like Abraham Bar Hiyya, Abraham Ibn Ezra, and others. Does it allow seeing mathematics as multicultural? The response reveals itself in Ethnomathematics, a discipline that gained attention since the 1980s, when Ubiratan D'Ambrosio, a Brazilian mathematician, defined the term as an intersection of pure mathematics with its socio-cultural-historical roots. The term Ethnomathematics constitutes a combination of the *ethno* and *mathematics*, with the prefix *ethno* endowing socio-cultural context, and *mathematics* relating to knowledge such as counting, weighing and measuring, comparing, sorting, classifying, designing and playing.

Since then, mathematicians and educators from around the world discussed at conferences and in the research literature various links between ethnomathematics and mathematics. Their proposals and conclusions led to acknowledging the field of ethnomathematics as an academic discipline and to the building of an ethnomathematical program relating to six dimensions: Cognitive, Conceptual, Educational, Epistemological, Historical, and Political, reflecting the agenda of extensive view of mathematics, and attitudes that may direct the learning process in the classrooms (Rosa, D'Ambrosio and coworkers, 2016). D'Ambrosio criticized "the mainstream image of mathematics education focused mainly on the transmission of mathematical content, with less attention given to cultural issues", and looked for "a new view of the history and philosophy of mathematics and their pedagogical implications" (p. 6). The mathematics community treads well on the new discipline, which over the past three decades has embarked on a course that guaranteed its approval by a line of mathematicians and scholars who explored the phenomenon. From studies, articles and books we can learn about practical mathematics, which finds its expression in the world of numbers and measurements, as well as in areas of architecture, ornamentation, weaving, living and working activities, and spiritual and religious practices. These ethnomathematical expressions cover a wide range of countries on the world's map: Africa, Alaska, North and South America, Portugal and Brazil, Australia, and so on. In Israel, the first book on the theme, titled "Ethnomathematics of Negev Bedouins: Existence in Forms, Symbols and Geometric Patterns", was published by Katsap and Silverman (2016). However, we could accept the perspective of multicultural mathematics as "a human situated in space and time that imply different knowledge and different practices to live by" (Domite and Pais, 2010, p. 1473), and can be comprehended by different populations around the world.

2. When Mathematics, Ethnomathematics and Pedagogy Meet

Let's proceed by taking a look at the discourse calling to implement ethnomathematics perspectives within mathematics education. The enterprise invites collaboration between mathematicians, educational researchers, mathematics educators, and teachers, regarding ideas exploit at the advantages and disadvantages of the intended collaboration. Shirley (2001) reviewed the professional literature debating educational implications of Ethnomathematics and summarized that it is central to the process of teaching; ethnomathematics should be included throughout preservice mathematics education programs, because teachers must have instructional skills to accommodate different backgrounds, and give rise to considerations that are crucial to mathematics teacher preparation. Many scholars (Adam, 2004; Orey and Rosa, 2005) asserted that it is necessary for ethnomathematics-based programs to identify and seek teaching practices directed towards pedagogical action, and to employ ethnomathematics as an educational tool that can help teachers and students make it a part of their own knowledge.

Nevertheless, there are scholars and researchers endorsing opposite opinions regarding ethnomathematics. Pais (2011) divided the objections to ethnomathematics into two categories of criticism: epistemological — related to mathematical knowledge; and pedagogical — related to the way ethnomathematics is implicated in education. For instance, in connection with pedagogical criticism, Horsthemke and Schäfer (2007) focused on ethnomathematics' encroachment into mathematical education curriculum in South Africa, and argued that there should be a basic agreement on some sort of mathematical standard, which should be transcultural. Even so, this critical position invites us to exercise caution and consideration in each individual case, including considering whether the integration of ethnomathematics into the mathematics school curriculum somehow justifies this move. In other words, "mathematics teachers should weigh the pros and cons, assess at which level of studies, in which school year, and for which pupil populations such a move would bear fruit" (Katsap and Silverman, 2016, p. 291).

3. Mathematics Education with an Emphasis on Socio-Cultural Environments: An Israeli Look

The Israeli education system comprises three sectors with different curricula, programs, and textbooks: Jewish sector, divided in turn into secular and religious-traditional, Arab sector, including Bedouin as well, and ultra-Orthodox Jewish (OJ) sector. This segmentation into sectors deals with socio-cultural dilemmas in mathematics education and requires appropriate activities for preservice teachers and school mathematics programs. Besides these, in Israel, as a country unceasingly absorbing incoming Jewish immigration, there is a recurring debate about integrating repatriates in the education system in general and in the mathematical education system in particular. One of these groups consists of students of Ethiopian origin (EO), who are struggling to express their abilities and getting good educational results. Low percent of EO students attempts to join the tracks leading to the secondary school graduation exams. Mulat and Arcavi (2010) found difficulties in learning and academic underachieving in mathematics among EO students, and formulated recommendations for an education system that promotes ethnic minorities to succeed academically. According to these scholars, the education system needs to learn about the roles of identities, self-regulation, and enhancement of motivation, and must build collaborative support programs for this group of students. The fusion of teachers and students with different cultural backgrounds and knowledge occurs instantly at different levels of the education system in Israel. In this context, immigrant and native Israeli teachers gathered during mathematical-social interactions that took place at the Kidumatica Club at Ben Gurion University, and created a reciprocal background that reflected constructive influence on students from diverse ethnic origins. The teachers shared an effort that brought all of them into a composite culture, and has proven that integration is possible (Neria *et al.*, 2008).

Several Israeli studies emphasize ways to make mathematics more significant by enriching subject matter understanding with knowledge rooted in the learners' cultural heritage. Amit (2015) describes integration of traditional units of measurement of the Bedouin folk into mathematics lessons, a step influencing students' motivation and achievements. A course taught in Orot College in Elkanah to OJ students, sought to alleviate the difficulties of solving mathematics problems by imparting algorithms and pedagogical techniques developed by Jewish Sages over the generations, which in the author's opinion can motivate new Torah-oriented mathematics teachers (Fendel, 2009). In a precollege course at Ben Gurion University, OJ students surprised the Algebra lecturer by using the word "re'aya" (evidence), which is known as a notion in Talmudic studies and debates. In turn, the inclusion of this term in the discourse triggered a classroom dialogue over mathematical claims (Ehrenfeld, 2016). The same target population (OJ) is featured in Sandik's doctoral study (in preparation). Sandik's research is examining mathematics teaching in a preacademic program via mathematical issues from students' cultural sphere. Furthermore, Zalmanson (2015) looked at mathematics lessons with varied contents in the setting of social equality. She demonstrated how texts of problems and exercises can be altered in order to consider different socio-cultural population, and thus enabling them a fair chance to understand the text.

4. Israeli Projects Within Ethnomathematics-Oriented Curriculum

Preservice teachers from both Jewish (secular and religious) and Bedouin sectors with different socio-cultural background attended the "History of Mathematics" course at Kaye Educational College in the Negev (Southern Israel). This course gave rise to several studies that examined the influences of interlaced ethnomathematical approach on mathematics preservice teachers. The first study began when a fixed chapter of "Mathematics in Judaism and Islam" was introduced in 2003 into the course curriculum. It was unique in that the participants had to choose and present in the classroom a school mathematics unit incorporating mathematical ideas from their respective cultures: either original and sacred texts, mathematics practices, or discoveries of mathematicians. The studies (Katsap, 2004; Katsap and Silverman, 2008) revealed that ethnomathematics in mathematics teacher education could influence the teachers' professional development. Teachers could incorporate humanistic and social aspects of mathematics, awakening sentiment of kinship to the cultural mathematical heritage and to ethnomathematics problems (see broader examples in Katsap and Silverman, 2016). The last study followed a project which aimed to introduce the ethnomathematical approach in school mathematics curricula through integration of adapted subjects, identified from the socio-cultural lives of students. Three different topics were taught by preservice teachers in two Bedouin towns in the Negev (e.g., Knowing a way to use graphs of functions in connection with writing poetry and playing Bedouin music on Bedouin instruments). According to a student that experienced it: "The graph helped me to understand what a fixed tune pace is, and how to build a stanza for a certain Bedouin song by adjusting between the melody and the number of words in the stanza." The findings show that combined factors, such as achieving the objectives of mathematics lessons on functions, and exposure to traditional Bedouin music, could indeed help to promote two goals: enhance pupils understanding of mathematical concepts by enjoying lessons, and to create a pride among the pupils caused by dealing with folk music of their culture (Katsap and Silverman, 2016).

In Northern Israel a teacher education course "Teaching geometry in cultural context" was taught at the Technion to preservice and in-service Arab and Jewish teachers, who gathered to learn and explore geometric ornaments drawn from diverse cultures. The ethnomathematical approach emphasized the structures and reduced the feeling of dissatisfaction among the participants, as reported by Verner, Massarwe and Bshouty (2013). The authors describe that they have started the inquiries by examining an approach to teaching geometry by means of constructing and exploring culturally meaningful ornaments among students from Arab high schools in Northern Israel. This case study shows how such activity induced students and teachers alike to increase positive emotions and learning motivation. Based on the analysis of motivating desires arising from multicultural interactions, Verner, Massarwe and Bshouty (2013) introduced a new engagement structure offering the potential to foster mathematics study, named "*acknowledge my culture*".

5. Concluding Remarks

We examined the potential value of incorporating an ethnomathematical perspective into mathematics education in Israel. This perspective can positively affect students and teachers from various social-cultural backgrounds. Our wish is to add ethnomathematics-oriented topics into Israeli mathematics curricula by using culturally significant examples.We hope that such an effort can change the attitude towards mathematics in a multi-culture population, and to decrease the achievement gaps. Consequently, school mathematics might become more meaningful and valuable, while remaining culturally relevant and adapted to mathematics classroom practices.

References

Adam, S. (2004). Ethnomathematical ideas in the curriculum, *Mathematics Education Research Journal*, **16**(2), 49–68.

Amit, M. (2015). Ethnomathematics: weaving culture, education and mathematics, Multiculturalism, Migration, Mathematics Education and Language M^3EaL Project, *Final International Workshop*, Lucca, 11–12 September 2015.

Domite, M. C. and Pais, A. (2010). Understanding ethnomathematics from its criticisms and contradictions, *Proceedings of CERME*, Lyon, France, No. 6, pp. 1473–1482.

Ehrenfeld, N. (2016). Between Mathematics and Talmud — The construction of a hybrid discourse in an ultra-Orthodox classroom, Thesis for the Master of Sciences Degree. Ben-Gurion University of the Negev the Faculty of Natural Sciences department of Mathematics, p. 32.

Fendel, H. (2009). 'New math' based on 2,000-year-old talmudic methods, *Arutz Sheva, Israelnationalnews.com.* Retrieved from: http://www.israelnationalnews.com/News/News.aspx/133252.

Horsthemke, K. and Schäfer, M. (2007). Does 'African mathematics' facilitate access to mathematics? Towards an ongoing critical analysis of ethnomathematics in a South African context, *Pythagoras* **65**, pp. 2–9.

Katsap, A. (2004). One mathematics, two cultures, and a history of Mathematics College course as a starting point for exploring ethnomathematics, in F. Furinghetti, S. Kaijser and C. Tzanakis (eds.), *Proc. HPM* 2004, Uppsala, Sweden, pp. 262–270.

Katsap, A. and Silverman F. L. (2008). A case study of the role of ethnomathematics among teacher education students from highly diverse cultural backgrounds, *The Journal of Mathematics and Culture*, **3**(1).

Katsap, A. and Silverman, F. L. (2016). *Ethnomathematics of Negev Bedouins' Existence in Forms, Symbols, and Geometric Patterns*. Rotterdam: Sense Publishers.

Mulat, T. and Arcavi, A. (2010). Mathematical behaviors of successful students from a challenged ethnic minority, *Proceedings of CERME 6*, Lyon, France, January 28th–February 1st 2009, Working group 13, pp. 2473–2482.

Neria, D., Amit, M., Abu-Naja, M. and Abo-Ras, Y. (2008). Ethnic and gender gaps in mathematical self-concept: The case of Bedouin and Jewish students, in O. Figueras, J. L. Cortina, S. Alatorre, T. Rojano and A. Sepúlveda (eds.). *Proceedings of the Joint Meeting of PME 32 and PME-NA XXX.*, Vol. 4, pp. 33–40, México: Cinvestav-UMSNH. Retrieved from: http://www.acri.org.il/education/2015/06/08/a-life-lesson-anti-racism-education-full-version.

Orey, D. and Rosa, M. (2005). Current trends in ethnomathematics as a program: Towards pedagogical action (Tendências atuais da Etnomatemática Como um programa: Rumo à Ação Pedagógica), *Revista ZETETIKÉ: CEMPEM — Faculdade de Educação — UNICAMP*. Campinas, Brazil. English Translation Retrieved from: http://www.csus.edu/indiv/o/oreyd/papers/etnotendenciasatuais.translation.doc.

Pais, A. (2011). Criticisms and contradictions of ethnomathematics, *Educational Studies in Mathematics*, **76**(2), 209–230.

Rosa, M., D'Ambrosio, U., Orey, D., Shirley, L., Alangui, W., Palhares, P. and Gavarrete M. (2016). Current and future perspectives of ethnomathematics as a program, in Kaiser, G. (ed.), *ICME-13 Topical Surveys*, Published by Springer International Publishing AG 2016-06-23, (eBook).

Shirley, L. (2001). Ethnomathematics as a fundamental of instructional methodology, *ZDM — International Journal on Mathematics Education*, **33**(3), 85–87.

Verner, I., Massarwe, K. and Bshouty, D. (2013). Constructs of engagement emerging in an ethnomathematically-based, *The Journal of Mathematical Behavior*, **32**, 494–507.

Zalmanson, G. (2015), $X = Y$: Teaching of mathematics as an opportunity for educating to equality, in Rivlin, N. (ed.), "Lesson for Life — Education against Racism from Kindergarten to High School", The Center for Critical Pedagogy, Seminar Ha-Kibbutzim, Education Department of the Society for Civil Rights. Internet version published on 08/06/2015, pp. 55–71.

About the Author

Ada Katsap is a retired senior lecturer at the Kaye Academic College of Education with 35 years of experience in the Israeli mathematics education system. She is the author of the book "Wisdom and Magic in Humanistic Mathematics" (Mofet Institute Publication, 2007), and the first author of the book "Ethnomathematics of Negev Bedouins' Existence in Forms, Symbols, and Geometric Patterns" (Katsap, A. and Silverman, F. L., Sense Publishers, 2016).

Chapter 43

Mathematical Learning Disability, Diagnosis and Intervention: An Examination of the Israeli Education System

Sarit Silverman[*,‡], **Nitza Mark-Zigdon**[#,§] **and Sarit Ashkenazi**[*,¶]

[*]The Seymour Fox School of Education, The Hebrew University, Jerusalem, Israel
[#]Levinsky Academic College of Education, Tel Aviv, Israel
[‡]sarit.silk@gmail.com
[§]nitzamark@gmail.com
[¶]sarit.ashkenazi@mail.huji.ac.il

Understanding the etiology of mathematical learning disability (MLD) has implications for assessments and remedial mathematics education. Diverging opinions regarding the source of the difficulty underlying MLD have slowed progress for new interventions and diagnostic testing in Israel and around the world. Traditional approaches suggest that the source is impaired quantity representation, while others propose skills that support mathematical learning, such as working memory. Currently, there is no normed assessment for MLD in Israel. We propose the establishment of a new test that includes basic numerical processing, mathematics curriculum-based tasks and cognitive tasks. In addition to assessment, educational and cognitive interventions are reviewed. Further progress is needed to establish a normed diagnostic tool, create personalized interventions and greater inclusion of children with MLD in mathematics classrooms.

1. Introduction

Mathematics is an important school subject that has significant impact on daily life skills, such as making financial and medical decisions (Agarwal and Mazumder, 2013; Gerardi, Goette and Meier, 2013). Despite this, research on typical and atypical mathematics development is limited, especially compared to reading (Szucs and Goswami, 2013). There is a debate in numerical cognition research on the nature of the cognitive weakness underlying mathematical learning disability (MLD), also known as developmental dyscalculia. According to the core deficit approach, the source of MLD is impaired quantity representation (Butterworth, Varma and Laurillard, 2011; Halberda, Mazzocco and Feigenson, 2008). People are born with number sense, a preverbal ability to intuitively understand and approximate quantity and the relationship between quantities (Cantlon *et al.*, 2009; Cohen Kadosh, Lammertyn and Izard, 2008). The core deficit approach proposes that this inherent, born skill is the foundation of later mathematical skills and its impairment is the source of MLD.

In contrast, the multiple components approach proposes that general cognitive skills not specific to mathematics contribute to MLD (Fias, Menon and Szucs, 2013). For example, working memory is essential for solving mathematics problems to store and update partial results or steps in the solution process (Ashkenazi *et al.*, 2013a,b; Metcalfe *et al.*, 2013). Similarly, a child with low inhibition can have difficulty suppressing inappropriate strategies (Szucs *et al.*, 2013).

The multiple components approach is a more accurate clinical description of how MLD manifests in the classroom. There are high comorbidity rates between MLD, reading disabilities and ADHD. The rate varies between studies, ranging from 17–66% (Landerl and Moll, 2010; von Aster and Shalev, 2007). The multiple deficit approach attributes this to the common cognitive skills required for mathematics and reading; both require mapping symbols, procedural knowledge, and executive functions (Ashkenazi *et al.*, 2013a,b). Even the lowest reported comorbidity rates are higher than expected if the disorders are assumed to be independent.

von Aster and Shalev (2007) proposed that different types of MLD can emerge over the course of development based on the cognitive skills required at different stages in formal mathematics education; skills specific to mathematics and other general cognitive skills can relate to MLD, as well as the comorbidity with reading and ADHD. For example, a preschooler with verbal difficulty would have trouble learning how to count and thus make arithmetic errors, and would likely have a reading difficulty. Another neurocognitive account for the heterogeneity of MLD was proposed by Rubinstein and Henik (2009). They made a distinction between: (1) developmental dyscalculia: a core deficit in number sense that originates from intraparietal sulcus impairment and (2) MLD: poor mathematics performance related to cognitive deficits such as working memory, visual-spatial processing or attention due to impairments in the angular gyrus, frontal regions, fusiform gyrus or intraparietal sulcus.

Therefore, MLD assessments, beyond mathematical skills such as quantity representation and school-like mathematics tasks, should include a comprehensive cognitive battery. Testing only for quantitative processing, as the core deficit approach suggests, overlooks large groups of children with MLD who have other difficulties. This outlook proposes a broader examination of risk factors for MLD and can obtain a clearer picture of the underlying cognitive weaknesses.

The shift in discussing MLD from a core quantity weakness to multiple cognitive weaknesses is also seen in the changes in its clinical definition in the diagnostic and statistical manual of mental disorders (DSM) V (2013), from the older version DSM IV (1994), which has ramifications on MLD diagnosis and remedial education design. The DSM IV defined 3 distinct learning disabilities each with its own core deficit. The updated definition of MLD is included under "Specific Learning Disabilities" that can manifest in difficulties with reading, writing or mathematics; each disability can result from several cognitive weaknesses. The flexibility in the DSM V can increase diagnosis of multiple learning disabilities, and recognizes the underlying etiological complexities.

The multiple definitions of MLD have slowed progress in advancing mathematics testing in Israel and around the world. Currently, there is no normed diagnostic tool for assessing MLD for Israeli children. There are different tests available without norms, or normed tests from other countries have been applied. Both methods are flawed and a normed tool is urgently needed to correctly diagnose MLD. Israel's population is highly heterogeneous, the mathematics curriculum serves as a guideline for each grade level, yet its application differs from school to school. A normed test is needed that will allow for comparison to children across the country. Application of norms from other countries is problematic because the results are based on different cultures and school curriculum, which have different requirements for each grade level.

2. The Curriculum-Based Measurement (CBM)

Due to a lack of a normed mathematics assessment in Israel, Ashkenazi, Mark-Zigdon and Henik (2009, 2013) created the CBM, which included curriculum based and number sense tasks. The curriculum based portion was built off the Israeli primary school mathematics curriculum (2006) and included oral administered, pen paper and computerized tasks. The computerized tasks included: quantity naming, symbolic number comparison and arithmetical facts (simple operations such as 4×2 that are typically retrieved from long term memory). The number sense tasks were computerized. For a summary of the subtests and examples please see Table 1. As the tool is being developed, cognitive tests such as working memory will be added in future stages.

A pilot was conducted with 3rd and 4th graders. Grade level expectancy templates were constructed for all the tasks for grades 1–4. Each child began testing at his/her current grade level, if s/he failed a task s/he was then tested 1 grade below in the same task. If s/he failed at 1 grade level below s/he was tested at 2 grade levels below in the same task (a 3rd grader could be tested on the 3rd, 2nd, and 1st grade levels). This method was used for all the tasks. For example, in the counting task, at the 1st grade level, children were asked to count aloud from 7 to 15, 2nd grade level from 326 to 334, 3rd grade level from 4493 to 4501, and 4th grade level from 9996 to 10,004. Accuracy was measured for each subtest: participants who scored 2 grade levels below their chronological age in most of the subtests were defined as MLD. 29 children were selected for the pilot from 600 3rd and 4th graders from 5 public primary schools in Beer-Sheva, Israel. 13 (11 female) were diagnosed with MLD. The mean age of the MLD group was 9.5-years-old ($SD = 0.9$ years).

To date, we have preliminary results from the pilot and a team of investigators from several Israeli universities, including Mark-Zigdon and Ashkenazi, are creating a new diagnostic tool. The tool will be computerized, adaptive and will test school-like arithmetic, number sense and general cognitive abilities. The creation of norms for the new tool will include 250 children grades 1–6, from different locations.

Table 1. Subtests and examples from the CBM.

Subtest	Content	Example
Part A. Knowledge of numbers		
Number-word sequence	Counting forward Counting backwards	Count from 793 to 801 Count from 506 to 498
Numerical system	Understanding of the base-ten system	Build biggest/smallest number from a given set of written digits (e.g., 3, 7, 4, 8)
	Equation transformation from horizontal to vertical position	Write: $340 + 3 + 5706 = 6049$
	Recognition of a numeral's place value in a written number	What is the value of the digit 5 in the number: 1252?
Sequences of numbers	Sequence completion	Complete the following sequence: 463, 473, 483, ___, ___, ___
Part B. Knowledge of number operations		
Equation	Addition Subtraction	$200 + \_\_\_ = 550$ $\_\_\_\_ - 100 = 600$
Simple multi-digit arithmetic	Addition Subtraction Multiplication Division	$20 + 50 =$ $70 - 30 =$ $40 \times 30 =$ $90/30 =$
Written word problems	Arithmetical operation presented in verbal format	In each class there are 25 children. How many children are there in 5 classes?
Arithmetic algorithms	Addition	$\begin{array}{r} 24 \\ \underline{+37} \end{array}$
	Subtraction	$\begin{array}{r} 56 \\ \underline{-43} \end{array}$
	Multiplication	$\begin{array}{r} 45 \\ \underline{\times 3} \end{array}$
	Division	$\frac{94}{6}$
Estimation	Multiplication	Is 32×19 bigger or smaller than 400?
Part C. Number sense measurements		
Quantity Naming	Enumerate configuration of 1–9 dots	3 4
Symbolic number comparison	Select the larger numeral	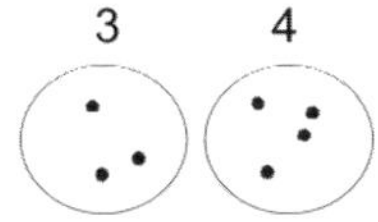

3. Interventions for Remediation of MLD

A byproduct of the various definitions of MLD is that many different approaches to intervention have emerged. In line with the core deficit theory, number sense interventions were created to improve the connection between Arabic numerals and internal magnitude representation, comparisons between magnitudes, arithmetic facts and representation of quantities on the mental number line (Butterworth, Varma and Laurillard, 2011; Kucian *et al.*, 2011; Wilson *et al.*, 2006). Number sense interventions focus on improving mathematical skills in young children, exclusively for deficits in quantitative understanding. This focus limits the scope of the intervention: it does not consider children's individual needs or more complex, formal mathematical skills.

A personalized intervention program, on the other hand, simultaneously takes into account the students' abilities based on assessment results, and on the mathematics curriculum. This highlights the advantage of curriculum-based assessments, which also include number sense and cognitive tasks, because it provides educators with comprehensive information on the students' strengths and weaknesses. Based on assessment results, teachers can create a personalized intervention program by modifying the class material; instead of repeating basic mathematics lessons with the expectation that more abstract content cannot be taught, the class material can be modified according to the students' strengths and weaknesses.

4. Summary

Due to the long-lasting effects of MLD (Agarwal and Mazumder, 2013; Gerardi, Goette and Meier, 2013), the Israeli education system should strive to advance progress. The highest priority should be the establishment of a normed, diagnostic MLD assessment for the Israeli school system, which should include quantitative understanding, mathematics curriculum-based tasks and cognitive tasks, such as the tool described above. Diagnosis is only the first step, results should be applied to create personalized intervention programs to improve mathematics performance. Further research is needed on Israeli children's mathematical abilities to improve instruction, MLD diagnosis and intervention, and help as many children as possible across the spectrum of mathematical abilities.

References

Agarwal, S. and Mazumder, B. (2013). Cognitive abilities and household financial decision making, *American Economic Journal: Applied Economics*, **5**(1), 193–207.

Ashkenazi, S., Black, J. M., Abrams, D. A., Hoeft, F. and Menon, V. (2013a). Neurobiological underpinnings of mathematical and reading learning disabilities, *Journal of Learning Disabilities*, **46**(6), 549–569; http://doi.org/10.1177/0022219413483174.

Ashkenazi, S., Mark-Zigdon, N. and Henik, A. (2009). Numerical distance effect in developmental dyscalculia, *Cognitive Development*, **24**(4), 387–400; http://doi.org/10.1016/j.cogdev.2009.09.006.

Ashkenazi, S., Mark-Zigdon, N. and Henik, A. (2013). Do subitizing deficits in developmental dyscalculia involve pattern recognition weakness? *Developmental Science*, **16**(1), 35–46.

Ashkenazi, S., Rosenberg-Lee, M., Metcalfe, A. W. S., Swigart, A. G. and Menon, V. (2013b). Visuo-spatial working memory is an important source of domain-general vulnerability in the development of arithmetic cognition, *Neuropsychologia*, **51**(11), 2305–2317; http://doi.org/10.1016/j.neuropsychologia.2013.06.031.

Butterworth, B., Varma, S. and Laurillard, D. (2011). Dyscalculia: From brain to education, *Science*, **332**(1049); http://doi.org/10.1126/science.1201536.

Cantlon, J. F., Brannon, E. M., Carter, E. J., Pelphrey, K. A. and Kanwisher, N. (2006). Functional imaging of numerical processing in adults and 4-y-old children, *PLoS Biology*, **4**(5), e125; http://doi.org/10.1371/journal.pbio.0040125.

Cantlon, J. F., Libertus, M. E., Pinel, P., Dehaene, S., Brannon, E. M. and Pelphrey, K. A. (2009). The neural development of an abstract concept of number, *Journal of Cognitive Neuroscience*, **21**(11), 2217–2229; http://doi.org/10.1162/jocn.2008.21159.

Cohen Kadosh, R., Lammertyn, J. and Izard, V. (2008). Are numbers special? An overview of chronometric, neuroimaging, developmental and comparative studies of magnitude representation, *Progress in Neurobiology*, **84**(2), 132–147; http://doi.org/10.1016/j.pneurobio.2007.11.001.

Education (2006). *Department for Curriculum Planning*, Jerusalem, Israel.

Fias, W., Menon, V. and Szucs, D. (2013). Multiple components of developmental dyscalculia. *Trends in Neuroscience and Education*, **2**(2), 43–47; http://doi.org/10.1016/j.tine.2013.06.006.

Gerardi, K., Goette, L. and Meier, S. (2013). Numerical ability predicts mortgage default, *Proceedings of the National Academy of Sciences of the United States of America*, **110**(28), 11267–11271; http://doi.org/10.1073/pnas.1220568110.

Halberda, J., Mazzocco, M. M. M. and Feigenson, L. (2008). Individual differences in non-verbal number acuity correlate with mathematicals achievement, *Nature*, **455**(October), 665–668; http://doi.org/10.1038/nature07246.

Kucian, K., Grond, U., Rotzer, S., Henzi, B., Schönmann, C., Plangger, F. and von Aster, M. (2011). Mental number line training in children with developmental dyscalculia, *NeuroImage*, **57**(3), 782–795; http://doi.org/10.1016/j.neuroimage.2011.01.070.

Landerl, K. and Moll, K. (2010). Comorbidity of learning disorders: prevalence and familial transmission, *Journal of Child Psychology and Psychiatry*, **51**(3), 287–294.

Metcalfe, A. W. S., Ashkenazi, S., Rosenberg-Lee, M. and Menon, V. (2013). Fractionating the neural correlates of individual working memory components underlying arithmetic problem solving skills in children, *Developmental Cognitive Neuroscience*, **6**, 162–175; http://doi.org/10.1016/j.dcn.2013.10.001.

Rubinsten, O. and Henik, A. (2009). Developmental dyscalculia: heterogeneity might not mean different mechanisms, *Trends in Cognitive Sciences*, **13**(2), 92–99.

Szucs, D., Devine, A., Soltesz, F., Nobes, A. and Gabriel, F. (2013). Developmental dyscalculia is related to visuo-spatial memory and inhibition impairment, *Cortex*, **49**(10), 2674–2688; http://doi.org/10.1016/j.cortex.2013.06.007.

Szucs, D. and Goswami, U. (2013). Developmental dyscalculia: Fresh perspectives, *Trends in Neuroscience and Education*, **2**(2), 33–37; http://doi.org/10.1016/j.tine.2013.06.004.

von Aster, M. G. and Shalev, R. S. (2007). Number development and developmental dyscalculia, *Developmental Medicine & Child Neurology*, **49**(11), 868–873; http://doi.org/10.1111/j.1469-8749.2007.00868.x.

Wilson, A. J., Revkin, S. K., Cohen, D., Cohen, L. and Dehaene, S. (2006). An open trial assessment of "The Number Race", an adaptive computer game for remediation of dyscalculia, *Behavioral and Brain Functions*, **2**, 20; http://doi.org/10.1186/1744-9081-2-20.

About the Authors

Sarit Silverman is a doctoral candidate at the Neurodevelopmental Learning Disabilities Lab at the Hebrew University of Jerusalem. Her dissertation topic is the unique impact of spatial working memory on mathematical development.

Nitza Mark-Zigdon is a lecturer and researcher at Levinsky Academic College of Education and Tel Aviv University. She was also the head of the Ministry of Education committee for special education mathematical curriculum accommodations.

Sarit Ashkenazi is the head of Learning Disabilities Program at the Seymour Fox School of Education at the Hebrew University.

Chapter 44

Through the Looking Glass: Research in Mathematics Education as Reflected at the Jerusalem Conference on Research in Mathematics Education

Talli Nachlieli[*,‡] **and Michal Tabach**[#,§]

**Levinsky Academic College of Education, Tel Aviv, Israel*
[#]Tel Aviv University, Tel Aviv, Israel
[‡]tallin@levinsky.ac.il
[§]tabachm@post.tau.ac.il

This chapter focuses on the first five years of the Jerusalem Conference for Research in Mathematics Education. Specifically, the paper examines the development of the conference during that period and presents findings regarding the span of research in mathematics education conducted in Israel as reflected at the first five conferences.

1. Introduction — A Conference is Born

The idea of establishing a conference for research in mathematics education in Israel was born in Seoul, South Korea during the ICME 12 conference. Several Israeli participants noted that while Israeli researchers in mathematics education often meet professionally and socially at conferences abroad, they rarely do so in Israel. In July of 2012 Professor Avi Berman, a mathematician from the Technion who is also deeply involved in mathematics education, took the initiative toward establishing a professional conference in Israel that would enable the local mathematics education research community to meet "at home" to discuss our work. In February 2013 the first

Jerusalem Conference on Research in Mathematics Education (JCRME) took place at the Jerusalem College of Technology (JCT).

In this chapter, we write about the development of this conference during its first five years and present the span of Israeli research in mathematics education as reflected in the programs of the first five conferences.

2. The First JCRME

After returning to Israel, three researchers met to plan the first JCRME: Professor Avi Berman from the Technion; Professor Noah (Thierry) Dana-Picard, former JCT president whose research as a mathematician also covers topics in mathematics education; and Dr. Talli Nachlieli from Levinsky Academic College of Education. The three decided upon a two-day conference that would take place in Jerusalem at JCT. Due to the short time and low budget, they decided that the program for the first conference would include invited talks only. Their third decision was to avoid parallel talks so as to initiate the development of a community, with all participants discussing studies in mathematics education together.

3. Goals of Mathematics Education Research Conferences

The goals that led to the establishment of the JCRME appear to be congruent with the goals of other national conferences on mathematics education research around the world. These goals include developing a research community, sharing and disseminating studies and research findings to promote and enhance collaboration between researchers, and thus promoting mathematics education at all levels. Unlike the JCRME, however, most such conferences are not stand-alone entities but rather one of the activities of an organization or association.

The Mathematics Education Research Group of Australasia Incorporated (MERGA) is one such association. It aims at promoting, sharing, disseminating and implementing its research and findings in mathematics education, particularly in Australia. One of the means to achieve these goals is the annual MERGA conference.

4. JCRME Changes and Developments

The program for the two-day JCRME1 included seven 90-minute sessions. Each session comprised two 30-minute invited talks and a 30-minute discussion led by a previously chosen discussant. That is, 14 research studies were presented and discussed during the two days. The presenters came from eight different universities, five colleges of education and the Center for Educational Technology (CET).

Over the course of the subsequent conferences the following changes and developments took place: (1) The sessions are now open to all Israeli mathematics education researchers who respond to a call for papers. Each proposal is reviewed according to clear and public guidelines by two invited reviewers. (2) The conference includes various types of lectures and invited talks, but mainly comprises peer-reviewed talks, symposia, research reports, posters and round table discussions. (3) Most talks are now given in parallel sessions, as the number of presenters and participants has grown. (4) The conference now charges a minimal registration fee to provide some degree of hospitality. (5) Two committees have been established: a program committee and an organizing committee. The program committee includes six participants. For continuity purposes, each member serves on the committee for two years. Every year three new members are elected and join three experienced members from the previous year. Each year the committee chooses one of its members to act as conference chair. The organizing committee includes at least three members. Since the second conference, Naama Adin has served as the administrative manager.

5. Categorization of the Conference Presentations

The data for this chapter incorporate abstracts and papers from plenary lectures (PL), symposia (SYM) and research reports (RR) as they appear in the proceedings of the first five conferences (Ayalon and Adin, 2017; Berman *et al.*, 2014; Naftaleiv and Adin, 2016; Tabach *et al.*, 2015).

We conducted a thematic analysis of the abstracts and papers published in the proceedings. The themes stem from the list of PME research categories found at IGPME.org (http://igpme.org/index.php/annual-conference/further-information/research-categories).

The abstracts presented at each conference are summarized in Table 1. As the table indicates, the number of RRs presented almost doubled from the second to the third conference, and increased by 50% at the fourth conference. Overall, the number of studies presented at each conference more than doubled from the first to the fifth conference.

Table 1. Number of PLs, SYMs and RRs given at each conference and analyzed for this study.

Conf. #	PL	SYM	RR	Total
1st	0	0	14	14
2nd	3	4	12	19
3rd	3	3	23	29
4th	3	3	32	38
5th	3	1	33	37

Table 2. Research methods used for the studies presented at JCRMEs.

Conf. #	Qualitative/ interpretative	Quantitative	Mixed methods	Theoretical/ philosophical
2nd	14 (74%)	1 (5%)	2 (21%)	1 (5%)
3rd	14 (48%)	7 (25%)	3 (10%)	4 (14%)
4th	23 (61%)	7 (18%)	7 (18%)	2 (5%)
5th	25 (68%)	6 (16%)	2 (5%)	2 (5%)

6. Findings

In this section we characterize the studies presented at each of the conferences according to type of research, research participants, topic of study and collaborations between researchers.

Type of research (Table 2). At each of the conferences, the majority of the studies presented used qualitative or other interpretative methods. We did not add data about the methodology of JCRME1 as most of the abstracts did not include this data.

Study participants (Table 3). We discovered some degree of stability regarding the research participants. Between a quarter and a third of the studies at the second through the fourth conferences focused on primary school students. At each conference one study was about preschool children. Over 40% of the studies presented at each conference involved secondary school students. The number of studies about preservice teacher preparation doubled from the second to the fourth conference, but decreased again at

Table 3. Distribution of research participants in each of the studies.

Conference	Preservice teacher preparation	Preschool	Primary	Middle school	High school	Adult ed. and out-of-school math
1st	1 (7%)	0	3 (21%)	5 (36%)	6 (43%)	1 (7%)
2nd	2 (11%)	1 (5%)	5 (26%)	7 (36%)	2 (11%)	2 (11%)
3rd	1 (4%)	1 (4%)	9 (33%)	8 (30%)	7 (25%)	1 (4%)
4th	8 (21%)	1 (3%)	9 (24%)	7 (18%)	9 (24%)	4 (10%)
5th	5 (14%)	1 (3%)	5 (14%)	10 (27%)	10 (27%)	4 (11%)

Table 4. Mathematics topics of the studies presented.

	Algebra and algebraic structures	Arithmetic, numbers and operations	Calculus and functions	Geometry, space and shape	Probability and statistics	Problem solving and problem posing	Proof, argumentation and reasoning
1st	3 (21%)	1 (7%)	2 (14%)	3 (21%)	2 (14%)	1 (7%)	2 (14%)
2nd	4 (21%)	1 (5%)	1 (5%)	2 (11%)	1 (5%)	5 (26%)	6 (32%)
3rd	2 (7%)	3 (11%)	4 (15%)	4 (15%)	2 (7%)	4 (15%)	5 (19%)
4th	7 (18%)	5 (13%)	4 (11%)	5 (13%)	2 (5%)	9 (24%)	5 (13%)
5th	3 (8%)	2 (5%)	9 (24%)	5 (14%)	2 (5%)	5 (14%)	5 (14%)

the fifth conference. We note that no studies about special education were reported.

Topics of study. Tables 4–6 reflect the diversity of the mathematics topics and general topics discussed in the studies. Note that a relatively high portion of the studies focused on three particular topics: *problem solving and problem posing*, as well as *proof and argumentation* (Table 4), and *teacher knowledge and practice* (Table 6). The topics of *culture* and of *equity and gender* are absent (Table 5).

Note that the numbers in each of the rows in Tables 4–6 do not add up to 100%: some of the studies did not focus on specific mathematical topics, e.g., studies about teachers' choices regarding tasks. Moreover, some studies referred to more than one topic, e.g., studies about geometrical proofs whose topic is both geometry and proof.

Table 5. Research topics of the studies presented at JCRME 2–5.

	Teaching and learning with technologies	Classroom assessment	Educational evaluation and policy	Classroom interaction and discourse	Equity and gender issues	Affect, emotions and attitudes	Culture, language and multilingualism	Curriculum development and task design
1st	5 (36%)	0	0	0	0	1 (7%)	0	2 (14%)
2nd	5 (26%)	0	1 (5%)	1 (5%)	0	1 (5%)	0	4 (21%)
3rd	4 (15%)	2 (7%)	3 (11%)	5 (19%)	0	4 (15%)	2 (7%)	4 (15%)
4th	5 (13%)	4 (11%)	0	3 (8%)	0	7 (18%)	0	2 (5%)
5th	9 (24%)	0	1 (3%)	6 (16%)	0	9 (24%)	2 (5%)	3 (8%)

Table 6. Studies that focus on teachers at JCRME.

	Teacher beliefs	Teacher knowledge and practice	Teacher professional development
1st	0	0	2 (14%)
2nd	1 (5%)	4 (21%)	3 (16%)
3rd	0	5 (19%)	4 (15%)
4th	5 (13%)	12 (32%)	8 (21%)
5th	3 (8%)	6 (16%)	8 (22%)

Table 7. Number of authors writing the papers.

	1 author	2 authors	3 authors	4 or more authors
1st	11 (79%)	2 (14%)	0	1 (7%)
2nd	5 (26%)	8 (42%)	5 (26%)	1 (6%)
3rd	6 (21%)	11 (38%)	9 (31%)	3 (10%)
4th	11 (29%)	16 (43%)	8 (21%)	3 (9%)
5th	10 (27%)	17 (46%)	7 (19%)	3 (8%)

Table 8. Affiliation of authors of studies presented at JCRME.

	University	College	University, college & JCT	Other
1st	8 (57%)	3 (21%)	1 (7%)	2 (14%)
2nd	13 (68%)	0	4 (21%)	2 (11%)
3rd	14 (48%)	4 (14%)	6 (21%)	5 (17%)
4th	17 (45%)	9 (23%)	11 (29%)	1 (3%)
5th	19 (51%)	8 (22%)	7 (19%)	3 (8%)

Presenters and collaborations. The presenters at JCRME come from universities, colleges of education, other academic centers (such as the Jerusalem Center of Technology), and centers for developing teaching and learning materials (such as the Center for Educational Technology). With respect to the number of researchers who authored papers (RRs, SYM or PLs) and their affiliation, most papers are authored by one to three authors (Table 7). Those authored by four to six researchers are usually symposia.

With respect to researcher collaborations (Table 8), we found that over 70% of the papers presented at each conference were co-authored. Most of the collaborations were between university researchers: at each of the conferences, most papers were authored by university researchers only (57%, 68%, 48%, 45%, 51%). Only a few papers were authored by college researchers only (21%, 0, 14%, 23%, 22%), yet the number of college researchers presenting their studies seems to be increasing. Starting with JCRME2, a bit more than 20% of the presented studies are the result of collaborations between researchers from universities and colleges of education (21%, 21%, 29%, 27%).

7. Conclusion

In this chapter we characterized the research in mathematics education conducted in Israel as reflected by the studies presented at the JCRME. The findings show that the number of studies presented more than doubled from the first to the fifth conference. Also, the number of studies presented by college researchers increased. This may suggest a shift in the activities of college educators towards doing research. Although it is often stated that research is mainly a solitary activity, most studies presented at JCRME were a result of collaborations between two or more researchers. Those collaborations are often between scholars from different types of institutions — universities and colleges of education. As for the topics of study, we found four main topics that are under-represented: (1) culture, language and multilingualism, (2) equity and gender, (3) special education, and (4) educational evaluation and policy. In a country marked by great cultural differences and diversity, there appears to be a need for such research. Finally, we suggest that the time has come to establish an Israeli research association in mathematics education.

References

Ayalon, M. and Adin, N. (eds.) (2017). *Jerusalem Conference for Research in Mathematics Education*, **5**, Jerusalem: JCT, Lev Academic Center.

Berman, A., Nachlieli, T., Adin, N. and Shpizer, L. (eds.) (2014). *Jerusalem Conference for Research in Mathematics Education*, **2**, Tel Aviv: Mofet.

Naftaleiv, Y. and Adin, N. (eds.) (2016). *Jerusalem Conference for Research in Mathematics Education*, **4**, Tel Aviv: Mofet.
Tabach, M., Levenson, E., Nachlieli, T. and Adin, N. (eds.) (2015). *Jerusalem Conference for Research in Mathematics Education*, **3**, Tel Aviv: Mofet.

About the Authors

Talli Nachlieli is a mathematics educator and researcher at Levinsky Academic College of Education. Her research interests include teacher preparation and education, and mathematics classroom discourse.

Michal Tabach is currently the head of the Department of Education in Mathematics, Science and Technology at the School of Education, Tel Aviv University.

Appendix

Israeli Mathematics Curricula (K-12): A Concise Form Based Upon the Hebrew Original Published by the Ministry of Education (See References Below)

Prepared by John Oberman

1. Preschool (Ages 3–6)

1.1. *The concept of number*

Topic	Sub-topics	Ability components	Expansion
Counting with and without objects	Counting	Count from 1 to 30	Count beyond 30 Skip count by 10s up to 100 Skip count by 2s and 5s
	Counting down	Count down from 10 to 1	
	One-to-one correspondence	Compare set sizes by One-to-one correspondence	

(*Continued*)

[a]The concise form does not include any pedagogical suggestions which exist in the original.

Preschool (Ages 3–6) (*Continued*)

Topic	Sub-topics	Ability components	Expansion
	Counting with objects	Count objects from 1 to 20	Extending the range beyond 20. Skip counting by 2,5, and 10
	Set size comparison	Compare set sizes by counting with objects	
	Ordinal numbers	Use ordinal numbers from first to sixth	Extending the range to tenth
Estimation	Estimating visually	Compare set sizes by visual estimation	
	Quantity estimation	Estimate the number of objects in a given set	
Recognition and display of quantities	Recognition and display of quanti- ties	Recognize and display the numbers from 0 to 10	Extending the display to 20
Operations with numbers	Composition/ decomposi- tion	Compose and decompose quantities up to 10 objects	
	Addition and Subtraction	Addition and subtraction to 10 using objects	Addition and subtraction of numbers over 10 with and without objects

(*Continued*)

Preschool (Ages 3–6) (*Continued*)

Topic	Sub-topics	Ability components	Expansion
	Divide a set of objects into equal quantities	Division of up to 10 objects into equal quantities	Extending the number of objects in the set
	Concept of half	Understanding the concept and use of half	
Repeating patterns	Repeating patterns	Identify, describe, and make repeating patterns	

1.2. ***Spatial sense and geometry***

Topic	Sub-topics	Ability components	Expansion
Spatial sense	Verbal description and visual representation of relationships in space	Use correct terms to describe basic relationships in space. Perform activities related to place and position in space	Make and read simple maps
	Visual perception of solids and shapes in space	Solve simple and realistic problems using visual perception of solids and shapes	

(*Continued*)

Preschool (Ages 3–6) (*Continued*)

Topic	Sub-topics	Ability components	Expansion
Shapes and solids	Recognition of basic solids and shapes	Identify and sort 3D solids. Identify the faces of 3D solids and shapes in the environment	
	Shapes and lines	Identify, name and draw 2D shapes. Identify, name and draw straight and curved lines	Classification
	Polygons	Distinguish different polygons by the number of sides and vertices. Name polygons.	
Solids and nets	Solids and nets	Identify and name 3D solids. Identify and name the faces in the nets of solids	
Symmetry	Reflective symmetry	Identify and form situations of reflective symmetry	

(*Continued*)

1.3. *Concept of measurement in everyday life*

Preschool (Ages 3–6) (*Continued*)

Topic	Sub-topics	Ability components	Expansion
Size relationships	Size relationships	Use the relative size concept. Arrange a group of objects according to a measured size	
Size measurement	Comparison of measurement — direct or using tools	Measure two objects by direct comparison or using something to measure which fits both objects	
	Arbitrary units of measurements and accepted tools for measurement	Measure using arbitrary units of measurement Recognize available tools for measurement	
Estimation	Visual estimation of measurement using arbitrary units	Estimate size in order to compare size of objects. Estimate size using arbitrary units	

(*Continued*)

Preschool (Ages 3–6) (*Continued*)

Topic	Sub-topics	Ability components	Expansion
Time measurement	Concepts of time	Use concepts of time in daily life and repeating events including comparison of time quantities	
	The clock	The clock as a measure of time Identifying the time on an analog clock relating to the time table in the kindergarten	
Money	Coin and bank note recognition	Recognize coins and notes used daily	
	Using money for purchasing and selling	Use "money" in purchasing and selling	
Collecting and presenting data	Sorting and arranging data	Sort and arrange objects according to criteria	
	Displaying data	Represent data with pictures, objects and diagrams	
	Data analysis	Answer simple questions relating to data	

2. Grades 1–6

2.1. *Natural numbers*: *Recognition*, *relationships*, *and operations*

Grade	Number recognition range	Relationships	Operations
1	100 range: Counting, reading and writing of numbers in this range, the number zero, even and odd, whole tens and the number line.	Equality and inequality	Addition and subtraction up to 20 (extending the range of these operations according to grade level). Addition and subtraction of whole tens. Use of rules for arithmetical operations. Multiplication and division up to 20
2	1000 range: Step counting, reading and writing of numbers in this range, place value, odd and even, the number line	Ordering numbers	Addition and subtraction up to 100 (extending the range of these operations according to grade level). Addition and subtraction of whole tens, simple equations. Multiplication and division: understanding of these operations, multiplication in the 100 range. The use of brackets with operations

(*Continued*)

Grades 1–6 (*Continued*)

Grade	Number recognition range	Relationships	Operations
3	10,000 range: Reading and writing of numbers in this range, the decimal place value system, different representations of numbers, the number line	Ordering numbers	Addition and subtraction up to 10,000. Multiplication and division in the 100 range, 2 digit by 1 digit multiplication, multiplication of whole tens and hundreds, simple equations, rounding numbers and number estimation, rules of operations, order of operations and use of brackets
4	Unlimited range: Reading and writing of numbers, the decimal place value system, different representations of numbers, infinity of natural numbers, the number line	Ordering numbers	The four operations with natural numbers and zero, relationships between operations, rules of operations, use of brackets, estimation

(*Continued*)

Grades 1–6 (*Continued*)

Grade	Number recognition range	Relationships	Operations
5	Unlimited range: Reading and writing of numbers, the decimal place value system, different representations of numbers, infinity of natural numbers	Ordering numbers	The four operations with natural numbers and zero, relationships between operations, rules of operations, use of brackets, estimation
6	Unlimited range: Comparison of number systems: natural numbers, whole numbers, rational numbers	Ordering numbers; inclusion of the different number systems	The four operations with natural numbers and zero, relationships between operations, rules of operations, use of brackets, estimation

(*Continued*)

2.2. *Simple fractions, decimal fractions, percentages recognition, relationships and operations*

Grades 1–6 (*Continued*)

Grade	Recognition	Relationships	Operations
1	Recognition of a half without formal symbol		At this stage understanding that two halves together make a whole
2	Recognition of a half, quarter as part of unit		At this stage understanding that in one whole there are two halves; in one whole there are four quarters; on subtracting one quarter from a whole three quarters of the whole remain
3	Recognition of the unit fractions $\frac{1}{2}, \frac{1}{3}, \frac{1}{4}, \frac{1}{5}, \frac{1}{6}, \frac{1}{7}, \frac{1}{8}, \frac{1}{9}, \frac{1}{10}, \frac{1}{12}, \frac{1}{60}, \frac{1}{100}$ demonstration using manipulative	Ordering fractions	Solving verbal problems such as: how many fifths are there in 3 wholes/ in 1 hour there are 60 minutes. What is one minute as a fraction of 1 hour? Expressing a part of a quantity as a unit fraction: how much is one third of 21? These verbal problems are solved without symbolic expressions

(*Continued*)

Grades 1–6 (*Continued*)

Grade	Recognition	Relationships	Operations
4	Recognition of a wide variety of simple fractions. Fraction as part of a whole. Fraction as part of a quantity	Ordering fractions, intuitive comparison fraction size	Addition and subtraction of simple fractions with denominators having a common factor. Addition and subtraction of mixed fractions with denominators having a common factor. Multiplication of a whole number and a fraction
5	Fraction as a point on a number line. Fraction as a quotient. Fractions larger than 1 and mixed fractions. **Decimal fraction** as simple fraction with denominators 10,100,1000 etc. **Percentage**- initial recognition.	Ordering fractions, comparison of fraction size. Comparison of decimal fraction size	Expanding and reducing (simplifying) fractions. Addition and subtraction of simple fractions. Addition and subtraction of decimal fractions. Size comparison of decimal fractions. Rounding decimal fractions. Decimal fractions as simple fractions. Simple fractions as decimal fractions — finite

(*Continued*)

Grades 1–6 (*Continued*)

Grade	Recognition	Relationships	Operations
6	Fraction as a quotient. Simple fractions and decimal fractions on the number line. Density of fractions. Repeating decimal fractions	Ordering fractions; inclusion of the different number systems	Multiplication of simple and mixed fractions with whole numbers and mixed fractions. Multiplication and division of decimal fractions by 10, 100, etc. Multiplication and division of decimal fractions. Part–Part comparison and Part–Whole comparison and calculations. Division of simple fractions. Percentages calculations, part–whole. Ratio division of a quantity

(*Continued*)

2.3. *Section 3 — Geometry*

Grades 1–6 (*Continued*)

Grade	Polygons	Solids	Transformations	Measurement
1	**Polygons** (recognition) **Quadrilaterals**		**Translation or Reflection**	**Length** (comparison). Informal measurement units. Commonly used units of measurement: cm. **Time** Whole hour measurement
2		**Solids** Recognition of different solids	**Reflection or Translation**	**Length** — cm and meter as units **Area** (comparison) Informal units of area measurement. Calculating area of a rectangle. **Weight** (comparison). **Time** Half and quarter hours
3	**Angles, Perpendicular lines, Parallel Lines, Triangles, Quadrilaterals** Properties of sides and angles		**Rotation**	**Length** in commonly used units. **Volume** (comparison). **Time and Weight** in commonly used units
4	**Diagonals, square and rectangle** Definition and properties. **Triangle** Properties of angles and sides	**Prisms** Vertices, edges, faces, nets and diagonals	**Symmetry**	**Area** in commonly used units. Formulas for calculating area and perimeter of rectangles. **Volume** of rectangular prism, surface area, connection between weight and volume. **Time** calculations.

(*Continued*)

Grades 1–6 (*Continued*)

Grade	Polygons	Solids	Transformations	Measurement
5	**Quadrilaterals** Properties, classification, inclusion relations of quadrilaterals. Tessellations, Height		**Symmetry of quadrilaterals**	Commonly used units of area. **Area Formulas**
6		Classification of **Solids**, nets of solids, regular solids		**Circle**. calculating volume of solids. Units of volume

3. Middle School (Grades 7–9)

3.1. *Seventh grade*

Algebra	Number	Geometry
	Phase 1:	
Variables, algebraic expressions and numeric and algebraic calculations	**Rules of Arithmetic calculations, squares and square roots of numbers**	**Rectangle, Prism Perpendicularity and parallelism**
Substitution in algebraic expressions and numerical solutions. Equal algebraic expressions with same numerical solution.	Rules of arithmetic operations. Commutative, associative and distributive laws of the four arithmetic operations.	Rectangle, perpendicularity of two lines. Parallel lines. Congruent shapes.

(*Continued*)

Middle School (Grades 7–9) (*Continued*)

Algebra	Number	Geometry
Combining like terms	Non-division by zero. Neutral terms of addition and multiplication. Reciprocal of numbers. Subtractions of the forms $a - -(b + c) = a - b - c$ $a-(b-c) = a-b+c$; Multiplication and division of the forms $a : (b \times c) = (a : b) : c$ $a : (b : c) = (a : b).c$. Powers with whole number exponents. Square root	Properties of a rectangle; perimeter and area of a rectangle; the square; Cuboid and surface area of a cuboid; volume of a cuboid; net of a cuboid

(*Continued*)

Middle School (Grades 7–9) (*Continued*)

Algebra	Number	Geometry
	Phase 2:	
Solution of equations and verbal problems	**Positive and negative numbers and zero**	**Area of polygons and circles. Angles and measurement of angles**
Solution of 1st degree equations with one variable including verbal problems	Negative numbers on a number line, order of numbers on the line, opposite numbers; four operations with integers; four operations expressed algebraically, powers with integer exponents, Cartesian coordinate system	Area of polygons, triangles, parallelism and angles of transversal, trapezoid, perimeter and area of a circle; angles-equal angles, comparison of angle sizes, opposite angles, straight line angles
	Phase 3:	
Functions. Equations and verbal problems		**Triangle and triangular prism**
Functions, solution of linear functions, graphs of functions		Triangle, angles, sides; quadrilaterals and angles; triangular prism, volume, faces

(*Continued*)

3.2. *Eighth grade*

Middle School (Grades 7–9) (*Continued*)

Algebra	Number	Geometry
	Phase 1:	
Linear equations and functions. Inequalities	**Ratio, proportion and scale**	**Congruent triangles, median of triangles and isosceles triangle**
Linear functions, inequalities, and verbal problems	Ratio between numbers, division of quantity according to given ratio, proportion, direct proportion, scale, inverse ratio	Congruent triangles, external angles of a triangle, median, isosceles triangle
	Phase 2:	
Solution of first degree equations, appropriate verbal problems and algebraic manipulations	**Percentages, statistics and probability**	**Similarity of polygons**
Solution of first degree equations, appropriate verbal problems and algebraic manipulations	Percentages, data handling and various representations-charts, graphs, diagrams, pictographs, and number line; frequency and relative frequency, range of a set of data, the three types of average — median, mode and mean; probabilities and properties of probabilities in symmetrical situations, estimation of probabilities using relative frequency	Similarity of triangles and polygons

(*Continued*)

Middle School (Grades 7–9) (*Continued*)

Algebra	Number	Geometry
	Phase 3:	
Solving simultaneous equations —1st degree. Appropriate verbal problems. Absolute value and inequalities	**Square roots. Irrational numbers**	**Pythagoras theorem in the plane and 3D**
Solving simultaneous equations — 1st degree. Appropriate verbal problems. Absolute value and inequalities-extension of inequalities	Square roots. Irrational numbers	Pythagoras theorem, Cylinder-straight cylinder, surface area, area of envelope, volume and net

3.3. *Ninth grade*

Algebra (including probability)	Geometry
Powers and roots	Kite and isosceles triangle
Integer Exponents, zero exponent.	Definitions.
Square roots	Concave and convex polygons.
	Diagonals of polygons.
	Form of geometrical proofs

(*Continued*)

Middle School (Grades 7–9) (*Continued*)

Algebra (including probability)	Geometry
Probability	**Geometrical constructions**
Probability of conditional events. Probability of dependent events. Probability of mutually exclusive events. Probability of compound events	Compass-and-straightedge construction. Line segment bisector. Angle bisector. Perpendicular to a point on a line. Perpendicular to a point NOT on a line. Same (congruent) angle. Construction of a triangle according to different constraints
Algebraic manipulations	**Parallel lines and trapezoid**
Multiplying binomials. Expanding (removing parentheses). Factoring. Solving quadratic equations by completing the square. Factoring a trinomial and solving quadratic equations	Definition of parallel lines. Pairs of angles on the transversal. Corresponding angles. Alternate interior angles. Consecutive interior angles. Definition and properties of trapezoid
Quadratic Functions	**Parallelogram and properties**
Definition and symbolic representation. Graphical representation. Different forms of quadratic function, vertex, factored and standard form. Solving verbal problems using the quadratic equation. Solving nonlinear set of equations	Definition of parallelogram. Construction of a parallelogram. Rotational symmetry of the parallelogram. Mid-points of the sides of a triangle and trapezoid, and the parallelogram properties

(*Continued*)

Middle School (Grades 7–9) (*Continued*)

Algebra (including probability)	Geometry
Algebraic skills in problem solving	**Rectangle and properties**
Choice of variable in problem solving.	Definition of rectangle as a quadrilateral with all angles right angles.
Simple problems solved algebraically.	Construction of a rectangle.
Use of algebra in geometrical problems	Rectangle, parallelogram and their diagonals
	Indirect proofs in geometry
	A variety of proofs presented indirectly
	Rhombus and square
	Definitions and inclusiveness
	Properties of rhombus and square

4. High School (Grades 10–12)

4.1. *Basic Level* (*"Three units"*)

Algebra
1st and 2nd degree equations; simultaneous equations 1st degree with a 2nd degree without parameters; quadratic equation; graphical representation of solution; factorizing algebraic expressions; changing the subject of a formula; verbal problems involving percentages; analysis of graphical solution to linear and square equations; powers with integers as exponents; scientific expansion of whole numbers with base 10
Analytic geometry
Distance between two points; line segment length using Pythagoras theorem; mid-point of line segment.
Equation for a line between two points; equation for a line according to slope; parallel lines; perpendicular lines; intersecting lines.

(*Continued*)

High School (Grades 10–12) (*Continued*)

Circle-representation $(x - a)^2 + (y - b)^2 = R^2$; intersection of line and circle; tangent to a circle

Sequences and series

Arithmetic sequences including verbal problems; constant differences between consecutive terms; rule for arithmetic series; summing an arithmetic series.

Geometric series; increasing and decreasing geometric series

Probability and statistics

Frequency and relative frequency; frequency charts; Pie charts; bar charts; analysis of diagrams; mean; measures of central value — mean, median and mode calculations.

Probability of independent events; complement of an event; normal distribution; probability of dependent events; not mutually exclusive events; standard deviation

Trigonometry

Right triangle definition of trigonometric function; Law of Sines, Cosines, and Tangents; solving various problems of polygons divided into right angle triangles; calculation of side lengths, perimeters and areas of polygons; use of formula

$$S = \frac{1}{2} a \times b \times \sin c$$

Solids in space, intuitive knowledge of spatial meanings and terms, length of edges of solids, volume, surface area-cuboid, pyramid with quadrilateral base

Differential calculus

Concept of derivative; point tangent; slope of a function at a point; concept of limits; derivatives as *dy/dx*; differentiation of polynomial functions with whole number exponents and zero; differentiation of functions $\sqrt{x}$, $1/x$, square root functions.

Differentiation of sum, difference, multiplicand, quotient; graphs of functions and derivatives

(*Continued*)

High School (Grades 10–12) (*Continued*)

Integral calculus
Integration of polynomials; integral as the reverse of finding a derivative; definite integration, calculation of area between the x-axes and the graph of a function; area between graphs of two functions

4.2. *Intermediate level ("four units")*

Analytic geometry
Distance between two points; line segment length; mid-point of line segment. Equation for a line between two points; equation for a line according to slope; parallel lines; perpendicular lines; intersecting lines. Equation of a circle in standard and canonical form; circle-line intersection; tangent lines to circles
Algebraic manipulations
Factorizing algebraic expressions by common factor, by completing the square; using factorizing with operations on algebraic fractions; factorizing for solving equations and inequalities; 1st and 2nd degree equations; simultaneous equations 2nd degree with two variables and one parameter, connection between value of parameter and solutions (unique, infinite, no solution); inequalities 1st and 2nd degree (without parameters)' irrational inequalities without parameters; graphical representation of solutions; solutions by substitution, powers with whole number exponents, roots — multiplication and division
Word problems
Problems involving distance, rate and time; buying, selling and discount; use of percentages in solutions; geometric word problems-area, compound polygons, volume of prisms; use of Pythagoras theorem
Sequences and series
Arithmetic sequences; geometric sequences; general sequences and rules

(*Continued*)

High School (Grades 10–12) (*Continued*)

Probability
Randomness; finite probability space; independent and dependent events; conditional probability; rules of probability; bias formula; Tree Diagram; binomial distribution (Bernoulli); sample spaces; combinations and permutations
Euclidean geometry
Polygons; computation of perimeters and areas; four laws of congruency of triangles; triangles and quadrilaterals-theorems proofs and properties; Pythagoras theorem; Thales theorem; three theorems of similarity in triangles; ratio of line segments in triangles and polygons; Altitude Rule and Leg rule of a right angle triangle; right angle triangles, altitudes, and the Geometric Mean; Circle-cords, arcs, diameter, sector, segment, tangent, intersection of circles-internal and exterior tangents; interior and exterior angles in circles; similar triangles in a circle; inscribed and circumscribed circles in triangles and polygons; centers of inscribed and circumscribed circles in a triangle; Locus–locus of points equidistant from the endpoints of a line segment is the perpendicular drawn through the midpoint of the segment; the angle bisector as a locus
Trigonometry
Right Triangle Definition; Unit Circle Definition and solution of trigonometric equations; Domains of the Trig Functions; Ranges of the Trig Functions; Periods of the Trig Functions; Identities and Formulas; Inverse Trig Functions; Law of Sines, Cosines, for triangle solutions; trigonometric functions of complementary angles; trigonometric formula for triangle area.
Applications in space using geometry theorems and trigonometric identities; calculation of angles, segment lengths, surface and envelope areas and volume of solids; cuboids and cube; prisms and pyramids; perpendicular to a plane; angle between a line and a plane; identifying a projection of a line on a plane; slope of a non-vertical line and the plane; angle between lines on a plane

(*Continued*)

High School (Grades 10–12) (*Continued*)

Differential calculus
Derivatives; slope of a function at a point; derivatives as dy/dx; limits; derivative rules; second derivative; finding maxima and minima using derivatives; inflection points; differentiation of polynomial functions, rational functions, trigonometric functions; square root functions; differentiation of sum, difference, multiplicand, quotient, absolute value, and compound functions; graphs of functions and derivatives
Integral calculus
Integration of polynomials, trigonometric functions(including use of identities), constant and variable functions, exponential functions, square and reciprocal functions, sums, multiplicands and quotients, multiplication by constant, rules of integration', indefinite and definite integration, integration by parts, integration by substitution; calculation of compound areas and volume of solids of revolution, integration of inflections

4.3. *Advanced level ("five units")*

Introduction to analytic geometry
Distance between two points; line segment length; mid-point of line segment.
Equation for a line between two points; equation for a line according to slope; parallel lines; perpendicular lines; intersecting lines.
Equation for a circle with the origin as its center.
Algebraic manipulations
Factorizing algebraic expressions; 1st and 2nd degree equations; simultaneous equations 2nd degree with parameters; inequalities 1st and 2nd degree; powers and roots
Sequences and series
Arithmetic sequences; geometric sequences; general sequences and rules

(*Continued*)

High School (Grades 10–12) (*Continued*)

Probability
Independent events; conditional probability; rules of probability; Tree Diagram. binomial distribution and necessary combinations and permutations
Euclidean geometry
Polygons; triangle theorems and properties; Pythagoras theorem; Thales theorem; theorems of similarity in triangles; ratio of line segments; Circle-cords, arcs, diameter, sector, segment, tangent, inscribed circles, intersection of circles; Locus
Trigonometry
Right Triangle Definition; Unit Circle Definition; Domains of the Trig Functions; Ranges of the Trig Functions; Periods of the Trig Functions; Identities and Formulas; Inverse Trig Functions; Law of Sines, Cosines, and Tangents
Differential calculus
Derivatives; slope of a function at a point; derivatives as dy/dx; limits; derivative rules; second derivative; finding maxima and minima using derivatives; inflection points; differentiation of polynomial functions, rational functions, trigonometric functions; square root functions; differentiation of sum, difference, multiplicand, quotient, absolute value, and compound functions; graphs of functions and derivatives
Integral calculus
Integration of polynomials, trigonometric functions (including use of identities), constant and variable functions, square and reciprocal functions, sums, multiplicands and quotients, multiplication by constant, rules of integration', indefinite and definite integration, integration by parts, integration by substitution; calculation of compound areas and volume of solids of revolution, integration of inflections.

Bibliography

High School Curriculum: http://cms.education.gov.il/EducationCMS/Units/Mazkirut_Pedagogit/Matematika/ChativaElyona/.

Middle School Curriculum: http://cms.education.gov.il/EducationCMS/Units/Tochniyot_Limudim/Math_Chatav/TachnitLimudim/.

Preschool Curriculum: http://cms.education.gov.il/EducationCMS/Units/Tochniyot_Limudim/KdamYesodi/TochniyotLimudim/Math.htm

Primary School Curriculum: http://cms.education.gov.il/EducationCMS/Units/Tochniyot_Limudim/Math_Yesodi/PDF/.

Glossary

Boaz Silverman
Technion — Israel Institute of Technology, Haifa, Israel
boaz.s@technion.ac.il

This chapter provides a brief description of key terms and abbreviations which are used in this book.

1. Terminology from A to Z (words in UPPER CASE are in Hebrew)

Advanced-level mathematics: The highest of three levels of study in Israeli high school mathematics (Also Known As Five units).

Advanced-plus level: A proposed fourth level of study in Israeli high school mathematics (AKA Five Plus-Plus).

Basic-level mathematics: The lowest of three levels of study in Israeli high school mathematics (AKA Three Units).

Chief Superintendent (of school mathematics): A national-level supervisor of all aspects of mathematics education (AKA the Hebrew acronym MAFMAR).

Department chair: A school-level supervisor of all aspects of mathematics education (AKA subject coordinator, in Hebrew — MERAKEZ-MIKTSOA).

Facilitator: A professional development facilitator coordinates, develops, and leads courses and programs for teachers.

Intermediate-level mathematics: The middle of three *Levels of Study* in the Israeli high school mathematics curriculum (AKA Four units).

Israeli education system: Comprises four stages — Preschool (ages 3–6), Primary School (commonly grades 1–6, rarely grades 1–8), Middle School (grades 7–9), and High School (commonly grades 10–12, rarely 9–12).

Levels of study: The current Israeli secondary school graduation examination comprises three levels — *Basic*, *Intermediate*, and *Advanced*. A fourth higher level is in development — *Advanced-plus*.

Pre-academic studies: An educational program aimed at preparing high school graduates for study at an academic institution (AKA MECHINA).

Psychometric entrance test: A standardized test covering three areas — mathematics, verbal reasoning, and English proficiency. It is a prerequisite for higher education in Israel (AKA The psychometric exams).

Secondary school: A collective name for Middle and High School (i.e., grades 7–12).

Secondary school graduation certificate: A certificate attesting to a successful completion of the Israeli education system. It is typically a prerequisite for higher education in Israel (AKA Matriculation certificate, or BAGRUT certificate).

Secondary school graduation examination: A collection of matriculation exams covering various academic disciplines. A passing grade in these exams is required for eligibility to receive certification (AKA BAGRUT exams).

State comptroller's report: The State Comptroller of Israel supervises and inspects all government bodies, the military and security services, and corporations, focusing on legality, efficiency, economy, ethics, and regularity issues (AKA DUACH MEVAKER HAMEDINA).

State Education: Israeli public schools can be divided into four tracks — general State Education (AKA secular or MAMLACHTI), Religious State Education (AKA MAMLACHTI-DATI), independent religious (AKA ultra-Orthodox or HAREDI), and Arab.

Teacher preparation program: A pre-service course or program aimed at training prospective teachers (AKA HACHSHART/HASMACHAT MORIM).

Teacher instructor: A member of the supervision team at the Ministry of Education, either at the national, district, or school level (AKA MADRICH MORIM ARTZI/MEHOZI/BEIT-SIFRI).

Teachers' union: The trade union and profession association for teachers and educators in Israel (AKA IRGUN HAMORIM).

Teaching certification program for those holding an academic degree: A teacher preparation program aimed at retraining holders of academic degrees from accredited institutions (AKA HASSAVAT ACADEMAIM).

2. Acronyms and Abbreviations

CET: The Center for Educational Technology (Also Known As MATACH).

CHE: Council for Higher Education in Israel (AKA the Hebrew acronym MALAG).

KESHER CHAM: Israeli National Center for Mathematics Education.

MALAM: Israeli Center for Science & Technology Education.

RAMA: National Authority for Measurement and Evaluation in Education.

MEITZAV: School Growth and Efficiency Measures.

About the Author

Boaz Silverman recently earned a PhD in Mathematics Education from the Weizmann Institute of Science. His research interests include: (1) cognitive and epistemological aspects of learning mathematics in tertiary school; (2) the interplay of learning disabilities and mathematics; (3) the role of context in the learning of mathematics; and (4) the contribution of textbooks to learning mathematics.